CAMBRIDGE SOUTH ASIAN STUDIES

CANAL IRRIGATION IN BRITISH INDIA

Perspectives on technological change in a peasant economy

Plate 1. 'The Ganges Canal, Roorkee (Saharanpur District)', by William Simpson (1863). (By kind permission of the British Library, India Office Records)

CANAL IRRIGATION IN BRITISH INDIA

Perspectives on technological change in a peasant economy

IAN STONE

Senior Lecturer in Economics, Sunderland Polytechnic

The right of the
University of Cambridge
to print and sell
all manner of books
was granted by
Henry VIII in 1534.
The University has printed
and published continuously
since 1584.

CAMBRIDGE UNIVERSITY PRESS

CAMBRIDGE

LONDON   NEW YORK   NEW ROCHELLE

MELBOURNE   SYDNEY

PUBLISHED BY THE PRESS SYNDICATE OF THE UNIVERSITY OF CAMBRIDGE
The Pitt Building, Trumpington Street, Cambridge, United Kingdom

CAMBRIDGE UNIVERSITY PRESS
The Edinburgh Building, Cambridge CB2 2RU, UK
40 West 20th Street, New York NY 10011–4211, USA
477 Williamstown Road, Port Melbourne, VIC 3207, Australia
Ruiz de Alarcón 13, 28014 Madrid, Spain
Dock House, The Waterfront, Cape Town 8001, South Africa

http://www.cambridge.org

First published 1984
First paperback edition 2002

A catalogue record for this book is available from the British Library

Library of Congress catalogue card number: 84-3200

ISBN 0 521 25023 4   hardback
ISBN 0 521 52663 9   paperback

CONTENTS

ILLUSTRATIONS

TABLES

PREFACE

This study is a foray into an under-researched aspect of Indian colonial history. It attempts to achieve some understanding of the effect upon a peasant economy of the vast irrigation schemes introduced by the British, through focussing upon the experience of the western part of what is now Uttar Pradesh. The work is based upon my Cambridge doctoral thesis which was financed initially by the Social Science Research Council and completed during the period when I was lecturing at Victoria University of Wellington, New Zealand. In making it ready for publication, I have considerably revised and extended it since my return to England, and I am grateful to the Twenty-Seven Foundation for an historical award which enabled me to carry out the additional research during 1981, and to the Geography Department of the University of Newcastle upon Tyne for providing me with facilities to make further refinements during 1982.

Teachers, colleagues and friends have, over the years, provided comments on various draft sections, or assisted with points of information. Notable among those are the late Professor Eric Stokes, Dr W.J. Macpherson, Dr Neil Charlesworth, Dr Chris Bayly, Dr Peter Robb, Dr Peter Musgrave, Dr Phil Bradley, Dr Tony Hellen, and Basil Poff. My greatest intellectual debt, however, I owe to Dr Clive Dewey of Leicester University who, as supervisor of my thesis, posed key questions and provided countless comments which, together with his good-humoured support, were vital ingredients to the project overall. I am grateful also to Mr B.H. Farmer, Director of the Centre for South Asian Studies, University of Cambridge, for his assistance and encouragement in the bringing of this work to the publication stage.

ix

In India, too, where I carried out fieldwork during the winter and spring of 1976 – and again the following year when *en route* to New Zealand – I received help of different kinds from many people. I am particularly indebted to Manohar Singh, whose efforts enabled me to gain access in the limited time available to a range of archival sources in my search for specific reports and files. I am grateful for the considerable efforts expended by officers of the Meerut District Cooperative Bank, and Executive Engineer R.D. Sharma, in assisting me to examine current farming and irrigation practices around Meerut. Special thanks are due to D.K. Jain, Secretary of the Bank, and G.P. Pant, who did so much to make my stay in Meerut a thought-provoking and comfortable one. I fondly recall, also, the friendship and assistance extended by Ram and 'D' Advani of Lucknow.

My material was gathered from a large number of libraries and archives, including: UP State Archives, Lucknow; UP State Regional Archives, Allahabad; UP Board of Revenue Libraries in Lucknow and Allahabad; UP Secretariat Records, Lucknow; Records of the Meerut Commissioner, Meerut; Collectorate Records, Muzaffarnagar; Central Design Directorate (Irrigation Department) Library, Lucknow; and the University Library, Cambridge. To the librarians and other staff of these institutions I am grateful. I am especially indebted, though, to the staff of the India Office Library and Records, upon which I placed particular reliance. The staff of the University Library at Newcastle – and especially Helen McFarlane – have been particularly helpful, as well as resourceful, in helping me to tie up loose ends.

Robin Mita and Olive Teasdale have executed with great care my diagrams and maps, and Doreen Morrison has contrived to produce copies of photographs which are better than the originals. Various people have supplemented my own efforts at the typewriter, notably Lorna Guerin and Stella Daniell at Victoria University, Eileen Temperley of Sunderland Polytechnic Social Sciences Department, and Kathleen Quinn. I am grateful for their attention to detail, as I am to Doreen Jones of CUP, who skilfully identified errors and problems in the manuscript and advised on improvements.

Finally, special thanks must go to my wife, Judy, for her

invaluable help in reading and commenting with great patience upon various drafts. I should say also that none of those mentioned – nor, indeed, any of the others who have helped along the way – bear any responsibility for remaining errors and omissions.

Jesmond   IAN STONE
Newcastle upon Tyne

ABBREVIATIONS

AJI	*Agricultural Journal of India*
AR	Assessment Report
BOR	Board of Revenue
CE	Chief Engineer
DG	*District Gazetteer*
DNB	*Dictionary of National Biography*
EE	Executive Engineer
EJC	Eastern Jumna Canal
GI	Government of India
IIC	*Report of the Indian Irrigation Commission, 1901–03*
IIC(E)	*Indian Irrigation Commission, 1901–03, Minutes of Evidence, United Provinces of Agra and Oudh*
IOR	India Office Records
IRR	*Irrigation Revenue Report of the North-Western Provinces and Oudh*
LGC	Lower Ganges Canal
MCR	Records of the Commissioner of Meerut, Meerut
NWFP	North-Western Frontier Province
NWP	North-Western Provinces
NWP&O	North-Western Provinces and Oudh
PP	*British Parliamentary Papers*
PRRC	*Papers on the Revenue Returns of the Canals of the North-Western Provinces*
PWD	Public Works Department
PWD(I)	Public Works Department (Irrigation Branch) Proceedings
RAR	*Report of the Board of Revenue on the Revenue Administration of the North-Western Provinces and Oudh*
Rev.	Revenue Department Proceedings
RRR	Rent-Rate Report
RSC	*Report of the Sanitary Commissioner for the North-Western Provinces and Oudh*
San.	Sanitation Department Proceedings
S&C	*Report on the Seasons and Crops of the United Provinces*

Scarcity	Scarcity Department Proceedings
SE	Superintending Engineer
SR	*Settlement Report*
UGC	Upper Ganges Canal
UP	United Provinces of Agra and Oudh
UPA	Uttar Pradesh State Archives, Lucknow
UPA(R)	Uttar Pradesh State Regional Archives, Allahabad
UPBORL	Uttar Pradesh Board of Revenue Library, Lucknow
UPPBEC(E)	*United Provinces Provincial Banking Enquiry Committee, 1929–31, Evidence*
UPPBEC(R)	*United Provinces Provincial Banking Enquiry Committee, 1929–31, Report*
UP Sec.	Uttar Pradesh Secretariat Library and Records, Lucknow
WJC	Western Jumna Canal

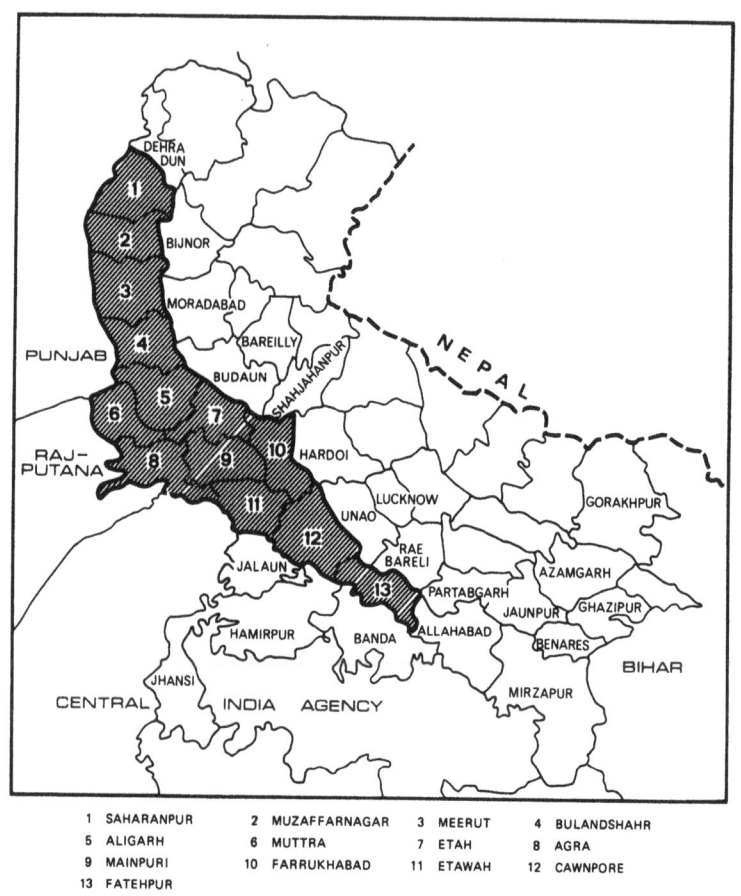

1	SAHARANPUR	2	MUZAFFARNAGAR	3	MEERUT	4	BULANDSHAHR
5	ALIGARH	6	MUTTRA	7	ETAH	8	AGRA
9	MAINPURI	10	FARRUKHABAD	11	ETAWAH	12	CAWNPORE
13	FATEHPUR						

Map 1. The United Provinces (with Doab districts shaded).

I

Introduction

Described recently by one economic historian as among the 'greatest monuments to British rule',[1] the harnessing of the waters of India's great rivers for irrigation purposes would appear one of the most positive ways in which the colonial regime contributed to Indian welfare. Despite the scale of this effort – one acre in six was irrigated from government schemes by the late 1930s – there is an almost total absence of modern historical studies into their impact. We are, in fact, very much in the dark as to the nature and distribution of the effects they had upon the peasant economy in general. The only study of this topic constitutes a single section of a book, published in 1972, on the Uttar Pradesh agrarian economy in the closing decades of the last century.[2] The western part of the UP,[3]

[1] W. J. Macpherson, 'Economic Development in India under the British Crown, 1858–1947', in A. J. Youngson (ed.), *Economic Development in the Long Run* (London, 1972), pp. 144–5. The other main 'monuments' identified were railways, law, and the civil service.

[2] E. Whitcombe, *Agrarian Conditions in Northern India*, vol. 1, *The United Provinces Under British Rule, 1860–1900* (Berkeley, Los Angeles, and London, 1972), pp. 64–91. Articles on the topic are rare and invariably undertaken mainly with a view to assisting with current policy formation. This is the case with E. Whitcombe, 'Development Projects and Environmental Disruption: The Case of Uttar Pradesh, India', *Social Science Information*, 11:1 (1972), pp. 29–49. Much of the work which does exist focusses upon pricing: B.D. Kanetkar, 'Pricing of Irrigation Service in India (1854–1959)', *Artha Vijnana*, 11:2 (1960), pp. 158–68. Additional material is available, by way of historical backdrop, in N. Ansari, *Economics and Irrigation Rates: A Study in Punjab and Uttar Pradesh* (London, 1968); and R.B. Reidinger, 'Institutional Rationing of Canal Water in North India: Conflict between Traditional Patterns and Modern Needs', *Economic Development and Cultural Change*, XXIII:1 (1974), pp. 79–104. Even historian D.R. Gadgil's *Economic Effects of Irrigation: Report of a Survey of the Direct and Indirect Effects of the Godavari and Pravara Canals* (Poona, 1948) is basically a technical cost-benefit analysis primarily oriented towards policy formation and the refining of project appraisal techniques.

[3] In 1901, the North-Western Provinces of the Bengal Presidency were combined with

1

specifically the Ganges–Jumna *doab*, was the focus for a substantial part of the canal-building activity during the nineteenth century,[4] and the study considers the impact of these schemes upon the ecology and rural economy of this tract. The conclusion its author, Dr E. Whitcombe, comes to is that 'the canals proved a costly experiment',[5] and it is the suggestion that the disadvantages actually outweighed the advantages that constitutes the starting point for this present study.

The basis for Dr Whitcombe's conclusion is the quite bewildering array of adverse effects which she associates with the introduction of the canals. It is argued that, due to the way irrigation was applied in the Doab, the policies over its use and the responses of the peasant cultivators themselves, the canals caused pronounced environmental and economic disruption. The very act of turning large quantities of water over an almost slopeless plains landscape with a comparatively high water-table inevitably led to waterlogging problems in low-lying land, and the problem was made worse by the widespread tendency for water channels to obstruct the existing natural drainage lines. Conditions of saturation were conducive to the spread of *reh* (saline deposits) over the land's surface as capillary attraction brought salt-laden moisture up from the subsoil; they contributed also to the more frequent incidence of malaria, and produced an 'unhealthy humidity'. The rise in the water-table made common earthen wells unstable, and as they fell in the cultivators were forced to rely on the canal's uncertain supply. The peasant afflicted by such calamities was confronted by a typically cumbersome bureaucratic structure which, even if liable to make compensation payments – and the grounds for compensation were elaborately circumscribed by

---

the Province of Oudh to form the United Provinces of Agra and Oudh (which in turn became 'Uttar Pradesh' after Independence). Except where it is necessary to be specific, the general term 'United Provinces' (abbreviated to 'UP') will be used, whatever the date. 'The Doab', throughout, refers to the Ganges–Jumna *doab*.

[4] While much of the early irrigation development was concentrated in the UP and Madras, the locus of attention later moved north-eastwards to the Punjab and Sind. By the late 1930s the areas (in millions of acres) irrigated by government works were as follows: Punjab 12.3, Madras 7.2, UP 5.3, Sind 4.7, Bihar 0.7, NWFP 0.5, Orissa 0.4, Bombay 0.4. V. Anstey, *The Economic Development of India*, 4th edn (London, 1957), p. 616. For more details, see Ch. 2 below.

[5] Whitcombe, *Agrarian Conditions*, p. 91.

the 1873 Canal Act – was indeed slow to yield up the requisite sums. Moreover, at the lower level of this structure, the cultivator came up against petty officials able to exploit their position of influence over canal supplies and add their various *haqs* to the 'already high' official charges for water from an unnecessarily costly system.[6]

The effect on cultivation itself, the argument continues, and on farm practices was no less damaging. The production of staple foods, particularly the coarse *kharif* staples, was downgraded in favour of commercial crops, a situation particularly urgent in drought years, when – as a consequence and against all expectations – the canal could do little to 'decrease the ravages of scarcity'. As well, the ease of canal irrigation compared with lifting water from wells gave encouragement to overcropping and a concomitant loss of soil fertility; and the disruption of fallowing cycles was aggravated by the attraction of pastoral castes into cultivation, which not only increased the tendency to imbalance and the curtailment of fuel and fodder supplies, but reduced the supply of cattle to agriculturalists. The latter were thus forced to rely on their own, 'often deteriorating', stock for their cattle supply, the inadequacy of which made for slovenly cultivation and constituted a yield-depressant in that farmers faced a reduced manure supply. Lured by the prospect of a less arduous life – and by that of a simultaneously increased value of product – the farmer involved himself in a system of production which 'disrupted [his] former pattern of work', while at the same time his 'techniques were not adapted to deal with such sudden and radical changes'.

Those familiar with the practical effects of large-scale canal schemes and with irrigation bureaucracies will probably find this unfortunate catalogue of effects perfectly plausible. Furthermore, given the character of the tract being studied, there is a distinct likelihood that such disadvantages could have substantially eroded the net benefits arising out of the Doab canals. Canals taken through the virtual deserts of the Punjab and Sind clearly produced large gains in terms of net output. The UP Doab may have been 'before the introduction of canal

---

[6] 'Unnecessarily costly' because of design faults which either increased construction costs or involved heavy maintenance outlays. See Ch. 2 below.

irrigation . . . among the most insecure in Northern India',[7] but it was also an area already widely cultivated and with extensive facilities for well irrigation. The margin of benefit, therefore, between the value of produce raised before the canal and that made possible by its construction was clearly narrower than the benefit arising from making settled agriculture possible in places previously sparsely populated by pastoralists.

There are problems, however, with this new interpretation. First, and perhaps most strikingly, there is the fact that for most of this period the western districts – and in particular the heavily irrigated northern districts – enjoyed a degree of broadly based material prosperity matched by few areas in India. There are problems also in that some of the responses attributed to peasant cultivators appear to be in conflict with current notions of 'peasant rationality'. Further, although contemporaries were well aware of the unfortunate side-effects of canals, this did not diminish their continued advocacy of canal construction and expansion. Successive Famine Commissions and an Irrigation Commission urged the extension of canal-building programmes, while the Government of India wrestled with the cautious and tight-fisted London Office to obtain the resources to execute such works. This, of course, might be dismissed as simply a reflection of Imperial perceptions as to what form of development was in the rulers' interests, but Indian nationalists, also, have been favourably disposed towards extending canal irrigation, at least to the extent that they have favoured a diversion of government spending in that direction and away from railways (which they consider drained away India's wealth rather than adding to it).[8]

[7] *British Parliamentary Papers* (hereafter *PP*) 1904, LXVI, *Report of the Indian Irrigation Commission, 1901–03*, p. 185. (This report will hereafter be abbreviated to *IIC*.)

[8] Historian Bipan Chandra, for example, is critical of the 'uniformly low' expenditure on irrigation alongside that on railways: see B. Chandra, 'Reinterpretation of Nineteenth Century Indian Economic History', *Indian Economic and Social History Review*, v:1 (1968), pp. 67–8. See also, for 'nationalist' views, R.C. Dutt, *Speeches and Papers on Indian Questions* (Calcutta, 1902), p. 77; and B.M. Bhatia, *Famines in India: A Study in Some Aspects of the Economic History of India (1860–1965)* (London, 1967), p. 198. Irrigation by no means rivalled the railways in terms of resources expended by the state. In fact, the proportion of total government outlays upon 'creative investment' (public buildings, irrigation, agriculture, transport, and communications) taken by railways was up to four times greater than that devoted to irrigation during the 1900s. Bhatia, *Famines in India*, p. 198.

The verdict on the impact of the canals, then, would appear to be very much an open one. Elizabeth Whitcombe assesses its effects upon the peasant community from a position which, very reasonably, regards with suspicion any disruption to what she sees as a viable and well-established social, economic, and technological balance by the incompetent meddling of the British with a peasant system and an ecology neither of which they understood. She accepts the possibility of increased production, but considers that this took place only in a context where a 'depressed peasantry laboured in a distorted environment'.[9] There is certainly – in what is, after all, only a short study – no attempt to weigh the qualitative and quantitative significance of the evidence used to support this position. Nonetheless, the accumulated material highlights the problem areas and creates a strong impression that in the canal divisions all was not well down on the farm. And she is joined in this view by Professor A.K. Bagchi, who, taking a long-term perspective, argues that the canals assisted in the creation of structures within the rural sector which were exploitative and obstructive to efforts to promote growth.[10] Thus, the cash requirements associated with canal irrigation, along with continual rent enhancement in expectation of higher yields, produced, according to Professor Bagchi, a shift in product mix away from millets – the staples of the poor – and, through making extensive cultivation by large landowners profitable, made 'their exploitation of the landless strata of poor peasants more intense'. Although this essentially neo-Marxist approach could – by bringing in the possibility of class alliances and sectional interests – successfully reconcile a situation where further canal schemes were prosecuted in the face of associated disadvantages to specific groups, it does rely heavily upon Dr Whitcombe's UP findings restated within a different framework.

This study[11] of the impact on a peasant economy of an

[9] Whitcombe, *Agrarian Conditions*, p. xi.
[10] A.K. Bagchi, 'Foreign Capital and Economic Development in India: A Schematic View', in K. Gough and H.P. Sharma (eds.), *Imperialism and Revolution in South Asia* (New York, 1973), pp. 49–50.
[11] The pilot study out of which this full-length investigation grew focussed upon eastern Muzaffarnagar, and the preliminary findings were contained in a paper presented at a Conference on Indian Economic and Social History at St John's

exogenous and extensively applied technical change within a colonial setting takes up in detail many of the important questions raised in the Whitcombe study. It is concerned particularly with the way the peasant community adjusted its activities in relation to this new irrigation source. Indeed, if it were necessary to isolate one fundamental issue on which it departs from the Whitcombe view, it would be the character of the peasant's response to the canal. According to Dr Whitcombe, the peasant took to the canal either because he had no choice (his wells fell in), or because he saw the prospect of relatively effortless (but possibly short-term) commercial gain. He thus accepted a technology and a system of farming with significant direct and indirect disadvantages in comparison with the traditional irrigation techniques to which agricultural practices, and society in general, were well adjusted. If it can be demonstrated that the decision to adopt the new mode of irrigation was an altogether more soundly based and rational decision on the peasant's part, then a whole set of alternative interpretations of the canal's impact is possible.

Chapter 3, indeed, does demonstrate that in many (though not all) circumstances, the canal was a more appropriate technology than the traditional methods of irrigation, given the priorities normally exhibited by peasant households and the economic, institutional, and physical conditions in the Doab. Chapter 4 then proceeds to analyse, with the aid of village data obtained from settlement handbooks, how this decision was translated in terms of production and the organisation of agricultural activity. In doing so it also shows how many of the supposed disruptive effects and disadvantages simply do not stand up to close empirical examination; some can only be seen as 'problems' when taken out of context, while the importance of others can easily be exaggerated. Thus, it was not the case – as contemporaries sometimes alleged – that the peasant left the canal to earn his living for him and retired to recount his good fortune from his *charpoy*; in most cases he spent, not less, but a

College, Cambridge in July 1975, subsequently published under the title 'Canal Irrigation and Agrarian Change: The Experience of the Ganges Canal Tract, Muzaffarnagar District (U.P.), 1840–1900', in K.N. Chaudhuri and C.J. Dewey (eds.), *Economy and Society: Essays in Indian Economic and Social History* (Delhi, 1979), pp. 86–112.

good deal more time at work once he received canal water. Moreover, the evidence that some yields were lower on canal-irrigated than well-irrigated land can be shown to be perfectly consistent with an increase in overall output (by value and quantity). Some cultivators rationalised their cattle resources, but the evidence is against the decline in cattle numbers which some officers fancied had taken place, and emphatically against a deterioration in the quality of stock. Similarly, the suggestion of a fall in staple food production is not valid in the context of the crop adjustments across the board induced by the canal, and taking into account the changes in consumption patterns accompanying rising incomes. Much the same can be said of the external diseconomies such as waterlogging and the spread of salinity. It cannot be disputed that some estates were seriously damaged and many holdings incurred production losses on account of these side-effects. But how extensive or persistent were these problems? Did they have a serious effect upon production? The effects, in fact, were variable over time, and there are good reasons for arguing that they may well have been exaggerated. The canal does, however, appear to have had a significant effect upon the incidence of malaria.

A major aim of colonial policy on canals was to decrease the 'ravages of scarcity', and Chapter 7 examines closely the protective role of the canal, re-examining the suggestion that its effectiveness in this regard was restricted by its apparent inability to protect the *kharif* food staples from drought. It can be demonstrated that the canal was far from impotent in meeting the challenge of climatic disruptions. To do this, however, it is necessary to trace its role in terms of the complexities of the peasant production system, involving both stocks and flows (of cash and food crops) and the disruption of exchange values resulting from famine-induced patterns of behaviour. The problem cannot be analysed solely in terms of *kharif* food staples.

To understand more fully, however, many of the patterns produced by the canal – including the types of benefits to which it gave rise and the way these were distributed among areas and within society – it is necessary to know more about the nature of canal irrigation and the way the innovation was applied. As with any technology, canal irrigation was not 'neutral' in its effects. It was intended to serve the perceived interests of its masters, in much the same way as the earlier irrigation works were. In its design, modes of operation, and intended effects, canal irrigation was ultimately a cultural expression, representing the priorities and aspirations of its western architects, and was inextricably bound up with some of the most vital aspects of colonial rule. It was related directly to the concern exhibited for the spread of commercial crops – and was thus tied up with official efforts at 'agricultural improvement' – as well as having an important bearing upon land revenue (frequently regarded as the point of most contact between rulers and ruled), political security, and famine prevention. It existed, in fact, at the very interface between the colonial presence and peasant society: the water trickling into the field *kiaris* was as tangible as any of the western innovations which filtered down to village level – and more tangible and more pervasive than most. Here, therefore, one might be tempted to think, was a vital agricultural input (under centralised European control) which could be used to bring about change along defined channels. The authorities, after all, determined the nature of the water supply and the terms on which it was made available, and the levers of control and manipulation were potentially influential in affecting crop and irrigation patterns and practices.

As Chapters 5 and 6 in particular show, however, the degree of control exercised by the central irrigation authorities, and their manipulative powers with respect to agricultural improvement, were effectively very limited. Why? The answer has to do with the technical aspects of the canal schemes and the broader administrative framework in which they operated.

More and more, researchers now agree that colonial power was frequently more apparent than real: a gloss that touched society but which did not overturn its fundamental structures. Colonial rule was always a juggling act: there were too few hands, not enough money, and too many considerations to attend to at once to permit policy-makers to realise – or, for that matter, even clearly define – their intentions in any specific area. The multifarious aims of different branches and levels of government, of departments, and of personalities within the administration spawned compromise (or inertia) as policies were interlocked into the overriding and all-embracing strategy of maintaining stability and British stewardship. As far as irrigation alone was concerned, on a policy level it was simultaneously linked with famine prevention, revenue stability, the settling of unruly tribes, expansion of cultivation, extended cultivation of cash crops, enhanced taxable capacity, improved cultivation practices, and political stability. Thus, as Chapter 5 shows, when it came to pricing canal water, it was difficult even to establish clearly the irrigation priorities that pricing was to reflect. Indeed, it was so difficult to separate the strands of canal policy from those of the broader administrative context that that element of control, of actually directing the way water was used through the price mechanism, was never really tried (though that is not to say that the pricing system was irrelevant to the determination of the way water was used).

For these and the technical reasons discussed in Chapters 5 and 6, the way canal water was deployed was far too dependent upon indigenous structures – within both the bureaucracy and the village – to avoid the blurring of policy intentions at the lower levels, where it was applied. Any institution serves itself to a certain extent, and this was true in irrigation, not only among those at the top, but particularly so at the lower level, where instead of western concepts of efficiency and economy of water use, the prevailing concerns were those of the peasant society to which the subordinate bureaucracy itself belonged, and thus bribery, status, reciprocity, and trading upon petty monopolies asserted themselves.

Technically, the canals proved difficult to regulate with any pretence at precision, though, as Chapter 6 shows, things did improve out of all proportion over the period. Administratively, there was a great reliance upon native staff within the canal bureaucracy, over which supervision was minimal. For these reasons alone – leaving aside the indecision over policy priorities at the state level – manipulative control was not possible. Effectively, canal water at ground level was surrendered to the society itself, and was deployed in accordance with local social, economic, and political priorities. The use patterns, the division of benefits, and the way water was used during drought years were thus largely determined independently of government interference. Only when disputes over water and its benefits occurred between parties whose respective resources showed a semblance of balance was even arbitration called for.

Just as the peasant socio-cultural system has mechanisms for adjusting to internal forces disturbing its balance (such as a gradual rise of numbers) and exogenous forces (such as a climatic change or a new overlord), and is flexible enough to incorporate change and new forms of intrusions without losing its long-term stability and viability, so, because of the way it was grafted onto society, canal irrigation was substantially 'absorbed' by that society rather than bringing about fundamental changes in its underlying structure. In common with other elements in the 'modernisation package', it does not deviate significantly from Professor Eric Stokes' assessment that 'the institutional and economic "inputs" of modernity were too feeble to blow apart the structure of Indian society . . . the elastic, accommodating nature of the latter was adequate to contain them'.[12] Indeed, Basil Poff has gone as far as arguing that, even where an entire framework was set up which suited – and was even designed to encourage – innovation and change, the Punjab canal colonists, despite undergoing significant upheaval, 'adapted the colony to themselves and not themselves to the colony'.[13]

If the intentions of British policy-makers – like those of their Indian counterparts in the post-colonial period – were, due to their limited control over an imperfect canal system, destined like other innovations to be moulded according to the specific needs, structures, and technical considerations which are the ubiquitous concerns of peasant society, this is not to dismiss the possibility of change and material progress. Peasant society exhibits continuity of form rather than changelessness, and while it might not constitute change on the scale of the British agrarian revolution – which, for at least some of the time, provided a crude policy model – given the right combination of

---

[12] E.T. Stokes, 'The First Century of British Colonial Rule in India: Social Revolution or Social Stagnation?', *Past and Present*, LVIII (1973), p. 151.

[13] B.J. Poff, 'Land Management and Modernisation in a Punjab Canal Colony: Lyallpur, 1890–1939', paper presented at the Institute of Commonwealth Studies, London, Feb. 1974, p. 6.

circumstances peasant society can display marked dynamism; it can expand production and improve productivity by the adoption of new crops and investment in technical and organisational innovations.[14] The absence of such trends is frequently interpreted in cultural terms (e.g. 'limited wants'), but is more often a reflection of specific institutional structures or technical constraints rather than a lack of concern for increasing output *per se*. Canal irrigation could play a key role in providing the appropriate technical environment for the release of expansionary forces, though it was, by itself, limited in its potential to increase output. It was vital that the institutional as well as the technical environment was conducive to such an influence through providing the incentive for investment and growth, as well as permitting the necessary related organisational and technical responses.

It follows, therefore, that a crucial focus of this study must be on the interaction between the precise form of the local irrigation facilities (reliability, terms on which it is available, and so forth) and the specific recipient environments (physical, economic, technical, social, and cultural). This point of interface is effectively the source of what Professor Eric Stokes referred to as 'differential impact', for it is this which determines not only the potential of the input, but the degree to which (and manner in which) it will be realised. Thus it is possible, in Chapter 8, by examining this area of interface, to show why the canal had different effects in different parts of the Doab; why growth rates and growth patterns differed. The canal's impact is not simply a reflection of the specific environment into which it was introduced. This final chapter examines how the pattern of distribution of benefits it created, in the Meerut division at least, was itself a factor in perpetuating the cycle of growth and prosperity which had been established. The comparative dynamism and vitality of the western districts, with their growth and noticeable diversification, were in striking contrast to the eastern districts of the UP particularly, and a far cry from the gloomy impression given in *Agrarian Conditions*. Ironically, it is the alleged villain of the piece, the canal, which is

---

[14] This is clearly demonstrated for the Punjab in C.J. Dewey, 'The Agricultural Output of an Indian Province: The Punjab, 1870–1940', paper presented at the Institute of Commonwealth Studies, London, April 1972.

finally shown primarily to explain the dynamism and prosperity.

The study begins, however, with a chapter which sets the development of the Doab canals in the broader context of the growth of irrigation in India as a whole, and which then traces the history of the significant UP works. It outlines the main physical difficulties confronting the pioneering engineers, and the experiments and controversies which marked the evolution in design and construction towards a technically effective basic system of irrigation.

2

Development of the canal system

Historical outline

Restoration and improvement of pre-British works (1817–36)

British canal-building activity in India commenced in 1817 and was initially confined largely to the plains area to the north of Delhi and the deltaic regions of Madras. Most of the early British schemes were, in fact, rehabilitated and extended versions of indigenous works found in various parts of the country. The original version of the Western Jumna Canal (WJC),[1] in the east Punjab, has been attributed to Firoz Shah Tughlak. This fourteenth-century excavation subsequently fell into disuse, but was apparently renovated during Akbar's reign, and then remodelled and extended to Delhi by Shahjahan's engineer adviser Ali Murdan Khan.[2] The Doab Canal (later Eastern Jumna Canal) also had its forerunner. It is thought to have been built in the early eighteenth century, though it was abandoned fairly soon afterwards. It appears to have been partially restored, down to just below Saharanpur, by the Rohilla Chiefs in 1784.[3] The Bari Doab Canal was preceded by the 130-mile-long Hasli Canal, which conveyed water from the Ravi to Lahore, and which was still functioning

---

[1] So called because it took off from the western bank of the Jumna River.

[2] Government of India (hereafter GI) Public Works Department (hereafter PWD), *Triennial Review of Irrigation in India, 1918–21* (Calcutta, 1922), p. 24. The historical details contained in this section have also been obtained from G.W. McGeorge, *Works and Ways in India* (London, 1894), pp. 121–30; G. Kuriyan, 'Irrigation in India', *Madras University Journal*, xv (1943), pp. 55–6; and P.T. Cautley, *Report on the Ganges Canal Works*, vol. 1 (London, 1860), pp. 7–10. It should be noted that there are numerous discrepancies between these accounts of earlier works.

[3] *IIC*, p. 9. The Doab/Eastern Jumna Canal will be referred to hereafter as the EJC.

as an irrigation work when the Punjab came under British rule; as indeed were several inundation canals in Multan, Muzaffargarh, and Dera Ghazi Khan districts.[4] Also still functioning when Tanjore was ceded to the British in 1801 was the work which was perhaps the most impressive of all the indigenous schemes – and which, when rehabilitated, was to form part of Sir Arthur Cotton's extensive irrigation works in Madras – the 1,080-foot 'Grand Anicut'. This work, which spanned the Cauvery, has been ascribed to the Chola Kings, though opinions as to the date of original construction vary from the second to the eleventh century. At one time, clearly it irrigated a large area – one estimate puts the figure at 600,000 acres.[5]

Not all the pre-British works were built primarily as works of irrigation. According to at least one source, the prime aim of the Moghul versions of the WJC was for the supply of water to the Emperor's hunting grounds at Hissar, while Ali Murdan's extension to Delhi was intended to feed the fountains and gardens of the Imperial palace. Similarly, the less ambitious twelve-mile cut linking the Kali Nadi with Meerut, traces of which remained when the British took over, was excavated with a view to supplying Meerut's groves and gardens, rather than for irrigation.

Many of these indigenous works did not achieve what was expected of them, or at least did so for a comparatively short period. Engineering capabilities understandably fell some way short of what was shown subsequently to be necessary for the design and construction of large-scale canal systems, and the history of the indigenous versions of both the WJC and the EJC is one of a succession of setbacks. Pre-British works made little use of masonry structures and those taking off from the Jumna, and others in and around the Doab, shared the characteristic of avoiding high ground and following the courses of rivers and drainage lines wherever possible. These crude works were thus vulnerable to the effects of floods and silt deposits, and, as well as causing swamps and damage to low-lying land (as did the

---

[4] *Ibid*. 'Inundation canals' are those which carry water only when the river from which they obtain their supply is in flood.

[5] *Triennial Review of Irrigation*, p. 27. The irrigated area, and the prosperity of Tanjore, had considerably diminished by 1801, as a result of the main stream of the Cauvery having begun to flow down a northern channel know as the Coleroon.

Rohilla version of the EJC around Saharanpur), were of limited use for purposes of irrigation. Even the short West Kali–Meerut Canal, though it functioned for a time through the construction of a crude embankment across the river, must have caused considerable damage to the river valley. More spectacularly, the town of Lalpur was reputedly destroyed when one of the embankments of Ali Murdan Khan's canal gave way.[6]

The earliest important work of renovation undertaken by British military engineers involved the WJC, which was reopened in 1820. In the two decades which followed – and while Cotton was establishing his considerable reputation as an irrigation engineer in Madras[7] – a group of Bengal Engineers was busy, in between military operations such as those against the Sikhs, restoring the Moghul works in the North-Western Provinces. Rehabilitation work on the 129-mile Doab Canal (EJC), which reopened in 1830 and served the western parts of the districts of Saharanpur, Muzaffarnagar, and Meerut, involved the considerable task of embanking the main channel to an average level of nine feet above the land surface for almost forty miles. In the light of the subsequent operating and financial success of the Jumna Canals, extensive improvements were made during the 1830s and after to the numerous dams and canal networks in both Rohilkhand and the Dehra Dun Valley behind the Siwaliks.

The nature of the work carried out was very much experimental. Funds were severely restricted and the original alignment of these works was in consequence largely adhered to for reasons of economy. Depressions were crossed on earthen banks which intercepted drainage and were prone to collapse and cause damage. Numerous masonry works were introduced, but both the early Jumna Canals, successful as they were in times of drought and in terms of financial returns (and important as they were in acting as objects of hydraulic

---

[6] A. Deakin, *Irrigated India: An Australian View of India and Ceylon, and their Irrigation and Agriculture* (London, 1893), p. 291.

[7] He constructed an impressive (and profitable) system of *anicuts* (weirs), irrigation channels, and navigation channels, first on the Cauvery and Coleroon Rivers (where his 'Upper Anicut' across the Coleroon ensured a water level sufficient for the full utilisation of the 'Grand Anicut', which he improved), and then on the Godavery and Krishna Rivers during the 1840s and 50s.

experimentation), eventually required extensive remodelling of main and distributing lines. Much of this work was considerably delayed by government reluctance to grant the requisite funding, so that much of the corrective work, particularly on the WJC, had to wait until funds for such projects became more available in the 1870s.[8]

Wholly British projects (1836–56)

One of the officers involved in the rehabilitation of these schemes was Sir Proby Thomas Cautley,[9] the architect of the grand and first purely British work, the Ganges Canal. This work was originally contemplated by Colonel John Colvin, the then NWP Superintendent-General of Canals, who envisaged in 1836 a scheme which would serve the north-east section of the Doab. Cautley examined the project and was discouraged by the physical obstacles presented by the low level of the land between the Ganges and the upland plain requiring irrigation. The idea was to be revived, however, after the severe famine of 1837–8. In the face, particularly, of the 'sacrifice of money' occasioned by the famine, the Ganges project suddenly assumed urgent importance, and in 1839 Lord Auckland granted funds for a full survey to be undertaken by Cautley.[10] Cautley's subsequent proposals were approved by Auckland and by the Court of Directors, who noted in a September 1841 despatch that

apart from the consideration of financial results, which we are far from contemplating with indifference, there are few measures connected with our revenue administration in India more calculated to contribute to the general improvement of the country, the amelioration of the condition of the people, and to raise the character of the Government, than those of the nature now under consideration. We concur in opinion with the Government of Agra that a higher

---

[8] *Triennial Review of Irrigation*, pp. 24–6. 'In nothing has the spirit of false economy more perniciously affected the prosperity of existing canals than by preventing the executive officers from submitting any comprehensive plans for remedying the evils of the original designs of the works', complained Cautley in 1849. P.T. Cautley, 'Canals of Irrigation in the North-Western Provinces', *Calcutta Review*, xii (1849), pp. 79–164.

[9] Cautley was the Superintendent of the EJC over the period 1831–43.

[10] Cautley, *Ganges Canal Works*, i, Ch. 1.

ground for advocating these works is found in the security they afford against famine and its attendant horrors.[11]

Moreover, the court initiated a considerable extension of the project's territorial scope. In the light of the severity of the famine in the central and lower parts of the Doab, it was directed that the projected Canal should be 'constructed on such a scale as would admit of irrigation being supplied to the whole of the Doab'.

The official enthusiasm which had greeted Cautley's proposals – and had prompted the decision to award him with a donation of Rs 10,000 for his efforts[12] – waned, however, even before construction got under way in 1843. This is partly because the incoming Governor-General, Lord Ellenborough, had a passion for the army and displays of pageantry, and clearly not for public works schemes; but there were also at the time doubts over whether such a large canal would not reduce the flow in the Ganges so as to limit its navigability, and fears that the serious fever outbreaks experienced along the Jumna Canals would similarly occur along the new canal. Whatever the precise mixture of reasoning behind his action, Ellenborough vigorously discouraged the project by withholding sufficient officers' assistance, by limiting funds, and by modifying the original design, directing that it should be constructed 'in the first instance [as] a canal of navigation, and all the water not required for that purpose may be distributed for the purposes of irrigation'.[13]

Until 1844, when Lord Harding replaced Ellenborough, Cautley's efforts were often diverted to surveying and levelling for want of sufficient staff to perform these mundane tasks. His health had suffered in consequence, and he took furlough from early 1845, his place being taken by Major (later General) W.E. Baker. When he returned from leave in Europe – during which time he had studied hydraulic works in England and

---

[11] Quoted in *ibid.*, p. 25. For an alternative to Cautley's account of the history of the Ganges Canal construction, see R.B. Buckley, *The Irrigation Works of India and their Financial Results*, 2nd edn (London, 1905), pp. 95–101.

[12] Nearly half of which, at the recommendation of the Military Board, was to be deducted to cover the full cost of supplying a temporary officer to assist Cautley in surveying the Ganges *khadir*. Cautley, 'Canals of Irrigation', p. 151.

[13] Cautley, *Ganges Canal Works*, I, pp. 29–33.

Italy and (while *en route* back to India) those in progress on the Nile – he encountered once more an official attitude favourable to resuming the project on the original basis.[14] With the active encouragement of Lieutenant-Governor James Thomason[15] and the development-oriented Governor-General, Lord Dalhousie, the vast and unique work, stretching the 350 miles from Hardwar to Cawnpore and incorporating altogether 650 miles of main and branch lines at a total capital outlay of £2.15 million, was completed for official opening in 1854 and became fully operational in 1857. It was the largest canal ever attempted in the world, five times greater in its length than all the main irrigating lines of Lombardy and Egypt put together; and longer by a third than even the largest USA navigation canal, the Pennsylvania Canal.[16]

In an effort to expand public works activities, Dalhousie proposed – in the year the Ganges Canal was opened – to reform the system under which such schemes were constructed and financed.[17] Up to 1854, public works of all kinds were carried on by the Engineer Department of the army, under the supervision of the Military Board in each of the three Presidencies. If the work was of a civil nature, then engineers were lent to the civil authorities, and the financial responsibility was that of either the local government or a civil department of the Government of India. Thus the Ganges and Jumna Canals, like the works in the deltas of Madras – and the Great Trunk Road – were all constructed under this system for the Government of India, and all expenditure incurred was treated as ordinary expenditure and charged against the revenues of the year. Dalhousie abolished the Military Boards, substituting in their place a Chief Engineer in each of the local governments, and at the same time formed a Central Public Works Secretariat at Calcutta for general supervision of the

[14] In the interim, a committee appointed to report on the likely implications of the Ganges Canal for the healthiness of the tract involved had concluded that the problem could be minimised through making allowances at the design stage. See Ch. 4 below for details.

[15] After whom Roorkee Engineering College, opened in 1848 for the training of PWD staff, both European and Indian, was renamed.

[16] *The Times*, 2 Nov. 1864, p. 4, quoting Lord Dalhousie.

[17] For further details on this section, see *PP* 1878–9, IX, *Report from the Select Committee on East India (Public Works)*, pp. iii–vi.

local establishment. In addition to establishing the basic framework from which were to develop the Public Works Departments, Dalhousie outlined important changes in the financing of public works. He proposed that expenditure on the construction and maintenance of works necessary for the administration of the colony should be classified as 'ordinary expenditure' and set against the revenue of the year; outlays on projects 'calculated to increase the wealth and promote the prosperity of the country', such as irrigation, harbours, and railways, should be met mainly with borrowed money. As far as irrigation was concerned, Lord Dalhousie wished to see the annual allocation of a specific sum, raised through loans, to promoting canal schemes; and in setting up the PWD he designated £2 million for this purpose.

Irrigation development through private companies (1858–66)

Dalhousie's retirement and the events of 1857 postponed the adoption of these financing proposals. In the post-Mutiny period there was considerable pressure on the Government of India to promote irrigation by the same indirect agency as that by which they were then extending railways – by guaranteeing a return to private companies – and the period 1858–66 was marked by what turned out to be a remarkable experiment in irrigation development through private irrigation companies.

The driving force behind both the Madras Irrigation Company and the East India (Orissa) Irrigation and Canal Company was Sir Arthur Cotton. He was at the height of his powers as an hydraulic engineer. His track record inspired the confidence necessary to overcome opposition and actually launch such an enterprise. Cotton had for some time been dissatisfied with the restrictive system of constructing public works out of current revenue. He criticised particularly the fact that it meant having to 'grind out of the present generation the whole cost of works in which they only have a life interest',[18] and in his book, *Public Works in India*, published in 1854, he showed himself to be a supporter of Dalhousie's plan for financing public works through raising loans. Certainly, if he was to be

---

[18] A. Cotton, *Public Works in India* (London, 1854), p. 25.

able to carry out some of the grand schemes he had in mind, freedom from the existing financing arrangements was a prerequisite.

Such a scheme was the one he devised during the late 1850s for peninsular India. A response, perhaps, to Cautley's Ganges Canal, its main elements consisted of, in the eastern section: a series of weirs on the Tungabhadra (one of the largest tributaries of the Krishna) and one on the Krishna itself; storage reservoirs to impound 50,000 million cubic feet of water to supply navigation canals; the diversion of part of the water of the Tungabhadra into the Penner; 600 miles of river improvement; and a network of five large canals for navigation and irrigation, including a fifty-mile east coast canal connecting the Krishna delta with Madras itself. To the west, a canal was to be taken 600 miles to Poona and the west coast, with subsidiary lines extending from Ahmadnagar (100 miles north-east of Poona) southwards to the coast at Mangalore (on the same latitude as Bangalore). Altogether, this system would traverse the peninsula, open up 150,000 square miles of country, give rise to an increased trade of around £5 million annually, and provide extensive irrigation in Hyderabad and Raichore – all for an estimated £2 million.[19]

The initial stage of the project, the eastern section, was estimated in 1859 to involve an outlay of £1.38 million. Despite the anticipated 22% yield on these works, the government felt unable at the time to sanction such an outlay, and it may indeed have been Cotton himself who was responsible for an (unattributed) 1859 *Calcutta Review* article arguing the case for allowing private companies to carry out such schemes and outlining the kind of safeguards which would be necessary to make the proposal acceptable in principle to the government.[20] Officials in India seriously discussed whether so remunerative

---

[19] For details, see Deakin, *Irrigated India*, p. 264 *et seq.*; and *Triennial Review of Irrigation*, pp. 47–51. This peninsular system was actually part of an even more comprehensive scheme envisaged by Cotton at the time. He dreamed of a huge irrigation and navigation scheme, with a navigable line 4,000 miles long, stretching from Karachi via Cawnpore, Calcutta, and Cuttack to Madras and across to Poona and the west coast. 'There is not a single obstacle to this', he wrote in 1858, 'and the results would be far beyond calculation.' *Triennial Review of Irrigation*, p. 48.

[20] Anonymous, 'English Capital and Indian Irrigation', *Calcutta Review* xxii (1859), pp. 172–85.

a scheme should be allowed to pass out of the Crown's hands; Madras officials, particularly, were unhappy about its doing so. In the end, though, in spite of all the objections, and without challenging any of the estimates, the Secretary of State for India determined that the need to introduce British capital into the colony – and particularly to speed up the progress of irrigation works – was so important that the scheme should be permitted to proceed. The Madras Irrigation Company was formed, and in June 1863 a 5% guarantee was given by the government on £1 million, along with the necessary grants of land. The government was to receive half the profits over 5% and retained the right of purchase of all shares after 25 years at the fair market price.

Within three years the Secretary of State was actually offering to buy the shares, for by 1866 nearly the whole of the capital had been expended and the planned sections were far from complete. Understandably, the shareholders declined his offer. An advance of £600,000 was made to the company and more stringent conditions laid down as to its use on a further section of the project, but the company's programme soon ground to a halt in the face of a succession of cost overruns.[21] Despite devoting to the purchase of boats a further £80,000 (apparently claimed from the government for surveying costs on those portions of the scheme abandoned for want of funds), navigation operations failed to make a profit. As one observer put it, 'The canal runs from nowhere, to nowhere in particular, and consequently there is nothing and nobody to carry.'[22] Irrigation was no better an investment, and showed a deficit on working expenses,[23] in spite of the fact that the government made large remissions to the peasants for crop failure caused by famine while still paying full rates to the company on their behalf. Although the completed sections had the capacity to irrigate 300,000 acres, the local agricultural conditions were

---

[21] For example, Col. Hugh Cotton (Sir Arthur's brother) had spent more than a third of the estimated cost of the Tungabhadra *anicut* before detailed surveys showed the site to be impracticable. The alternative site exceeded its estimated cost by a factor of five. [22] Deakin, *Irrigated India*, p. 267.

[23] Noted the Famine Commission: 'The establishment charges have been very high in comparison with what they would have been on a Government work of like nature, and the London Board of Directors and its office have been an additional heavy item of expense.' *PP* 1880, LII, *Report of the Indian Famine Commission*, pt II, p. 162.

such that irrigation was required only occasionally. The situation was made worse by the heavy irrigation charges, and barely 20,000 acres were watered on average. 'The extraordinary oversight which led to the unhesitating construction of these great works, without regard to the character of the soil to be watered, of the people who own it, or the results to be obtained by its execution, is a remarkable incident in the history of Indian irrigation', summed up one observer. 'Even had the works been well and cheaply built, they must have failed.'[24]

It was not until 1882, after having paid out 5% annually on the initial £1 million and having advanced £680,000 for which it received no interest, that the government was able to buy the company out for £1.18 million. With the additional loss due to debts abandoned of £1.5 million, the government finished up paying more for an isolated fragment of the eastern part of the design than was originally estimated for the entire gigantic scheme.

A similar conclusion – also involving terms favourable to the shareholders – had attended the private irrigation initiative in Orissa. The company, which was incorporated in 1861 with a capital of £2 million,[25] was set up to execute – though not with the aid of a 5% guarantee – another of Cotton's grand schemes: this time in the deltaic region of the Mahanadi and Brahmani Rivers and extending from Cuttack to Calcutta. River and flood controls, irrigation canals with a projected irrigation of over 1.5 million acres, and a navigation channel linking Cuttack with Calcutta were proposed for an area undeniably in need of effective land and coastal communications, and subject to serious drought and increasingly destructive floods. Even though the cost estimate was subsequently raised to £3 million, when the government took them over, in 1869, the actual cost was on course for a final total of around £6 million.[26] The

[24] Deakin, *Irrigated India*, p. 267.
[25] The Madras Irrigation Company wished to add this enterprise to its existing structure, but was refused permission by the Secretary of State. *Ibid.*, p. 283.
[26] The government paid the company just under £1 million for the canal, as well as cancelling a debt of £150,000 and paying £50,000 compensation to company servants. *Triennial Review of Irrigation*, p. 53. The company had also, at the time when the government was seriously doubting the wisdom of continuing to allow private enterprise to develop canals, begun work on the Son and Midnapur Canals,

scheme not only shared with its Madras Company counterpart a serious problem of overrunning cost estimates, it similarly suffered from an overestimation of potential demand for irrigation. With an average of sixty inches of rain in the Mahanadi delta, rate reductions and other concessions failed to raise ordinary demand for water to more than a few thousand acres. But if the scheme failed to realise the profits initially envisaged, in a non-commercial sense the Orissa scheme was not entirely a failure, since it effectively reduced flooding, gave some protection against famine, and did help to overcome a difficult local communications problem.

State construction financed by loans (from 1866)

By the mid-1860s, there was widespread disenchantment in the government concerning the wisdom of allowing private companies to construct and manage works of irrigation, and when the Government of India determined in 1866 upon a course of major extension of irrigation works through public agency, it revived Dalhousie's general plan for financing such works through loans. The Secretary of State for India approved of the scheme of raising loans in London specifically for irrigation works, subject to the condition that such schemes could be expected to be 'remunerative'. The Government of India was given the discretion to authorise the commencement of projects which met the necessary conditions, and it was decided that around £500,000 per year would be earmarked for irrigation out of ordinary revenue, with an addition of roughly £2 million obtained through loans.[27] From 1867, for the first time, an adequate and reliable flow of funds went into canal development, and a large number of 'productive' works, involving an outlay of £15 million, were sanctioned over the next decade. This period was important from the point of view of UP

modified versions of which were eventually completed by the governments of Bihar and Bengal respectively.

[27] Government of the North-Western Provinces, Public Works Department (Irrigation Branch) Proceedings (hereafter NWP PWD(I)), Nov. 1867, (proceeding no.) 22, GI Circular 91, 'Construction with Loans', 18 Sept. 1867. (In the India Office Records (hereafter IOR).) For further details relating to this section, see N.D. Gulhati, *Administration and Financing of Irrigation Works in India* (New Delhi, 1965), pp. 58–63.

irrigation in that both the Agra Canal and the Lower Ganges Canal were initiated under these new arrangements. Other major works included the Sirhind Canal in the Punjab and the Mutha Canals in Bombay.

The refusal on the part of the government to sanction an 1875 proposal for a canal to serve virtually the whole of the Chenab–Ravi watershed, the Chenab Canal, was to mark the end of the initial boom in loan-financed canal development. This particular project was complicated by the fact that its financial prospects were dependent upon colonisation of the tract. But there were more general reasons contributing to the decision. Against a background of concern over the financial returns to canal irrigation (and thus the danger that servicing the loans might become a permanent charge on government revenues), a growing body of opinion wanted the government to curtail its public works programme, and particularly irrigation. 'To prosecute its irrigational operations on the scale at present contemplated', wrote William Thornton in 1875, 'and on the same principles as heretofore, may therefore seem to be a course that must inevitably involve the State in serious financial embarrassment.'[28] A budgeting squeeze, brought on, among other things, by famine costs and the rising cost of remittances to England as a result of the depreciation of silver, reinforced the arguments of those opposing the enhanced level of government involvement in public works. Moreover, a large proportion of the borrowing for 'Extraordinary Public Works', as schemes constructed from loans were termed from 1868, was intended for railway projects.[29] Naturally, advocates of rail expansion sought to divert the pressure for reduced government public works activity away from railways and onto irrigation. Much of the criticism of canals during this period – in terms of both their overall financial performance and their operational performance – took place against this backdrop.[30] The railway lobby was too powerful to be resisted, and –

---

[28] W.T. Thornton, *Indian Public Works* (London, 1875), pp. 116–17.

[29] It was decided in 1870 that further extension of railways should – as with irrigation – generally be undertaken by the state directly, rather than by companies, and that the funds should be obtained by borrowing.

[30] See, for example, J. Dacosta, *The Indian Budget for 1877–8: Remarks on the Financial Position of the Government of India* (London, 1877), pp. 20–7.

despite the tireless advocacy of Sir Arthur Cotton for navigation and irrigation canals as a more appropriate form of transport and famine protection than railways – it was irrigation outlays which were cut back sharply from 1875, while annual average expenditure on railways rose markedly over the levels of the early 1870s.[31]

The issue eventually led to the appointment in 1879 of a Parliamentary Select Committee to examine ways in which loan capital raised in London for Indian irrigation works might be safeguarded.[32] It was also to recommend a formal criterion for allocating loan funds among various canal projects. The committee was hampered by the fact that the revenue accounts of existing irrigation works were grossly unreliable; its investigations nonetheless revealed that for the normal year examined, 1875–6, gross receipts from all works covered aggregate working expenses plus interest on capital. They even yielded a small profit overall. The committee thus recommended that the practice of loan-financed construction be continued, subject to a 'productivity test'. Essentially, this test required that only such projects as would, ten years from their date of opening, yield an annual income equal to the interest on the capital expended on their construction (including, in such capital, accumulated interest arrears) should be accepted.

Up until this time, major irrigation works were considered as essentially commercial undertakings and were classified as 'productive works'. Such a designation did not mean that a work was constructed and operated without regard for its potential role in combating famine. On the contrary, in 1867 it was determined that certain design features intended to increase the protective capabilities of the irrigation works should be incorporated into north Indian canal schemes.[33] This, of course, did not help those areas which, though in need of irrigation to provide defence against bad seasons, were

---

[31] Annual outlays on irrigation works from borrowed funds fell steadily from the peak level of £1.23 million in 1875 to around £600,000 in 1880. The total for the decade 1871–80 was £9.2 million as against £23.9 million spent on railways. *PP* 1880, LII, *Report of the Indian Famine Commission*, pt II, p. 147.

[32] *PP* 1878–9, IX, *Report from the Select Committee*, pp. iii–vi.

[33] Details are given in Ch. 6 below. The Irrigation Commission noted, in fact, that it considered artificial the distinction between 'productive' and 'protective' works, since the former had always had an important protective function. *IIC*, p. 75.

unable to receive such works because either the construction costs or normal conditions of agriculture ruled out the possibility that the prospective works would be formally 'productive'. The 1880 *Report of the Indian Famine Commission* drew attention to the *indirect* returns to irrigation works (including savings on famine relief expenditure) and it was with such calculations in mind that the Government of India decided to set apart from the revenues of each year a sum of Rs 150 *lakhs* (£1.5 million)[34] to be held in a Famine Fund. Half of this fund (Rs 75 *lakhs*) was to be allotted to 'protective' works (including railways as well as canals) in the construction of which direct returns were to be a secondary concern of the government. Subsequently it was decided that the full Rs 75 *lakhs* should be devoted to protective irrigation works.

Changes in the funding methods for public works schemes were accompanied by a restructuring of the administrative framework. The PWD was divided, in 1866, into military, civil (including irrigation), and railway branches. The former system was one in which revenues from all provinces were treated as belonging to a single fund, and expenditures from which were controlled by the Government of India, with provincial governments having no discretion in sanctioning fresh charges. This situation led to considerable tension between central and local government, and gave little incentive for local economy in the use of funds. It gave way to Sir Richard Strachey's 1867 reform proposals, which were to render the execution and management of irrigation works a state government function.[35] This was done through provincial PWDs, with the Government of India and Secretary of State exercising powers of superintendence, direction, and control over local technical and administrative activities through an Inspector-General of Irrigation and the PW Secretariat in Calcutta. Strachey's system was operative for virtually the rest of the period, until the first round of constitutional reforms,

---

[34] A *lakh* of rupees connotes Rs 100,000 (nominally £10,000); a *crore* of rupees connotes Rs 100 *lakhs* (nominally £1 million).

[35] J. Strachey, *India: Its Administration and Progress*, 4th edn (London, 1911); Gulhati, *Adminstration and Financing of Irrigation*, pp. 15–17; *Report of the Irrigation Commission of India*, vol. 1 (New Delhi, 1972), p. 257. The first appointee to the post of Inspector-General was Strachey himself.

which made irrigation a provincial but 'reserved' subject from 1921. The firm control of the central government and the Secretary of State was then removed, except in the sense that provincial governments had to seek prior approval of the latter (through the Government of India) before sanctioning projects costing in excess of Rs 5 million. Irrigation projects were then financed by local governments, either from the general revenues of the province or with loans raised on the security of the Indian Government, with occasional grants from the central government for minor or unproductive works.[36]

Confidence in the soundness of irrigation returned after 1880, and, within the administrative and financial framework described, construction of irrigation works proceeded throughout the country. The area irrigated by major productive works increased from 4.6 to 10.9 million acres between 1878–9 and 1900–1.[37] New schemes included the Lower Chenab and Lower Jhelum Canals (the Punjab), the Lower Swat Canal (NWFP), the Jamrao and Nira Canals (Bombay), the Son Canals (Bihar), and the Periyar and Kumool–Cuddapah Canals in Madras. In the UP, the period saw the coming on stream in 1874 of the 109-mile-long Agra Canal, which was initially sanctioned to provide famine relief work in 1868 (see Map 2). With its headworks on the Jumna just below Delhi, it took water to the drier parts of Agra and Muttra which lay south of the river.[38] The construction of what came to be called the Lower Ganges Canal (LGC) also commenced in 1868 and it was opened ten years later. This canal, with its independent headworks at Narora in Bulandshahr district, and an intake of 4–5,000 cubic feet per second (cu-secs), intersected the Etawah and Cawnpore branches of the original Ganges Canal and thenceforth these became part of the LGC system. Three new major channels – the Deoband, Hathras and Mat branches – took up for what became known as the Upper Ganges Canal

---

[36] 'Minor works', productive or protective, were constructed out of general revenues.

[37] *IIC*, p. 25. Increases in the irrigated areas of other classes of state irrigation works were small in comparison. Minor works showed an increase of barely 20% over the period and the area covered by protective works was, in total, less than one-twentieth of the *increase* in irrigated area on major productive works. *Ibid.*

[38] Strictly speaking, therefore, this canal lies just outside the Doab, though since the districts served were administratively part of the Doab, the Agra Canal will be similarly treated for the purposes of this study.

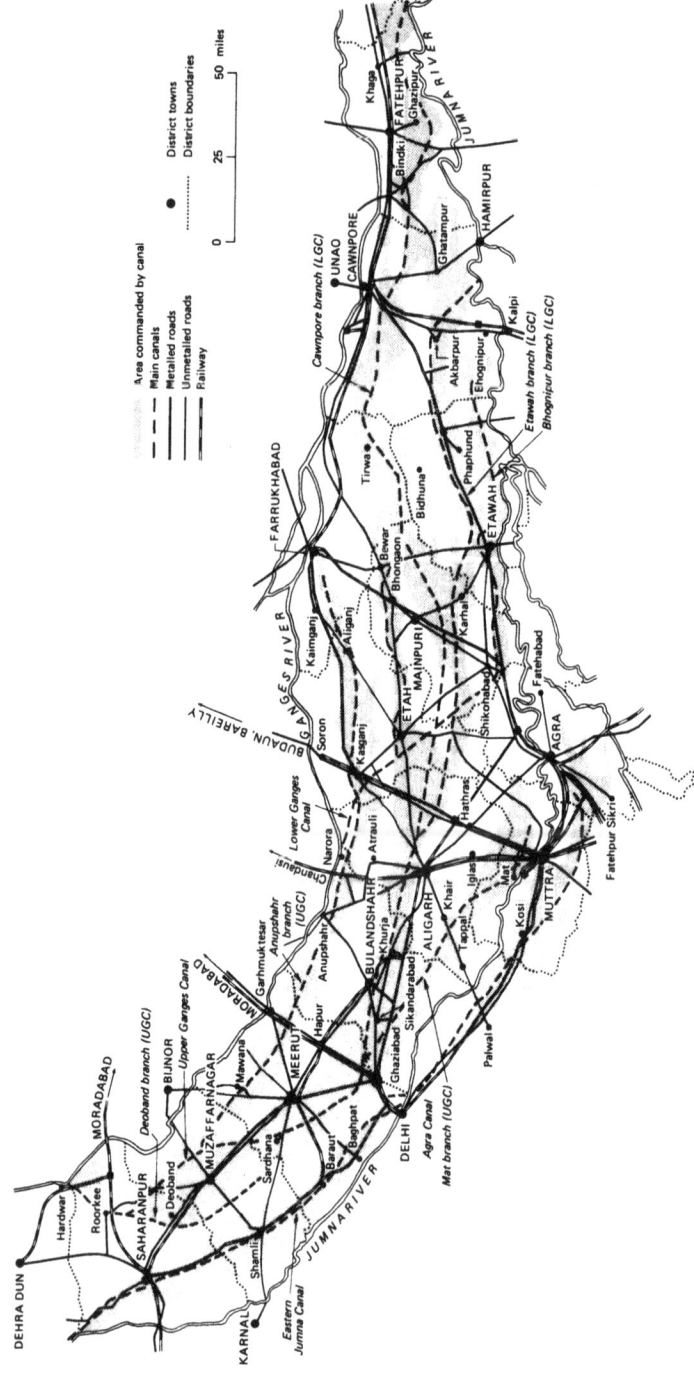

Map 2. Districts of the Ganges–Jumna *doab, c.* 1908.

(UGC) system the water no longer required in the lower reaches.

The Agra Canal and the EJC were not especially large systems in terms of capacity, annually watering on average 240,000 and 300,000 acres respectively around 1900, but the Ganges systems were very substantial indeed. Even in its truncated form the (Upper) Ganges Canal, in the mileage of its channels, was still the largest in India. It had a capacity of 8,000 cu-secs and irrigated during the 1890s 900,000 acres in normal years, rising to 1.25 million acres in the 1899–1900 drought. In comparison, the other giant, the LGC, watered 800,000 and just over one million acres respectively. The two canals together consisted of 1,230 miles of main channels and branches and 6,400 miles of distributaries, at a total cost by 1920 of £8.2 million.[39]

By 1900, when the construction of the basic systems tracing down the Doab had been completed, along with their remodelling and major extensions, 2.5 out of the 11.5 million acres normally cultivated in the tract were irrigated by the canals during dry years, and steady improvements in the operation of the system increased this figure to around 3.5 million acres by 1920, thus clearly exceeding the considerable area of well irrigation.[40] The total cost of these Doab schemes up to 1920 was just under £10 million.[41]

Apart from the earlier works in the Dun Valley, Bijnor, and Rohilkhand north of Bareilly (which together did not typically water more than 150,000 acres and were classified as 'minor' works), UP canal-building outside the Doab was very limited up until the opening of the Sarda Canal, serving the Ganges–Gogra *doab*, in 1928. The Sarda Canal project was initially examined, along with two other projects, the Eastern Ganges and Ramganga Canals, in 1870. Projects prepared for the latter canals were both abandoned because of uncertainty over the likely demand for water and the low winter levels in the rivers concerned. The Ganges schemes in the Doab left no water to spare and the Ramganga flow in winter was tiny.[42] The

---

[39] *Triennial Review of Irrigation*, pp. 64–5, 204–5.

[40] *IIC*, Ch. xix. For an indication of the area under irrigation by source, for each district, see Map 4; for long-term trends in total canal-irrigated area, see Figure 4.

[41] *Triennial Review of Irrigation*, pp. 204–5. [42] *IIC*, pp. 189–90.

Sarda Canal project was less straightforward, and the possibility of constructing a significantly scaled-down version of Captain (later Colonel) J.G. Forbes' 1870 project for a scheme watering 2.4 million acres was repeatedly re-examined over the period. Doubts among local officials about the project's ecological impact, the likelihood of limited returns in an area with a normal average rainfall of forty inches, and the consistently strong opposition of the *talukdars* to a canal held back its acceptance, although the urgent need to provide famine relief work in 1880 and 1896–7 almost led to the commencement of one or other of the schemes.[43]

The Sarda Canal was one of the two schemes recommended by the President of the Famine Commission of 1878–80, Richard Strachey, for immediate and special enquiry. The other involved a system of canals supplied from the Rivers Betwa and Ken in Bundelkhand. The Betwa Canal, sanctioned in 1881, was India's first explicitly 'protective' work and was constructed, initially as a temporary relief employment scheme, at a cost of Rs 42 *lakhs* for the protection of Jalaun. Its expected capacity was 120,000 acres, but it averaged less than 40,000,[44] partly because much of the soil in the area served was retentive of moisture. This, plus a costly storage system to supplement the winter river supplies in a dry year, meant that the system ran at an accounting loss. The government was, in fact, aware that, except in Sind, the Punjab, and Oudh (the Sarda Canal), its construction programme of works likely to prove directly remunerative was approaching completion. Any considerable expansion of irrigation to famine-prone tracts would involve costly storage works and thus be less remunerative than the existing works (which tapped perennial supplies of large rivers). An important task of the 1901–3 Irrigation Commission, then, was an examination not so much of whether the new generation of works would prove directly remunerative, but of 'whether the net financial burden which they may impose on the state in the form of charges for interest and maintenance will be too high a price to pay for the protection against famine they may be relied upon to afford'.[45] The commission's view was that whether or not a canal was

[43] *Ibid.*, pp. 190–5.   [44] *Ibid.*, p. 188.

[45] *Ibid.*, p. 2; see also Kanetkar, 'Pricing of Irrigation Service in India', pp. 163–4.

considered 'unproductive' depended crucially on the definition of the cost of famine, and following its recommendations the 'productivity test' applied to such works was de-emphasised so as to allow for indirect returns, including the saving of famine relief expenditure.[46] Most of its strong recommendations for large works focussed on Madras and the Bombay Deccan. For the UP, the commissioners were generally against the Sarda Canal, suggesting various schemes for using Sarda waters in the Doab system.[47] The Ken Canal, serving Banda district, was strongly recommended, and this relatively small work, held in abeyance since 1876, opened in 1909, followed in 1911 by another small work in Bundelkhand, the Dhasan Canal.

Financially, the irrigation works programme was a clear success. By 1921, £57 million had been spent on 'productive' works (irrigating an average of 17.3 million acres) and £12 million on protective works (irrigating an average of $\frac{3}{4}$ million acres) for India as a whole. Net revenue on this outlay was about 9% and 1% respectively, with a combined net annual profit of £3.1 million.[48] For the UP, the comparative overall figure was 7.5% return on a total capital outlay of Rs 1,328 *lakhs* (see Table 2.1), with a net annual profit on all the works of Rs 58 *lakhs*. Of the four large Doab Canals, which dominated these returns, the EJC was spectacularly successful, annually returning around 25% on total capital outlay,[49] with the UGC returning around 10%. These two canals, with relatively cheap headworks and a high proportion of high-rated crops, generated returns which, by 1901, not only exceeded accumulated interest, but produced a running surplus amounting to almost half the total capital outlay on all four schemes (see Table 2.2).

Aspects of canal engineering in the Doab

Physical conditions and canal construction

Cautley described the northern plains of India as a 'region

[46] *IIC*, Ch. IV.
[47] The commissioners suggested that this might permit the release of a larger supply for the WJC to allow increased irrigation in Hissar and Rohtak.
[48] Gulhati, *Administration and Financing of Irrigation*, p. 63.
[49] Rising to 32% on a capital outlay of Rs 53 *lakhs* in 1921. *Triennial Review of Irrigation*, p. 29.

Table 2.1. *Details of physical and financial development of canal irrigation in the UP, 1831–1921*

Year to 31 March	Mileage of channels					Financial details (in Rs *lakhs*)						Area irrigated (in millions of acres)	Name of canal or branch, and date completed
	Main channels and branches	Distrib-utaries	Drainage cuts	Escape, navigation, etc.	Total	Capital outlay to end of year	Gross revenue during year	Working expenses during year	Net revenue	Simple interest charges	Net profit		
1831	–	–	–	–	–	4	0.1	0.3	0.3	0.2	0.1	–	EJC (Doab) 1830
1841	–	–	–	–	–	7	0.9	0.7	0.3	0.2	0.1	–	Dun and Rohilkhand systems (improved and extended) 1824–1840s
1851	–	–	–	–	–	12	1.8	0.9	0.9	0.4	0.5	–	
1861	–	–	–	–	–	215	7.8	5.2	2.6	8.2	–5.6	0.70	Ganges Canal 1854 Bijnor Canals 1859
1871	784	3,967	–	–	4,751	287	28.3	11.3	17.0	10.9	6.1	1.05	Dun Canals 1864
1881	1,209	5,424	–	–	6,633	634	49.3	19.0	30.3	22.8	7.5	1.73	Agra Canal 1874 LGC 1878
1891	1,400	6,726	1,820	227	10,173	800	67.7	29.2	38.5	29.3	9.2	2.01	Betwa Canal 1886
1901	1,509	7,659	3,311	311	12,790	910	99.7	34.5	65.2	32.7	32.5	2.00	
1911	1,778	9,371	3,605	329	15,083	1,144	105.4	38.3	67.1	34.9	32.2	2.27	Fatehpur branch 1902 Mat branch 1905 Ken Canal 1909
1921	1,888	10,151	3,730	367	16,136	1,328	153.3	54.2	99.1	40.7	58.4	3.40	Dhasan Canal 1911 Hathras branch 1913 Ghagar Canal 1918

Table 2.2. *Financial details of major works in the Doab, 1895–6/1900–1*

Canal	Number of years in operation	Annual average for period 1895–6/1900–1				Excess revenue over interest charges to end of 1900–1 (in Rs *lakhs*)
		Total capital outlay to end of year (in Rs *lakhs*)	Excess annual net revenue over interest charges (in Rs *lakhs*)	Percentage of net revenue on capital	Area irrigated (in acres)	
Upper Ganges	46	298.9	17.1	9.47	949,000	127.6
Lower Ganges	22	353.2	0.9	3.97	774,100	−26.2
Agra	26	95.4	1.8	5.60	228,700	−14.1
Eastern Jumna	70	38.6	8.8	26.55	286,000	220.9
Total		786.1	28.6	7.37	2,237,800	308.2

Source: Based on table in *IIC*, p. 185.

designed by nature as a great field for artificial irrigation'.[50] Outwardly it would appear a comparatively straightforward task to divert water from the perennial rivers flowing out of the Himalayas across the intervening plains, or *doabs*, via canals which could, with the aid of a network of channels, distribute the water for the irrigation of the flat expanse of farmland. However, each of the stages of this exercise involved confronting the Doab's specific geomorphological characteristics and active physical processes. These posed a considerable challenge to hydraulic engineering, particularly since its exponents were so little experienced in dealing with Indian conditions and were additionally hampered by a tight budgetary constraint. A lengthy period of experimentation and adaptation was necessary before a system was devised which could provide what, in the context of local agricultural conditions, was a useful and workable irrigation facility and one which was not attended by serious environmental disruption.

Understanding the physical characteristics and problems of works of hydraulic engineering in the Doab therefore involves taking account of the processes and form of the tract's geomorphology. In broad terms, the flat alluvial Doab landscape was formed through the medium of the major rivers from debris washed down from the mountains. Near the foot of the hills, the soil contains boulders and stones brought down by the mountain torrents, but thirty miles from the hills these disappear and the soil is a fine river deposit which varies from sand to stiff clay, the different types of which are distributed in patches over the surface and at different levels in the subsoil. This is an ongoing process: heavy rainfall causes floods as the rivers break out of the tortuous channels which confine their flow for most of the year and fill the broad flood valleys. Where the slope declines, the *khadir*, or valley, widens and the velocity of flow is reduced, and the water deposits some of its suspended material: first the coarser sand particles, and then – in the backwaters and depressions – the lighter particles which make up clay. As the floods subside, the river begins to carve a new dry-season channel (or channels) through the soft deposited

---

[50] Cautley, *Ganges Canal Works*, I, p. 3.

soil, and the less swampy parts of the broader valley become available for winter cultivation. In this way the geomorphological detail, if not the general pattern of the landscape, is continually subject to change. The pattern is one of generally light soil mixed with discontinuous clay beds found at various levels, which are often important in well irrigation (in that they check subsoil drainage flows) but which may interfere with drainage sufficiently to render the soil uncultivable, particularly where salts are present in the soil which can in their dissolved state be brought up to the root zone or surface through capillary attraction. This problem is especially likely to be encountered in the central and lower sections of the Doab, where the general gradient of the land slopes off to six to twelve inches per mile and long shallow depressions occur, along which the receding floodwaters wind with almost imperceptible slowness. Numerous beds of an alluvial formation of limestone, called *kunkar*, one or two feet thick in places, are also a cause of barren stretches scattered over the Doab.

The builders of canals needed, therefore, to devise economical ways of tapping the great rivers, with their tendency to wander and their very variable flow over the year, and to convey that supply to and along the watersheds of the tract via a network of canals and distributaries, taking care to avoid interruption to the complex and often subtle drainage lines. They had also to avoid the danger of damage to the works posed by the boulders carried by the flow of the rivers emerging from the hills, the serious scouring effect of water travelling at excess velocity in unlined alluvial soil channels, and the problem of the channels filling with silt where the flow speed was inadequate to keep the material suspended in the water. These were the fundamental engineering problems which had to be solved before an effective modern system of irrigation could be established. Moreover, they were essentially new problems. The experience of other (smaller) works in existence in Europe was not readily transferable to the northern Indian context, and the development of the basic irrigation infrastructure was essentially based on trial and error.

Nowhere is this more in evidence than in the construction of canal headworks, and particularly the Ganges Canal head at Hardwar. The advantage of choosing this site for the headworks was that it took off at a high point (where the river emerged from the hills), and thus avoided the problems of constructing a dam on the sandy bed of the river (the bed was still rocky at Hardwar) and of 'lifting', via embankments, the canal out of the *khadir*, several miles wide, through which the Ganges River flowed at a level lower than the Doab plains. The river itself had carved a main channel along the left bank of the valley, some distance from the high bank and the arable upland plain (*bangar*) of the Doab. Moreover, in the upper sections of the valley it was separated from the *bangar* by an old bed of the Ganges, the Budhganga, which was by then a long depression filled with reeds and swamps and inhabited by wild pigs and tigers (see Map 3).

One problem avoided, however, did not mean that there were not grave engineering difficulties still to be solved. The river at (and for some miles below) Hardwar has the character of a torrent, and is studded with islands and sandbanks, between which the stream runs in a series of pools and rapids. A system evolved which used an elaborate series of temporary and permanent works to channel the supply – and in particular the December–January supply, which could be as low as 4–5,000 cu-secs – so that it could be diverted into the canal. It was an intricate operation which involved directing the whole of the water flow past the sacred Hardwar *ghats* during the low season, but keeping the main channel of the river away from the head of the canal during floods. The system's main features are shown in Map 3A: the Chillawala weir, the Hardwar dam, and a system of three temporary bunds and numerous groynes, spurs, and revetments reaching five miles upstream were used to control the river and to channel the water past Hardwar. The Myapur dam (see Plate 2) served to regulate the depth of water immediately above the canal, and the lift gates of the Myapur regulator[51] controlled the quantity of water entering the canal.[52] When the river was in flood, or no water was required for the canal, the Hardwar dam could be used as an escape; the

---

[51] Regulator 1 on Map 3A. Construction of regulator number 2 commenced in the early 1880s.

[52] For more details, see Buckley, *Irrigation Works*, pp. 132–6; and Deakin, *Irrigated India*, pp. 179–85.

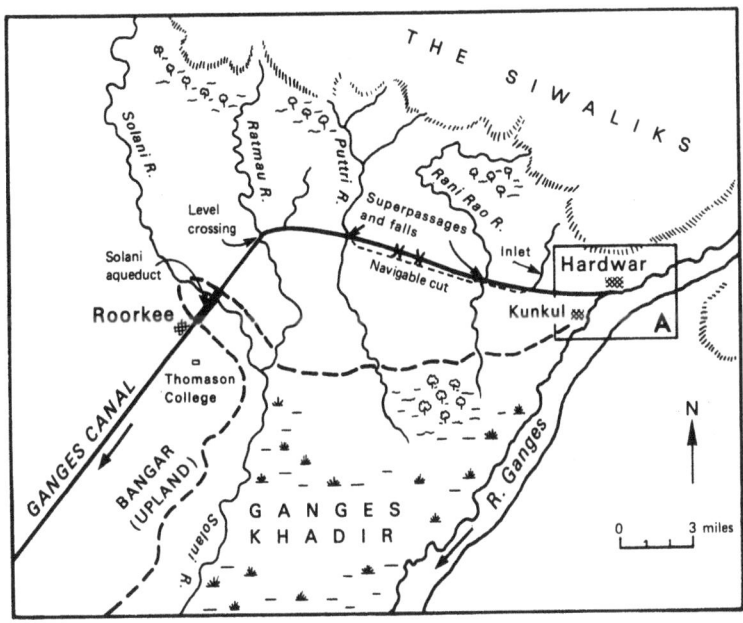

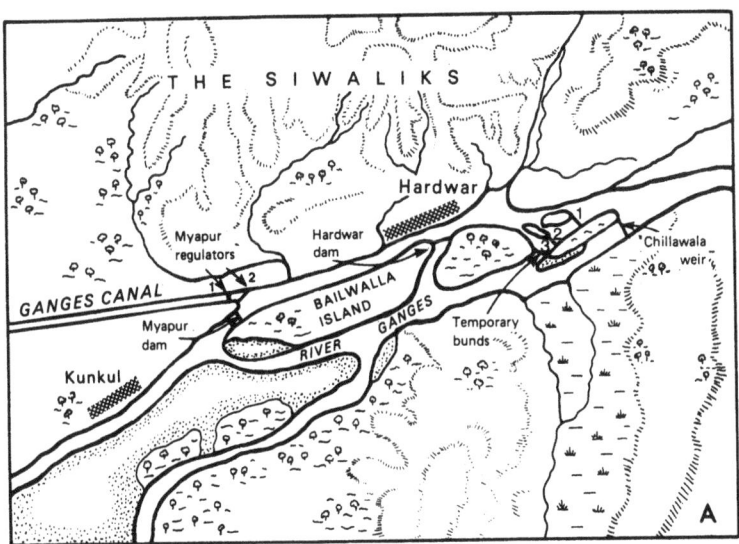

Map 3. Upper and headworks sections of the Ganges Canal, *c.* 1880.

Plate 2. Myapur regulator and dam, Upper Ganges Canal. (R.B. Buckley)

May–June floods usually washed away the temporary bunds in any case.

Constructing a temporary dam with bunds was used on both the UGC and the EJC as an inexpensive means of diverting the low-season river supply into the canal. Work on the construction of new bunds commenced each year at the close of the rains. Bund number 1 had to be laid in fast-running water ten to twenty feet deep.[53] To cope with the current, a fourteen-inch-diameter coir hawser was, with the aid of literally hundreds of men, stretched across the river and propped up on piles. It was possible, then, with the use of pulleys, to work boats in the fast current. These boats with the aid of derricks, transported from the banks large wooden cribs, part-filled with boulders. One by one, these crates were sunk in a line across the line of flow, lashed to the next crate, and then filled to the top with boulders loaded from the boats (see Plate 3). To make the bund waterproof, grass mats (*tattis*) were placed on the upstream side and kept in place with a layer of shingle. The second and third bunds, which were less elaborate (the third one was little more than a shingle bank, though the second was often made of

[53] The slope at this point being around ten feet per mile.

cribs), had the purpose of holding back the water which leaked through the bund(s) ahead.

This whole annual operation, involving an outlay of Rs 35,000, was supervised by an executive engineer, who, because ordinary boats were prone to be damaged when forced into boulders by the current, used a somewhat bizarre craft called a *surnai*, consisting of two inflated buffalo skins, their legs pointing upwards and tied up with string, supporting a *charpoy* on which one or two officers could sit. The craft was piloted by two *surnaiwallas* swimming behind, whose duty it was also to keep the skin inflated by occasionally untying one of the legs and blowing into it.[54]

For the most part, this mixture of permanent and temporary works was an effective control over the headwater supply to the canal. With the aid of level readings from gauges placed several miles up river – conveyed by men swimming down to the Myapur station on *tumris* (elongated gourds forming a crude lifebelt placed under the chest) – the level of water in the canal could be regulated to within one-half of an inch.[55] Problems did

Plate 3. Temporary bunds, Ganges Canal headworks. (R.B. Buckley)

[54] L.R. Nicolls, 'Agricultural Engineering in India (VI)', *Engineering*, XLVI (6 July 1888), p. 16. Under this general title Nicolls wrote a series of twelve articles covering many aspects of canal engineering and administration which appeared in vols. XLV and XLVI of *Engineering*.   [55] *Ibid.*

occur from time to time, however. Occasionally a late-season flood would occur in September and wash away the bunds under construction, and even endanger the canal itself, since it was difficult to work the gates on the Myapur regulator against the full force of a flood. On one occasion considerable damage to the canal was only avoided by chance, when some barges and trees lodged in the regulator entrance and restricted the flow into the main canal.[56] This kind of problem, and the tendency of the pocket just above the regulator to become filled with silt and debris, thus obstructing the flow into the canal, led to adjustments to the headworks in the 1880s, including the construction of a new regulator (number 2 on Map 3A). When the bunds under construction were washed away by September floods, it took an additional two months to complete the task, during which period only a limited supply was available in the canal at a time when rainfall was often inadequate to maintain the crops. But it was not until 1922, after a failure of the bunds coincided with a prolonged break in the rains causing a 'preventable loss' in crops and revenue exceeding the projected cost of new headworks, that the bunds were replaced by a permanent headworks facility.[57]

The decision to site the head at Hardwar also involved considerable engineering difficulties over the first twenty miles to Roorkee. The difficult terrain and the problem of coping with cross-drainage, which spilled in torrents from the catchment basins in the hills during the wet season, were the reasons why Cautley – who had encountered such problems, though on a much smaller scale, in the upper sections of the Doab Canal – was initially discouraged when he made a preliminary survey in 1836. Deep excavations were required, up to thirty feet in places, since most of this section followed an east–west line round the bottom of the hills. Due to the slope, four of the fourteen falls were needed in the first ten miles to keep the bed near to the level of the ground and at a level which enabled the works to accommodate the series of torrents which lay between the head and the Doab watershed (see Map 3). The first of

[56] *Ibid.*, p. 15.

[57] Government of the United Provinces of Agra and Oudh, Public Works Department (Irrigation Branch) Proceedings (hereafter UP PWD(I)), June 1922, (proceeding no.) 6. (In the IOR.)

these torrents, the 100-foot-wide Ranipur *rao*, occurred just five miles from the head, and was dealt with by means of a superpassage, 200 feet wide, fourteen feet deep, and supported over its length of 450 feet by eight arches of 25-foot span, which carried the seasonal torrent across the canal (see Plate 4). A similar, though wider (300 feet), superpassage was constructed at mile 9 to convey the Puttri torrent over the canal. A very extensive system of training works, together with an excavated channel, conveyed the torrent (usually three to four feet of water, with waves of similar height, and accompanied by a continuous flow of two to three feet of sand and debris) over the canal and away by means of a cleared channel, to be subsequently lost in the swamps of the Ganges *khadir*. Cautley dealt with the (perennial) Ratmau River, which intersected the canal at mile 13, by means of a 'level crossing'. A mile of revetment walls, plus a forest of pines and extensive cribwork, protected the canal as it permitted the Ratmau water to enter the canal from the side over a dam, and to pass out immediately opposite through a weir with a system of sluices which could be regulated by the forty workmen on guard according to the flow of the Ratmau. A regulating bridge with ten twenty-foot

Plate 4. Ranipur superpassage, Ganges Canal. (R.B. Buckley)

openings on the downstream side of the crossing prevented an undue supply rushing down the canal.[58]

The largest seasonal torrent, the Solani River, in a valley approaching three miles in width, still barred the way to the upland plain, and an enormous aqueduct was constructed by Cautley in the seven years following 1845. This included over 2¼ miles of embanked, earthen-brick-lined channel which linked the masonry section to the levels on either side of the valley. The Solani aqueduct proper was 1,170 feet long and 172 feet in width. Fifteen arches of fifty-foot span carried the 6,500 cu-secs at a height of 24 feet above the river (see Plate 5). It was the largest aqueduct in India, and indeed the world, until the completion of the Nadrai aqueduct, which conveyed the LGC across the Kali Nadi.[59] The cost of the aqueduct was not far short of a quarter of the total cost of the canal at the time of its opening.[60] Indeed the aqueduct itself, embellished with huge pairs of sculptured lions and marking the point at which the canal reached the irrigable plan, was chosen as the location for

Plate 5. Solani aqueduct, Ganges Canal. (R.B. Buckley)

[58] Detail for this section from Nicolls, 'Agricultural Engineering (IX)', *Engineering*, XLVI (7 Sept. 1888), p. 224; and Deakin, *Irrigated India*, pp. 183–4.

[59] The Nadrai aqueduct was 1,310 feet in length, carrying the canal over the river at a height of 88 feet. [60] Cautley, *Ganges Canal Works*, III, p. 193.

the splendid opening ceremony, which was performed by NWP Lieutenant-Governor John Colvin in the early morning of Saturday 8 April 1854.[61]

Cautley left India the following month, his departure being marked by the rare honour of a special thirteen-gun salute fired from Fort William on the orders of the Governor-General. It was not until some time after Cautley had left – when the distribution system was more developed and the full supply was admitted into the canal following the completion of repairs to sections damaged during the Mutiny – that the principal design defect, that of excessive slope, became apparent. Cautley had been concerned to fix upon a slope which would give the water sufficient velocity to reach all parts of the system, but which would not cause damage to the banks and the bed of the channel. Using experience gained on the Doab and Delhi Canals, Cautley calculated the intake required at Hardwar (6,750 cu-secs) to carry out the projected irrigation along the 800 miles of main and branch lines, the cross-sections of which progressively diminished from a width of 140 feet on a depth of ten feet at the head down to twenty by four feet towards the termination of the Etawah and Cawnpore branches.[62] The slope chosen, eighteen inches per mile in the upper sections falling away to fourteen inches towards the tail, was similar to that of the Doab Canal. On discovering the sandiness of the subsoil, Cautley reduced the slope by three inches per mile in 1850, diminishing the mean and bed velocities, calculated from the formula of the French hydraulician Dubuat (1734–1809), to between 4.04 and 3.53 feet per second.[63] The problem was that little was known at the time about the flow of water in large earthen channels, and that what was an appropriate slope on the Jumna Canals was an excessive slope on a much larger one. Dubuat's formula for open flow channels was proved to be quite unreliable, particularly for large works.[64] As Joyce Brown

[61] For a description of the ceremony, which was attended by an estimated audience of 50,000, see J. Brown, 'Memoir of Colonel Sir Proby Cautley, F.R.S., 1802–71, Engineer and Palaeontologist', *Notes and Records of the Royal Society of London*, xxxiv (1979–80), pp. 207–8.

[62] More detail on this is included in J. Brown, 'Sir Proby Cautley (1802–71), A Pioneer of Indian Irrigation', *History of Technology*, iii (1978), pp. 65–7.

[63] Cautley, *Ganges Canal Works*, iii, pp. 142–6.

[64] Buckley, *Irrigation Works*, p. 217.

points out, nothing very reliable or comprehensive was in fact to be available until the work of Ganguillet and Kutter (1869) and Manning (1890). She concludes: 'In reality, Cautley, working as he was in the 1850s on large channels carrying high flows through sand, was better placed to add data to the search for empirical design formulae than the existing formulae were in a position to guide the design of so singular a project as the Ganges Canal.'[65]

The excessive slope given to the bed of the main channel was amplified in its effects by the design of the falls. This was based on the 'ogee' curve falls developed by Cautley on the Doab Canal, with the purpose of delivering the water gently to the lower level, thus reducing erosion. In practice, with such a large body of water, this system increased further the velocity of flow as the water ran down the ogee, causing serious damage to the flooring, scouring the bed, and setting up a wave which considerably damaged the banks.[66] This erosion, in turn, caused serious silting problems on the first few furlongs of each distributary (the bed-slopes of which were often only four inches per mile), which reduced the effective supply and led to the creation of substantial hillocks of silt, cleared from distributaries alongside the distributary heads.[67] It also raised the level of the beds of the terminal branches quite considerably – by four feet in the case of the Etawah branch – with the mixed effects of almost obscuring the bridge arches altogether while at the same time giving a better command over the country for irrigation purposes.[68]

It soon became patently clear that the design faults were sufficiently serious to endanger the canal itself should it run for a number of years with a full intake at Hardwar. The intake was thus reduced and considerable debate ensued as to the

---

[65] Brown, 'Sir Proby Cautley', p. 69.

[66] The actual mean velocity was found in 1864 to be 4.7 feet per second at the head and 4.0 at Nanoon (the point of bifurcation) even with only two-thirds of the full supply in the canal. J. Crofton, *Report on the Ganges Canal, with Estimates and Plans*, vol. 1 (Calcutta, 1865), p. 19. (Report dated 23 Nov. 1864.)

[67] The problem was made worse by the early practice of installing distributary gates which lifted up to allow (silt-laden) water from the bed of the canal to flow into the distributaries. Later practice involved the use of double sluice gates, the lower of which were generally left down. Nicolls, 'Agricultural Engineering (XII)', *Engineering*, XLVI (16 Nov. 1888), p. 472.   [68] *Ibid.*

appropriate course of action (see below). In the event, the main errors were corrected during the late 1860s, at a cost, including extensions, of Rs 55 *lakhs*. The remodelling consisted of adding protective works to the fourteen masonry falls. Their crests were raised by masonry weirs to reduce the slope of the canal, and the ogee curve was in some cases converted to a perpendicular fall on a design which had been proved successful on the Bari Doab Canal.[69] The reduced velocities allowed the formation in the channel of a layer of fine sand, which reduced percolation through the bed and sides, saving water and reducing waterlogging damage. Indeed the LGC – the slope of which was no more than six inches per mile – allowed specially in its design for a system of double banks incorporating silt traps, which served significantly to reduce percolation losses.[70]

Public dispute between Cautley and Cotton

In the winter of 1862–3, the Madras Irrigation Company – which then seemed, according to *The Times*, 'ambitious of monopolising every irrigation project in Bengal as well as Madras' – sent Cotton to reconnoitre the Ganges Valley.[71] It was part of his brief to report on the likely cost of correcting the Ganges Canal's operational defects with a view to purchasing it from the government, and he travelled over from Bihar to visit the works. Finding fault with such a large and path-breaking scheme was not at all difficult. On the strength of the report Cotton sent home, the East India Irrigation Company made an offer to the Indian Government to buy the Ganges Canal, forwarding with its offer a copy of Cotton's report.[72] Cotton's

---

[69] On the construction of which Crofton, the officer principally responsible for the remodelling of the Ganges Canal, had been involved during 1850–9. For details of adjustments to the falls, see Buckley, *Irrigation Works*, pp. 217–18.

[70] Nicolls, 'Agricultural Engineering (XII)', *Engineering*, XLVI (16 Nov. 1888), pp. 471–3.   [71] *The Times*, 2 Nov. 1864, p. 4.

[72] *Ibid.* Cotton's 'Private Memorandum upon the Ganges Canal' accompanied a communication from the East India Irrigation Company to the Government of India in July 1863. Crofton, *Report on the Ganges Canal*, I, p. 75. The 'Private Memorandum' is printed as an Appendix in P.T. Cautley, *A Reply to Statements made by Major-General Sir Arthur Cotton, on the projection of the Ganges Canal Works* (London, 1864).

'private Memorandum upon the Ganges Canal' found its way into print through the interest of the local authorities, but Cotton had already made the issue public when he outlined his findings in an address to the Calcutta Chamber of Commerce on 7 May 1863. Cautley felt constrained to reply to the charges – both on his own behalf and on behalf of the government and his engineer colleagues – and had a pamphlet printed for private circulation,[73] which was but the first of several brought out in the mid-1860s during the course of this famous and public professional controversy between two of India's most distinguished engineers.[74]

Cotton argued that there were five 'fundamental' and fourteen 'minor' mistakes in the projection of the canal. In summary, he considered that the former mistakes included the fact that the head of the canal was sited too high up the river, with the attendant problems of slope and the need to cross drainage from the Himalayan foothills; that unnecessary excavation had been incurred because the canal was carried below the surface of the country; that the brick used in masonry works was less suitable than the locally procurable stone; that the whole of the water supply was admitted at the head and thus some was conveyed 350 miles before being used rather than being obtained from a point closer at hand; and that there was no permanent and thus reliable headworks system.[75] These mistakes, and particularly the decision on the siting of the headworks, Cotton argued, had resulted in the canal's costing three times more than was necessary. Cotton's preference was for a weir to be sited nearly 100 miles below Hardwar at Sookertal, near the confluence of the Solani and Ganges Rivers, thus shortening the main channel and saving the expense of the Solani aqueduct and the other works in the first twenty miles.

Somewhat ironically, the excessive slope given to the canal was included by Cotton in his list of 'minor mistakes', most of which were concerned with the efficiency of the canal from the point of view of navigation. He criticised the lack of connecting navigating lines between branches, the decision to terminate the canal at Cawnpore rather than Allahabad, and the

---

[73] Cautley, *Reply to Statements*.
[74] A more detailed account of the argument is given in Brown, 'Sir Proby Cautley', pp. 77–81.   [75] Cotton, 'Private Memorandum', pp. 44–5.

Plate 6. Colonel Sir Proby Thomas Cautley. (By kind permission of Roorkee University)

Plate 7. Major General Sir Arthur Cotton (from Lady E.R. Hope, *General Sir Arthur Cotton: his life and works.* Hodder & Stoughton, London, 1900)

numerous inconveniences to boat traffic, including the limitations on boat size of low bridges, sharp curves in lock channels, and the breadth of the canal at the lower end.

In his *Reply to Statements*, Cautley dealt carefully with each of the criticisms in turn, defending himself on all the points except those relating to the slope.[76] He calmly pointed out that the

[76] Cautley, *Reply to Statements*, pp. 17–18.

depth and extent of excavation could be justified on the grounds that much of the soil was porous and thus unsuitable for embankments; that although there was indeed suitable stone in the Siwaliks which could have been quarried, it would have needed careful sorting and would not in the end have been cheaper than bricks; that the headworks were temporary rather than permanent for reasons of economy rather than design, and it had long been his view that a permanent intake facility might be necessary; and that to construct a dam further down the river – to function either as headworks or as a supplementary supply to save admitting the full supply at the existing head – would involve great practical difficulties. 'It will be found more easy to propose weirs and dams on the sandy tracts of the Ganges and Jumna than to execute them', Cautley concluded,[77] pointing out the differences in conditions between the northern rivers ('heavy slopes with large masses of water pouring down with overwhelming violence') and those on which Cotton had constructed works in Madras ('larger bodies of water, but on very much smaller slopes in connection with a true delta').[78]

Cotton was not the sort of man to be denied the last word. Cautley's hope that his *Reply to Statements* would put the matter at an end was almost immediately destroyed by Cotton's 'Observations . . . on the foregoing Reply',[79] and he was drawn into the series of exchanges with Cotton.[80] The latter continued to insist that the works in Madras were built on sand, and that the only difference between the rivers was that the Madras rivers were larger. As was his manner, Cotton interpreted Cautley's counterarguments as merely the product of 'inexperience'. Cautley, for his part, continued to insist that the character of the northern rivers differed significantly from those of the deltas, pointing in support of his argument to the fate of

[77] *Ibid.*, p. 17.   [78] *Ibid.*, p. 18.

[79] In A. Cotton and P.T. Cautley, *A Discussion, regarding the projection and present state of the Ganges Canal, and the measures required to make it reliably useful and profitable* (London, 1864).

[80] P.T. Cautley, *A Disquisition on the heads of the Ganges and Jumna Canals, North-Western Provinces, in reply to strictures by Major-General Sir Arthur Cotton* (London, 1864); A. Cotton, *Reply to Colonel Sir Proby Cautley's Disquisition on the Ganges Canal* (London, 1864); P.T. Cautley, *A Valedictory Note to Major-General Sir Arthur Cotton, respecting the Ganges Canal, with a postscript touching certain misrepresentations of a writer in the 'Times' on the same subject* (London, 1864); A. Cotton, *Reply to Sir Proby Cautley's Valedictory Note on the Ganges Canal* (London, 1865).

the attempt made in 1828 to establish a new head on the WJC, at a point lower down near Karnal: the works had been promptly swept away by floods.[81] In addition, he noted that Cotton's proposed Sookertal dam could be up to 1½ miles long and very costly.[82]

The respective arguments of Cautley and Cotton were, in fact, the subject of official scrutiny during 1864. In February of that year, a committee of PWD engineers convened to recommend remedial measures for the Ganges Canal invited one of its members, Captain (later General) James Crofton,[83] the Officiating Superintending Engineer of the EJC, to submit plans and estimates for remodelling the canal. Crofton's report, which was completed by November of the same year, and contained suggested measures for correcting the slope and protecting the masonry works, also considered Cotton's criticisms. The conclusions in the report were not dissimilar to Cautley's. On the central issue of the site of the headworks, Crofton argued that the alternative suggested by Cotton not only involved the foundation of a weir upon sand, but failed to avoid the need for extensive excavation; he estimated that a channel averaging over fifty feet in depth for 57 miles would be required, and that the total cost would be at least three times that expended on the first 23 miles of the existing work.[84]

Incensed at having his advice rejected, Cotton was deflected from the increasingly pointless and often unpleasant exchanges with his reluctant adversary, Cautley, and attacked Crofton and the Government of India instead. He assailed Crofton's report through the letters page of *The Times*,[85] and in a characteristically long and abusive letter to the Government PWD Secretary insisted that Crofton – many years his junior – was entirely out of his depth; never having even 'seen a weir on one of our great successful works in Madras', he was 'certainly not a man who could be safely entrusted with the reformation of a work of such magnitude'.[86]

The Government of India, for its part, considered reasonable

[81] Cautley, *Disquisition*, pp. 55–6.   [82] *Ibid.*, pp. 80–1.
[83] See career outline, Appendix 1 below.
[84] Crofton, *Report on the Ganges Canal*, 1, p. 78.
[85] *The Times*, 26 Oct. 1865 and 11 Nov. 1865.
[86] NWP PWD(I), April 1867, 50, Cotton to Sec. GI PWD, 9 Nov. 1865.

Crofton's view that there was no *prima facie* case for abandoning the upper 100 miles of the canal in favour of the Sookertal proposal. However, in the light of the publicity surrounding Crofton's report, and taking account of Cotton's distinguished record as an engineer, a committee of 'impartial' engineers was appointed to decide between the two proposals. It found that the construction of a weir at Sookertal, at an estimated cost of Rs 112 *lakhs*, could not be recommended, and it urged that Crofton's remodelling plans should proceed with only minor modifications.[87] The committee agreed that when full intake into the canal was resumed, a permanent weir system might be necessary to safeguard the cold-season supply. Giving some ground to Cotton, committee members suggested that an additional canal head might well be constructed at Rajghat (near the eventual headworks of the LGC at Narora) when it was economically feasible, and that the possibility of building a weir across the Jumna at a point below Aligarh should be further investigated. The report ended with a comparison of the financial results of the Madras and Ganges Canal works. Aside from the contribution of problems of slope, Mutiny damage, and limited development of the distribution system to the poor financial performance of the Ganges Canal, it was clear that differences in conditions between the two areas restricted the financial potential of the Ganges *vis-à-vis* the delta works. Differences in staple crops, the availability of wasteland, the pattern of demand for irrigation, and the expense involved in bringing the water to the fields – together with the availability of still-water navigation in Madras – accounted for a marked difference in expected returns.[88]

Cotton was angered by the committee's report, and in a lengthy communication to the Under-Secretary of State he launched a tirade against it, saying that the 'whole bias of the writers was to make their paper reflect the views taken of the matter by the Government of India'.[89] He alleged that his views had still not been given a fair hearing – 'the president was

[87] A. Swinton, *Report of the Committee on the Ganges Canal* (Roorkee, 1867); also printed as an Appendix to NWP PWD(I), April 1867. (For a summary, see NWP PWD(I), April 1867, 51.)

[88] Swinton, *Report of the Committee on the Ganges Canal*, pp. 47–8.

[89] NWP PWD(I), Sept. 1867, 1, Cotton to Under-Sec. of State for India, 15 Jan. 1867.

altogether opposed, both professionally and personally, to myself, and only one of the six could be an unbiased person' – and concluded that it was the intention of the committee to 'smooth over' the 'sad failure of the Ganges Canal'.

Following another lengthy letter on the subject, and on the general need to expand development of irrigation and navigation, directed to the Government of India in November 1867,[90] the latter concluded that Cotton was trying to 'enlist public opinion against our views and acts'. The government wearily pointed out that Cotton's views *had* been considered by the committee, restated the official commitment to extending canal works, and made it clear that it considered the matter closed. Cotton continued to press the case for more canals, both for irrigation and – since they were cheaper than railways – for transport. His cause was boosted by the support of John Bright, who was moved by the 1877–8 famine to campaign for canals, and Cotton became a key witness before the Select Committee on East Indian Public Works, pushing his case to the extent of proposing the summary and indefinite suspension of nearly all railway schemes and works in favour of navigation canals.[91] His views found little support, and even some hostility, among members of the Select Committee. The somewhat eccentric Cotton, whose earlier reputation had been diminished by involvement with the irrigation companies and the unsavoury clash with Cautley and government officials, faced a committee which, in irrigation matters, was concerned with narrow financial criteria. He could, by then, easily be dismissed as a crank with 'water on his brain'.[92]

How does this whole Cautley–Cotton episode relate to the development of irrigation in the NWPs? It would be easy to represent this as a clash between two personalities, with Cautley being reluctantly drawn into confrontation, partly through a sense of duty with respect to the government and his engineer colleagues, to defend himself against the implied charge of misuse of public funds. It is understandable that Cotton should have resented the emergence of a rival to his

---

[90] NWP PWD(I), Aug. 1868, 27, Cotton to Sec. GI PWD, 12 Nov. 1867.
[91] Reply to qn. 2722.
[92] R.C. Dutt, *The Economic History of India in the Victorian Age*, 3rd edn (London, 1908), p. 364.

claim as the pre-eminent engineer in the entire history of the subcontinent, particularly one whose reputation was based on a work which, however impressive in its scale and achievement, was patently flawed in design and unprofitable besides. With the air of a man supremely confident in his own abilities, the somewhat dogmatic Cotton campaigned tenaciously for the acceptance of his views. His personalised attacks on Cautley and on other individuals and bodies with whom he clashed naturally made him enemies, and inevitably had the effect of causing a closing of ranks among the north Indian PWD officers, with whom Cautley was popular and as an engineer very highly regarded (as Crofton's reference to 'our loved and honoured chief' in his *Report on the Ganges Canal* shows).[93] Joyce Brown's interesting account of the ceremony and celebrations to mark the opening of the canal includes a quotation from a witness, who referred to his 'standing modestly on the scene of his glories, and hurrying with the simplicity of a child from any egotistical reference that the occasion might seem to force on him, to the generous and therefore welcome office of praise and thanks to others'.[94] This clash, then, inevitably spread beyond the individuals themselves, and embraced departmental provincial loyalties. Even today canal engineers in the UP not only know the details of the dispute, but exhibit some resentment against Cotton on behalf of the person they regard as the father of canal irrigation in the north.

However, there was more to the conflict even than that. This confrontation took place against a background in which the private irrigation companies were increasing their involvement in irrigation development in India. It was in the interests of the East India Irrigation Company to use Cotton's 'Private Memorandum' in order, as Cautley pointed out, 'to depreciate, in the eyes of Government, the property for which they were bidding'.[95] Moreover, they were unintentionally aided in this by Colonel Richard Baird Smith's report on the 1860–1 famine, which showed that the canal had proved inadequate to prevent starvation, on account of the inefficiency of the temporary

---

[93] Crofton, *Report on the Ganges Canal*, I, p. 84.
[94] Brown, 'Memoir of Colonel Sir Proby Cautley', p. 208, quoting *Delhi Gazette, Supplement*, 12 April 1854.
[95] Cautley, *Disquisition*, pp. 98–9.

headworks in keeping up supplies when the river ran low.[96] Although this was nothing more than an attempt by the former Director of the Ganges Canal to encourage the government to provide resources for permanent headworks – the Ganges Canal Committee actually pointed out that the inefficiency of the temporary system in 1860–1 was exaggerated[97] – it was grist to the mill as far as Cotton and the private irrigation companies were concerned. The broader dimension to the episode is also suggested by Cautley's belief that the articles in *The Times* supporting his adversary's views were actually written, with Cotton's assistance, by R. A. Dickinson, Secretary of the Indian Reform Association.[98]

The tide of opinion which favoured private enterprise in irrigation coincided with the period when Cotton was at the pinnacle of his career. He was the man with the magic touch; no scheme, it appeared, was too ambitious given his powers. But the tide began to ebb around the mid-1860s, as the private company schemes degenerated into fiascos and Cotton's intemperate outbursts turned opinion against him. After his knighthood in 1861 and his 2nd Class KCSI in 1866, no further public acknowledgement of his services was forthcoming, though Cotton lived until 1899. Cautley, by contrast, remained high in the government's esteem, serving for ten years on the Council of India. In March 1865, the Governor-General wrote a despatch to the Secretary of State for India, stating:

Whatever be the present ascertained defects of the Ganges Canal, the claims of Sir Proby Cautley to the consideration of the Government of India for his eminent services are, in our estimation, in no way diminished, and his title to honour as an engineer still remains of the highest order.[99]

It was unfortunate that Cotton's energies in the 1860s and beyond were taken up in activities which antagonised so many involved in irrigation, since his undoubted talents as an engineer would have been valuable during the period of

---

[96] R. Baird Smith, *The Special Commissioner's Final Report on the Famine of 1860–61* (Calcutta, 1861), pp. 32–6.

[97] Swinton, *Report of the Committee on the Ganges Canal*, p. 36.

[98] See Brown, 'Sir Proby Cautley', p. 81; and Brown, 'Memoir of Colonel Sir Proby Cautley', p. 211.

[99] Quoted in *Dictionary of National Biography* (hereafter *DNB*), vol. III, p. 1243.

irrigation expansion. Certainly, from an engineering viewpoint, Cotton's views on the feasibility of dams across the sand beds of the Ganges River eventually were shown to be correct in principle by the success of the LGC and Agra Canal weirs at Narora and Okhla respectively, which were constructed 'generally in the manner proposed by Sir Arthur Cotton'. Richard Strachey, who sanctioned these works as Inspector-General of Irrigation, admitted in 1868 that the change of opinion on the matter of a weir at Narora was due to Cotton's involvement, although the bitter controversy had had the effect of delaying his acceptance of an idea which was considered 'visionary' when it was first suggested.[100] This did not mean that Cotton's initial advocacy of a lower weir was any more practical in the sense that the canal was already built, and Cotton's suggestion of a supplementary canal for additional supply to the Ganges Canal was patently a more expensive option than that of simply remodelling.[101] It was only subsequently, with the acceptance of the idea of extending irrigation wherever a reasonable return was in prospect – and this included virtually the entire Doab – that Cotton's views became 'most relevant'.[102]

Cotton's weakness was pronounced in respect of the non-engineering aspects of irrigation schemes, particularly in assessing aspects of demand for the water and navigation facilities and in estimating costs. This was amply demonstrated in relation to the private company schemes in which he was involved, where the scale of error in demand projection and cost estimation suggests that the profitability of his earlier

---

[100] NWP PWD(I), Sept. 1869, 61, Note by Inspector-General of Irrigation on Remodelling of the Ganges Canal and Extension of Irrigation in NWP Doab, 22 Oct. 1868. To be fair to Cautley, his main objection to Cotton's suggestion for lower headworks was the likely cost. In describing the weir plans as 'visionary' he elaborated thus: 'My meaning must be taken in the economic sense. With unlimited outlay a dam might be built across the Ganges sufficient for any purpose.' *The Times*, 2 Nov. 1864, p. 4.

[101] Cotton insisted that he did not intend to reach the upland via a deep cutting (as Crofton assumed, and on which his calculations were based), but to take the canal along the valley, rising gradually out onto the upland level. NWP PWD(I), April 1867, 50, Cotton to Sec. GI PWD, 9 Nov. 1865.

[102] NWP PWD(I), Sept. 1869, 61, Note by Inspector-General of Irrigation on Remodelling of Ganges Canal and Extension of Irrigation in NWP Doab, 22 Oct. 1868.

Madras works may have been substantially due to chance factors in the choice of location. Another example of Cotton's faulty judgement in commercial matters involved the potential of navigation canals. His view of the Ganges Canal when it opened was that 'as a work of communication (in addition to its value for irrigation) there can be no question that it will be of incalculable value'.[103] In point of fact, although considerable expense was incurred in making the Ganges Canal suitable for navigation, including the provision of a separate lock channel system in the northern sections (see Map 3), navigation in the Doab was always a secondary function of the canal. As a result, many of the design features of the canal actually hampered traffic and effectively undermined any attempt to increase its use for navigation purposes.

It did not help, for example, that the main routes were not designed to serve the large trading centres. Moreover, the velocity of the water made towing boats upstream a slow business, particularly as there was no towpath provision under the ordinary bridges (which were often designed, particularly in the lower branches, without sufficient headroom for loaded boats). Boats also had to cope with delays at the numerous locks, closures (unscheduled as well as scheduled), and various exactions made of boatmen by the canal *chaukidars* stationed *en route*.[104] The small number of boats which did manage to operate despite these impediments – mostly for local rather than through traffic – quickly diminished when railways were constructed alongside the roads and canals already aligned on the watersheds of the country. From a recorded total of 2,463 boats in 1861, the numbers fell to 432 in 1869 and 239 in 1872,[105] though it was not until the early 1900s that navigation was officially abolished. The amount of goods carried was always small and the tolls (for example, Rs 21,000 in 1868) were insufficient to cover the expenditure incurred on account of navigation.

[103] Cotton, *Public Works in India*, p. 114.
[104] Nicolls, 'Agricultural Engineering (III)', *Engineering*, XLV (18 May 1888), pp. 473–4; NWP PWD(I), Oct. 1869, 152, Memorandum by Traffic Manager on Ganges Canal Navigation, 1 June 1869.
[105] From NWP PWD(I), Oct. 1869, Memorandum by Traffic Manager on Ganges Canal Navigation, 1 June 1869; and Thornton, *Indian Public Works*, p. 120.

Problems of labour and material supplies

The construction and operation of large canal schemes entailed the mobilisation of substantial human and physical resources. During the construction phase, skilled workers had to be recruited, building materials gathered or directly manufactured, and armies of labourers assembled. By the 1920s, close on 2,000 miles of main channels and more than 14,000 miles of state-built distributaries and other channels had been excavated in the UP,[106] many of which, of course, were in cuttings or embanked for part of their length. Some of this excavation took place in relatively remote areas, such as the upper sections of the Ganges Canal, including the construction of the numerous headworks installations, falls, weirs, and aqueducts. An executive engineer, of which there were six (one for each division) on the Ganges Canal, might easily have 5,000 men and hundreds of carts working daily on his section of construction. The brick-making alone for the Solani aqueduct kept 3,500 labourers occupied for seven to eight months a year over five years. The actual construction work kept 2,750 employed for a full six years.[107]

Labour for the considerable excavations in the sparsely populated northern areas was supplied mainly by the nomadic Oade tribe, members of which had by tradition wandered the area between the Sutlej and the Jumna with their cattle and other livestock undertaking excavation of tanks and watercourses.[108] Distinguished from other excavators by their use of the inlaid and ornamental *phaora* (mattock), and by their use of donkeys rather than women and children for carrying earth, they gathered in large numbers in tented encampments along

---

[106] That is, all channels above the level of village water-courses (*guls*). See Table 2.1.

[107] Cautley, *Ganges Canal Works*, I, Ch. 2. The Nadrai aqueduct taking the LGC across the Kali Nadi was of similar dimensions and took 4,000 workmen, often working at night, four years to construct. Over a million cubic feet of concrete and 4.75 million cubic feet of brickwork were used in the second aqueduct on that site. The initial construction was swept away by floods in 1885. *Triennial Review of Irrigation*, pp. 63–4.

[108] Three-quarters of the work on the EJC was performed by Oades, who also carried out most of the digging of the first 110 miles of the Ganges Canal. Cautley, *Ganges Canal Works*, II, p. 547.

the lines of excavation. Only seldom, through an occasional donkey raid, did they come into conflict with local cultivators, and they displayed a convenient reluctance to move away from an area where further work was in prospect. They worked either in large parties under a leader (with the contract terms, advances, and distribution of payments organised through a *bania*), or in small groups on specific projects where they were paid daily or monthly and supervised by a European overseer.[109] Oade contractors took on sections of from 25 to 100 feet and were paid Rs 1½–6 per 1,000 cubic feet, depending on the nature of the soil and the depth of the work.[110] An able-bodied man typically managed to move 1,000 cubic feet of soil in around twelve days, thus earning a daily rate of around two *annas*.

Obtaining labour was, perversely, more difficult in the more populated parts of the Doab, below Muzaffarnagar. The more developed state of agriculture in the Middle and Lower Doab generally gave less opportunity for the Oades to graze their livestock, and they were reluctant to follow the work opportunities too far southwards. Although many parties applied for contracts prior to the commencement of digging, the relatively low rates offered – Rs 2 or less per 1,000 cubic feet – and the existence of alternative income-earning opportunities made progress on the works very slow, particularly in Bulandshahr and Cawnpore, and for months only small groups of workers could be found. Indeed, in the latter division, those contractors who could be found were so inexperienced as to take contracts on rates below the level at which labour could be hired, and for a time actually made losses. Ultimately the rates were raised, but the problem of securing a continuous supply of labour remained: the works became 'nearly deserted' at harvest time and whenever a day's rain had fallen, and canal officers complained that 'squads of workers' regularly departed for weddings and fairs.[111]

A significant part of the earthwork operations on the schemes carried out in the 1870s and 80s appears to have been done by

---

[109] *Ibid.*, p. 544.

[110] This figure was inclusive of carrying, levelling, and smoothing the banks and slopes, and of making a towpath, boundary ditch, and roadway along the canal.

[111] Cautley, *Ganges Canal Works*, II, p. 571.

local labourers employed on famine relief schemes and by gangs of 'Purbeahs', specialist excavators from Oudh, whose grass hut villages sprang up wherever work was available. Some of the 100-yard contracts on the Agra Canal excavations – which began in October 1868 to provide relief work – were taken by Purbeahs, though most of the workers were inhabitants of the villages through which the canal passed. Not surprisingly, with drought diminishing employment opportunities, the Chief Engineer's inspection report notes considerable competition for contracts. Most of these were for lengths of 100 feet, but one was for a mile, and another, from a European by the name of Robinson, was a Rs 70,000 contract for a twenty-mile section.[112]

In addition to the unskilled labourers required for construction, a substantial number of skilled tradesmen had to be assembled, and provision made for the supply of needed materials. While Muslim bricklayers and stone-masons were usually available from the locality, many of the fitters, carpenters, and divers came from the Punjab, and a significant number of *khallassis* (sailors) from Bombay were employed in high-level work including erecting scaffolding. Artisans from a distance were procured for the Ganges Canal masonry works from among the thousands of pilgrims visiting for the Hardwar fair, but many skills were passed on during the earlier phases of public works construction so as to provide a sizeable pool of tradesmen which could be drawn upon for subsequent works.[113]

For specialised supplies, including timberwork and metal components and fittings, a workshop and foundry were set up at Roorkee. Reliable contractors to manufacture or arrange for the supply of construction materials were non-existent in the area when the early canals were built. Thus, on the Ganges Canal, for example, one executive engineer and his two assistants were occupied solely in organising materials. This principally involved brick- and lime-making on a huge scale, and the purchase (and often the personal selection) of timber from local forests.[114] All engineers, in fact, who joined the

[112] NWP PWD(I), March 1869, 27, Agra Canal Inspection Report, 20 Feb. 1869.
[113] For details relating to this section, see Nicolls, 'Agricultural Engineering (II)', *Engineering*, XLV (4 May 1888), p. 424.
[114] Cautley, *Ganges Canal Works*, I, Ch. 2.

service in the early days were required to learn how to make bricks by means of a simple trench kiln system, and how to manufacture lime from the *kunkar* deposits found in the plains. This lime, when mixed with white lime and brick dust, conveniently possessed sufficient tensile strength to meet most irrigation requirements.[115] While it continued to be usual for engineers to make small numbers of bricks as required, contractors did come forward to supply bricks for later works. The brick-burning for the remodelling of the headworks at Myapur in 1882 was done, for example, by a large European firm. Not all contractors, however, had the capital and the expertise necessary for the job. In one instance, an urgent need for bricks led an engineer to use a host of petty contractors who apparently wasted their advances in producing substandard products and then hurriedly decamped to escape their debts, leaving the area 'littered with valueless underburned bricks'.[116]

Engineers and the irrigation service

The canal-builders of the 1820s to 60s carved, almost literally, an important place for themselves in Indian history. The group of men who worked on restoring and developing bold new schemes – Cautley, Baird Smith, Napier, Durand, Yule, W.E. Baker, Greathed, Richard Strachey[117] – are far from obscure figures. Their claim to fame, however, rests on an even firmer base than the substantial one of having laid the foundations of modern canal-building in northern India.

Some of the men gained prominence more through their military achievements than through their irrigation activities. The early phase of canal-building was carried out by military engineers, who made their way into the ranks of the Bengal (later Royal) Engineers via the East India Company's Artillery and Engineers Seminary at Addiscombe, from which the best cadets were selected for subsequent training in Trigonometrical Survey methods and a course on mines and explosives at

---

[115] Nicolls, 'Agricultural Engineering (VIII)', *Engineering*, XLVI (24 Aug. 1888), pp. 175–6; W.P. Thompson, *Punjab Irrigation* (Lahore, 1925), pp. 58–9.

[116] Nicolls, 'Agricultural Engineering (VIII)', *Engineering*, XLVI (24 Aug. 1888), p. 176.

[117] See Appendix 1 below for career outlines of the main officers involved.

Plate 8. Bengal (PWD Irrigation) Engineers, *c.* 1862.
Front (left to right): Hutchinson, Gulliver, Maclagen, Turnbull, Dyas, Home.
Middle: Innes, J.G. Forbes, Pollard, Sim, Taylor, Becher.
Back: Padday, Newmarch, Crofton.
(By kind permission of the British Library, India Office Records)

Chatham.[118] All of the officers saw action at some time, since this was still a period of active territorial expansion. Most were called away from civil projects to assist in the Sikh campaigns of the 1840s, and some were to distinguish themselves during the Mutiny. No one more so than W. Greathed, who carried despatches between Agra and Meerut, in between relieving beleaguered civilian stations, at the height of the uprising; and although he was wounded during the siege of Delhi, he subsequently acted as Directing Engineer during the attack on Lucknow.[119] Richard Baird Smith (as Chief Engineer at the assault on Delhi)[120] and Henry Brownlow also made notable contributions in retrieving the situation for the British in

---

[118] There is a certain irony in the fact that Cautley, unlike his later rival, Cotton (who also graduated in 1816), did insufficiently well in his examinations to be selected for the engineers. He was assigned instead to the artillery, and it was as a lieutenant in the Bengal Artillery that he was sent to assist on the reconstruction of the Doab Canal in 1825. Brown, 'Memoir of Colonel Sir Proby Cautley', p. 190.

[119] *DNB*, VIII, pp. 474–5.   [120] *Ibid.*, I, pp. 512–15.

1857–8.[121] Robert Napier, of course, who from the position of assistant on the Doab Canal went on to become Chief Engineer, Punjab (1849–56), eventually finished a spell of almost half a century in India as Commander-in-Chief (1870–6).

Napier went on to become Governor of Gibraltar, and in this was one of several officers who moved into areas of civilian administration. Henry Durand, a good friend of Cautley's who worked on the Delhi Canal in the 1830s, became Private Secretary to the Governor-General in 1842 and subsequently a member of the Governor-General's Council and Lieutenant-Governor of the Punjab. Richard Strachey, who worked under Cautley on the Ganges Canal, was the first person to hold the post of Inspector-General of Irrigation on its creation in 1866. Though other NWP officers were to hold this post, the highest in the irrigation service, none was able to match Strachey's contribution to the reorganisation and development of Indian public administration. Strachey played a major part in reorganising the PWD, in formulating the policy of constructing public works with the aid of loans, and – with his brother John – in helping to develop and consolidate the new system of government gradually adopted after 1858. He presided over the Famine Commission of 1878–80 and, like Yule and Cautley, served on the Council of India on his return to England.[122]

A number of engineers became well known for achievements outside the military, administrative, and hydraulic engineering spheres. This was, of course, the age of the gifted amateur, adding to the base knowledge of a whole range of fledgling fields of science and natural history. On leaving India in 1862, for example, Yule took up residence in Palermo, and from there published his famous *Marco Polo*.[123] He was also renowned for his work for the Hakluyt Society and subsequently produced the invaluable *Hobson Jobson*.[124] Strachey was also an important figure in both the Royal Geographical Society and the Royal Society, of which he was elected a Fellow in 1854. Fever attacks while he was working on the Ganges Canal in 1847 led him to

---

[121] *Who Was Who 1897–1916* (London, 1920), p. 95.
[122] *DNB, Second Supplement* (London, 1912), III, pp. 439–42.
[123] Published 1871; enlarged edn 1875.
[124] *DNB*, XXI, pp. 1320–2; H. Yule, *Hobson Jobson, A Glossary of Anglo-Indian Colloquial Words and Phrases* (London, 1886).

take refuge in Naini Tal, from where he collected a great number of botanical species (many of them previously unknown), contributed significantly to important geological debates relating to glacial and other Himalayan features, and wrote travel–descriptive and physical geographies. Though his researches in physical and botanical geography earned him the Royal Society medal in 1897, it was in meteorology that he made his most useful scientific contributions to knowledge, particularly the instruments he designed to give graphic expression to formulas he had devised for working out meteorological problems.

Cautley, too, achieved distinction outside the field of engineering: as a palaeontologist. At Saharanpur, while working on the Doab Canal, he met up with a young Scot named Hugh Falconer, who was Acting Superintendent of the Botanic Garden there. The two went on fossil-hunting expeditions with two other officers, Baker and Durand, into the Siwaliks, where they found a wealth of significant specimens. For their published work on their discoveries, Falconer and Cautley were awarded the Woollaston Medal of the Geological Society in 1837. In the early 1840s Cautley despatched a total of 256 chests of fossils, averaging 4 cwt each, to the British Museum.[125] For this work, and for his achievement in designing the Ganges Canal, Cautley was elected a Fellow of the Royal Society in 1846.

Of course, most of the engineers who served in the irrigation service over the period filled their time in far less spectacular ways than those of the officers described above. Most of the work to be done within a canal division was of a fairly routine nature, though that is not to say undemanding. The engineer's work involved the general administration of establishment and accounts, surveying to provide for record maps and to indicate possible developments in irrigation, the settling of local disputes and other judicial functions connected with the canal, the inspection of channels, masonry works, and plantations,[126]

---

[125] *DNB*, IV, pp. 1242–3; for more detail, see Brown, 'Memoir of Colonel Sir Proby Cautley', pp. 197–202.

[126] 'Plantations' refers to the extensive tree-planting carried out by the Canal Department along the embankments of canals, for the purposes of shading those who travelled along the towpath and providing a source of income from the sale of timber.

and the design and construction of new works as well as the repair and remodelling of existing ones. Finally, there was the day-to-day task of ensuring the efficient distribution of supplies and of regulating the flows within the division. The executive engineer would carefully monitor incoming gauge cards and (from the 1860s) telegrams from different parts of the system, and despatch runners (or telegraph responses) to ensure that supplies were kept at the appropriate level. When, due to rainfall or sudden demand, 'gauges showed a change of so much as an inch in an hour', recorded Australian observer Alfred Deakin, 'the engineer abandons his rest, and all night the tick tick of the telegraph key is heard in his office, as reports rush in and orders are sent out' for inlets and escapes to be modulated so as to maintain the required flow levels.[127]

The work was undoubtedly arduous. An executive engineer was typically in camp 24 days a month throughout the year,[128] and was virtually alone among Europeans in having to remain out in camp during the hot season. Inspection duties compelled the officers to ride often twenty miles in a day, moving from one canal resthouse to another, with their belongings and office following on by camel, bullock cart, or boat.[129] Apart from the fact that irrigation was carried on during the hot season, the likelihood of floods or heavy rain meant that the canal engineer had constantly to be on hand, as graphically described by Deakin: 'When the country [is] under water, and the panic-stricken people flying for shelter to the high ground, the engineer must stand his ground, see that all the gates are closed and all escapes open, and that gangs of men are ready at hand with crates and timber, and branches at any threatened point.'[130]

There were, of course, the diversions of shooting, hunting, and fishing to be set against the physical hardships and social deprivation experienced by the canal officers. Some men, like James Crofton, admitted a preference for the remote station life, where they were to be found casually dressed in simple Norfolk jackets, away from the oppressive formalities of

---

[127] Deakin, *Irrigated India*, p. 228.
[128] NWP PWD(I), Jan. 1869, 41.
[129] Nicolls, 'Agricultural Engineering (v)', *Engineering*, XLV (22 June 1888), p. 608.
[130] Deakin, *Irrigated India*, p. 229.

Plate 9. Engineers outside canal bungalow, *c.* 1862, including James Crofton (right) and Joseph Dyas (centre-right). (By kind permission of the British Library, India Office Records)

colonial society.[131] Nevertheless, conditions clearly took their toll. Few officers shared with Brownlow the experience of being badly mauled by a tiger in a towpath attack, but the irrigation records are littered with references to the sufferings – and frequently the deaths – of officers in the field, mainly from malaria, sunstroke, and 'overwork'.

Arduous conditions and poor remuneration were seen by the Indian Irrigation Commission as the reason why, towards the end of the 1800s, engineers were 'reluctant to enter the canal service and anxious to leave it'. This produced a situation in the 1890s when subdivisions were under the control of subordinates and temporary engineers; in stark contrast to the situation 'many years ago when it was the most popular branch of the PWD in Northern India, and consequently commanded the services of the most competent engineers'.[132] 'The nature of the

---

[131] 'Memoir on Lt.-General James Crofton', *Royal Engineers' Journal*, v (April 1909), pp. 174–84. [132] *IIC*, p. 103.

work has got the reputation of being more unhealthy, more isolated, and more expensive than any other branch of the PWD', explained the Chief Engineer H. Marsh, in giving evidence before the commission. 'Unless a man is a bachelor and very keen and very interested, he strains every point to get out of the Irrigation Branch.' No less than six officers had recently departed the UP service (three of them to Egypt) 'to avoid the isolation and expenses of working in the jungle'.[133]

Many aspects of the work, of course, were no less unpleasant forty years earlier. But it may have helped the popularity of the service in the earlier period that the engineers, as military men, were better paid than the growing civilian element in the service. Military engineers received 20% of the salary for their grade within the Irrigation Department, and full military pay and allowances in addition.[134] Overall, therefore, military engineers enjoyed not only the prestige attached to their regimental status, but conditions of pay which put them on a par with the ICS. Civilian engineers felt deeply the disparities in conditions, pay, and status between them and the ICS. Compared with the civilian engineers, editorialised the *Engineer* on 1 June 1888, the ICS enjoyed less exposure, less pecuniary responsibility, more pay, more leave, higher status, better pensions, and more frequent recognition in the honours lists.[135] The situation noticeably worsened during the 1880s, when promotion possibilities within the canal service – already reduced by a slowing in the rate of expansion of public works activity – were further diminished as the depreciation of the rupee so reduced the real value of pensions for those intending to retire to England that many officers put back their retirement plans.

The engineers' sense of grievance associated with disparity in pay and conditions was to continue through into this century, and to contribute to the ongoing rift between the Canal and Revenue Departments discussed in Chapter 6. There would, of course, have been no shortage of alternative job opportunities for civil engineers, either in Europe or further afield, at this time. It remains a strong possibility, however, that the popularity of the UP canal service among engineers

---

[133] *Ibid.*, p. 281.   [134] NWP PWD(I), Dec. 1869, 18.
[135] Vol. XLV, pp. 543–4.

faded in part because of the diminishing opportunities for creative activity within the department. The early engineers were engaged in the projection and execution of schemes which frequently involved the adaptation and elaboration of existing engineering methods and principles so as to cope with the specific physical and agricultural environment of the area. A total structure – both the physical works and the administrative system – had to be devised. Later engineers, reaching executive positions in the 1880s and after, served at a time when only a limited number of new works, almost all of them minor in comparison with the earlier schemes, were sanctioned, and they were largely involved in the relatively mundane tasks of administration and maintenance. Even the extension work in which they were occasionally engaged was conducted according to generally well-tried prescriptions. The job must have lost some of its appeal at this point, since it took on the character of a caretaker function.

3

Canal irrigation as an appropriate technology

Two quite different irrigation technologies – one, a technique developed and refined in a past age, the other, in its technical and administrative aspects, an unintentional mixture of old and new, indigenous and exotic – came into head-on collision in the UP Doab during the nineteenth century, as the area commanded by the rapidly expanding canal systems frequently encroached upon tracts with long-established systems of well irrigation. The overlapping, and its outcome, reveals much about the nature of existing techniques of irrigation, and their related agricultural systems; it provides a perspective on the intruding technology, and the associated agricultural possibilities; and, through the kinds of responses of the peasant community, it reveals important aspects of their decision-making processes relating to production.

The overlapping of facilities, and the subsequent widespread substitution of the canal for wells, has been condemned as undesirable by both contemporary and latter-day critics. While one writer has argued that such duplication clearly constituted an 'economic waste',[1] others have gone further, pointing to the inflexible cash outlays associated with the use of canal water, the uncertain nature of the water supply, and the observed tendency for the yields of sugarcane, opium, tobacco, and garden crops in general to be lower where grown with canal water.[2] Lower yields were obviously related to the uncertain

---

[1] B.N. Ganguli, *Trends of Agriculture and Population in the Ganges Valley* (London, 1938), p. 75.

[2] 'The people themselves admit', the Saharanpur Settlement Officer reported, 'that in the long run irrigation from wells is more beneficial to the soil than that from the canal.' L.A.S. Porter, *Final Report of the Settlement of the Saharanpur District* (Allahabad, 1891), p. 13. (Hereafter settlement report titles will be abbreviated to the district name followed by the letters *SR*. The several reports for each district are

supply, but they were also the result of the 'careless use' of canal water, and to overcultivation; even the Jats, renowned agriculturalists, were susceptible, it appears, to the slovenly practices which spread with the canal, and abandoned their *pakka* (masonry) wells to take advantage of an easier system of watering their fields.[3] Elizabeth Whitcombe's is the most complete critique, since it embraces the foregoing specific criticisms within its theme. Essentially it appears to be based on the supposed merits of a long-proven system of techniques and organisation of production under well systems, the entire structure of which had been drastically and rapidly transformed by the intrusion of a technology which, because of its apparent functional shortcomings and its overtly commercial orientation, possessed features in many respects 'inappropriate' to the context into which it came. It has been argued that an 'inappropriate' technology was so widely adopted and the old technology discarded because, in some cases, the canals rendered the subsoil unstable, thus causing wells to fall in, and because, in other cases, the prospect of an easier life appealed to the 'satisficing' element in the peasants' psychological make-up.[4] 'Far from firing him with the much heralded spirit of industriousness', concludes Dr Whitcombe, 'canal irrigation required less by way of labour than his well had demanded.'[5]

If the reasons for the overlapping in the first place were, given the building of a canal system in the Doab, partly

distinguished from each other by the date of publication.) 'Well water gives much heavier crops than canal water', remarked S.N. Martin of Muzaffarnagar. 'All the Zemindars agree on this point.' 'Settlement Report of the District of Muzaffarnagar', 1866, p. 59, in A. Cadell, *Muzaffarnagar SR* (1882). See also C.H.T. Crosthwaite, *Etawah SR* (1875), p. 15; and R. Misra, *Economic Survey of a Village in Cawnpore District* (Allahabad, 1932), p. 38.

[3] H. Le Poer Wynne, *Saharanpur SR* (1871), pp. 139–40. Observed the Muzaffarnagar Collector, W. Young: 'The ease with which the water can now be got has spoilt the peasant as an agriculturist; formerly where he and his family had to work hard for every drop of water, they became inured to labour and did not repine at it, while now the peasant turns on the canal water and goes home to sleep.' Government of the North-Western Provinces and Oudh, Public Works Department (Irrigation Branch) Proceedings (hereafter NWP&O PWD(I)), Jan. 1881, (proceeding no.) 8. (In the IOR.)

[4] The 'satisficing' peasant expends effort only up to the point where a 'target' income is attained. See H. Ezekiel and P. Mathur, 'Marketable Surplus of Food and Price Fluctuations in a Developing Economy', *Kyklos*, XIV (1961), pp. 396–408.

[5] Whitcombe, *Agrarian Conditions*, p. 80.

unavoidable for technical reasons and partly due to the internal needs of the department responsible, it can be clearly established that the indigenous mode of irrigation, though comparatively favoured by the conditions in the Doab, provided overall no solid foundation for agricultural improvement. As a source of irrigation it was markedly restricted and restricting. This is reflected by the fact that cultivators abandoned the most efficient of wells (the extent of the destruction by the canal was far from total) even where canal water was only available on the basis of being manually lifted from the channels. Emphatically, it is not necessary to explain this decision through reference to the notion of leisure preference. The plain fact is that, for all its disadvantages, the canal provided more nearly for the needs of most peasant family producing units. The canal was an innovation which met their requirements, and it did so because it slotted into the productive aspects of the peasant system in a way which made it generally more advantageous than even the most favourable well irrigation.

Geographical conditions, wells, and the siting of canal distributaries

Well irrigation in the Doab

Substantial parts of the UP were among the most suitable in the country for well irrigation, and the subsoil characteristics in many parts of the Doab itself made the tract 'peculiarly favourable' to the construction and operation of temporary (*kachha*) wells.[6] The Doab water-table was relatively close to the surface and the alluvium generally firm, with frequent (if not continuous) water-hoarding beds of clay (known as the *mota*) spread through the area. Frequently, simple earthen wells could tap these underground springs and a good supply of water could be obtained for as little as Rs 5, with the cost rising to Rs 50 if the well was deep and layers of wet sand necessitated a stabilising *garoli* (brushwood lining).[7] Temporary spring

---

[6] *IIC*, p. 185. For background to this section, see G. Kuriyan, 'Irrigation in India', 2 pts, *Madras University Journal*, xv (1943), pp. 46–58, and xvi (1944), pp. 161–85.

[7] The excavation of *kachha* wells entailed an 'almost universal' charge of Rs 1.25, irrespective, it seems, of depth. The specialists who cut a hole in the *mota* and

wells were thus found 'extensively' in the Doab, occasionally working more than a single water-bag and frequently watering four acres in a season.[8] Many of these *kachha* wells would last for years, though they generally required an annual expenditure of Rs 10 or so for a new supporting cylinder. In any case, where they fell in they could be replaced fairly readily, and their location suited to the needs of the season.

Where the subsoil strata were less firm, or where multiple lifts were required, masonry wells could be constructed. Such wells, in fact, were seldom operated in individual units. 'Few wells', it was observed, 'have less than two cultivators working at them', and while Sitapur village (Aligarh) possessed two eight-lift wells employing 28 and 32 head of cattle respectively, instances of sixteen and twenty lifts were not unknown.[9]

Above the ground, too, the Doab conditions were relatively favourable to irrigation. The cold weather of northern India, along with soils retentive of moisture, meant that few waterings were required to supplement fully a fairly reliable, if light, winter rainfall in bringing *rabi* crops to maturity. The best wells, with multiple lifts, could thus annually irrigate upwards of twenty acres.[10]

The earliest estimate available puts the number of wells in the Doab during 1860–1 at 70,000 *pakka* and 280,000 *kachha*. Just under 1.5 million of the ten million or so acres normally

prepared the well for the bucket charged from Rs 1 to Rs 3. See J. Clibborn, *Papers Relating to the Construction of Wells for Irrigation* (Roorkee, 1883), p. 20. The conditions in the neighbouring Punjab were generally less conducive to well irrigation, making the overlap between the systems of minor importance. The alluvial deposits there were coarser in structure, and clay was seldom encountered. As a result, Punjabi farmers placed greater reliance on *pakka* wells supplied by percolation. These wells were, for the most part, located near to the rivers, since in the central parts of the *doabs* the water-table was at too great a depth. Kuriyan, 'Irrigation in India'.

[8] *IIC*, pp. 200–1.

[9] Clibborn, *Papers Relating to the Construction of Wells*, pp. 1, 19. See also E.T. Atkinson, *Statistical, Descriptive and Historical Account of the North-Western Provinces of India*, vol. II, *Aligarh* (Allahabad, 1875), p. 381. (Subsequent references to this district gazetteer series will be abbreviated to the district name, followed by the letters *DG* and the date of publication. All volumes published at Allahabad.)

[10] This is in contrast to the situation prevailing throughout most of the Indian peninsula, where the subsoil water resources were insufficient to sustain well irrigation and the hotter climate made more frequent waterings necessary, thus limiting irrigation to high-value crops. (For distribution of rainfall by month in the western UP, see Figure 1.)

cultivated in the Doab around that time were under well irrigation.[11] Thus, even in an area where conditions overall were considered 'peculiarly favourable' to well irrigation, tracts unconducive to the construction and operation of wells were to be found interspersed between well villages or out on the Doab fringes. Eastern Muzaffarnagar, specifically the Ganges–Kali *doab*, contained just such a cross-section of local conditions. Prior to the sweep of the Ganges Canal through this tract, virtually all irrigation was confined to the fertile south-western corner, where masonry wells were numerous and earthen wells practicable. To the north, and east towards the Ganges high bank, conditions were similar to those in the eastern Doab fringes further south: the subsoil was too unstable for earthen wells, and water-level too low (60–110 feet) and soils too light to justify *pakka* wells costing in excess of Rs 2,000.[12] Such dry conditions favoured production options entailing low risk, low input, and low value output; agriculture thus impinged only superficially on the landscape, ebbing and flowing with the seasons. In the richer moisture-retaining soils, better crops could be grown – some wheat during the *rabi* and cotton, even sugarcane, in the summer – though soil preparation for such crops was arduous and the returns uncertain.[13] But in the poorer soils, the pulses and millets of the rainy *kharif* were prominent, and the *rabi* area was small.

While the dry farmer juggled his crop combinations and resources to suit his uncertain environment, his counterpart in well villages had the benefit of greater operational flexibility and control over aspects of the environment which made worthwhile the investment of time and effort in more intensive cultivation. Wells were not extensively used during the hot season – though they could facilitate some pre-monsoon sowings and assisted in a limited way when rains came late or were ill-distributed; they were of most use in the *rabi*, when they could ensure the sowing of a good part of the normal area under winter crops and improve the yield of the finer grains by two or three timely waterings during growth.

---

[11] *PP* 1862, XL, *Report on the Famine of 1860–1 in the North-Western Provinces of India*, paras. 98–100.

[12] For a description, see 'Report on the Settlement of the Ganges Canal Tract', 1878, p. 67, in Cadell, *Muzaffarnagar SR* (1882).   [13] *Ibid.*, p. 19.

Intrusion of canals into well tracts

As the Doab contained areas where conditions were especially favourable to the construction and operation of wells, and as it also contained substantial (and often contiguous) areas where conditions militated against their use, the question inevitably arises: why should the canals' architects have displayed a pronounced tendency to direct their water into well tracts, often ignoring the needs of nearby dry villages? The subsequent growth in wealth of the neighbouring Punjab was, after all, largely the result of pouring canal water over dry, lightly populated zones outside the well tracts. Such areas in the UP were time capsules of untapped productiveness, awaiting the release which canal irrigation could bring through extending production possibilities and altering the studied balance between risk reduction and output expansion as a lure to entice resources on to hitherto marginal land. Large potential returns could be made even on the land already farmed, as incremental gross output came to reflect the switch from coarse millets to higher-yielding cereals such as wheat, the average yield of which could be doubled by irrigation. The 'enormous gains' resulting from the canal's irrigation of 48,000 acres in the dry and sandy *parganas* of eastern Meerut serve to illustrate this potential.[14]

Despite the combined need and potential of the dry tracts, the Director of Agriculture was still able to observe that the one-fourth of the Upper Doab watered by the Ganges and Jumna Canals in the late 1870s 'was less in want of canals than the three-fourths which they do not irrigate'.[15] Alan Cadell, Muzaffarnagar Settlement Officer, was especially alert to this recurring theme in canal distribution policy: in the trans-Hindan portion of *pargana* Shikarpur, irrigation from the Kalarpar branch (EJC) supplied an area of eighteen estates replete with *pakka* and earthen wells, but provided only a

---

[14] R.W. Gillan, *Meerut SR* (1901), p. 5.
[15] Government of the North-Western Provinces and Oudh, Revenue Department Proceedings (hereafter NWP&O Rev.), Aug. 1879, (proceeding no.) 37, Note by E.C. Buck. (In the IOR.) Buck went on to observe: 'Similar mistakes have, I understand, been made in Bengal and Madras.'

miserly service to the dry, and accessible, estates further south; in Thana Bhawan the department had 'carefully avoided' the Rajput tract on the east bank of the Kirsani (where the sinking of *kachha* wells was difficult), supplying instead the *kachha* well tract west of the river and the *pakka* well tract further to the east.[16] Throughout the district, Cadell charged, 'canal water has been wasted in fertile neighbourhoods which do not require it, while close by there are still arid tracts'.[17]

Purely technical factors relating to canal irrigation clearly played a role in the pattern of widespread duplication. However, it is clear that, particularly in the early stages, commercial policies pursued by the Canal Department were also a powerful influence contributing to the distribution bias.

The technical constraint was basically that which governed the routes which were feasible for the main channels and distributaries, and the extent to which this naturally coincided with favourable physical conditions for wells. From his experience in Meerut, Settlement Officer E.C. Buck concluded that wells could be dug 'with greater facility and of a more lasting character in proportion to their distance from drainage lines of magnitude'. Consequently, 'the best well tracts were on the watersheds', and when, therefore, the canals were introduced (of necessity along the watersheds, due to the need to command the largest area by taking distributaries off each side), 'it was a natural consequence that they should traverse those tracts which were most fully irrigated by wells and avoid those in which the loose and broken condition of the surface formed as great an impediment to irrigation channels as to wells'.[18] It was, in fact, often difficult to take water to the dry tracts of each *doab* (which lay beyond the stiff loam, or *dumat*, of the watersheds, along the line of the rivers), without 'leading a water-course along the tracts where wells are plentiful'.[19] To

---

[16] 'Report on the Permanent Settlement of the Western Pergunnahs of the District Muzaffarnagar', Feb. 1870, p. 32, in Cadell, *Muzaffarnagar SR* (1882).

[17] *Ibid.*, p. 44. The phenomenon was not confined to the Upper Doab: M.A. McConaghey commented with regard to *pargana* Karhal (Mainpuri) that the result of the introduction of the canal was 'more to substitute from that source for well and *jhil* (pond) irrigation than to bring land formerly dry under its influence'. M.A. McConaghey, *Mainpuri SR* (1875), p. 40.

[18] E.C. Buck, *Meerut SR* (1874), pp. 7–8.

[19] E.C. Buck, 'Note on the Effect of the Canal on the Agricultural Conditions of the

ignore the importance of watershed alignment throughout the entire distributary system was to court severe waterlogging problems – something which had quickly become apparent on early versions of the older canals, on which a herringbone pattern of distributaries was adhered to almost regardless of the configurations of the land.[20]

There is clearly enough in what Cadell alone observed to suggest that the technical constraints only supply part of the explanation: even where the distributaries (*rajbahas*) traversed partially dry tracts, as in Shamli, east of the Kirsani, the well-irrigated estates were favoured; and 'many dry estates at no great distance from the line of *rajbuha*' were 'left unaided'.[21]

Superimposed upon and reinforcing the dictates of the technical constraints was an irrigation policy which in effect favoured the intrusion into areas of established well irrigation. Although inspired by a desire to counter the effects of periodic drought, the early works especially were nonetheless still looked upon as commercial ventures: only those schemes which could pay for the annual expenditure on maintenance and operation, and meet the interest charges on the capital, were sanctioned.[22] Given the commercial orientation, and the not infrequently expressed view during the formative phases of canal development that 'revenue should be the end and aim of all canal administration',[23] it was natural that canal water should be taken to the areas readily yielding high returns to capital. The short time frame was a natural outcome of the reliance on current revenues for funding the earlier works, and

Pergunnah of Tirwa, Zilla Farruckhabad', 1 May 1871, *Revenue Reporter*, v (n.d.), pp. 212–27. See also NWP&O Rev., Aug. 1879, 38; and W.H. Smith, *Aligarh SR* (1882), pp. 3–4. (Copies of the Revenue Reporter are available at the Uttar Pradesh Board of Revenue Library, Allahabad.)   [20] *Meerut DG* (1875), p. 241.

[21] 'Report on the Permanent Settlement', p. 48, in Cadell, *Muzaffarnagar SR* (1882).

[22] Cautley candidly underscored this concern: 'It is very certain that if the restoration and extension of these [indigenous] works had not promised an increase in revenue to the British Government they would never have been undertaken. To suppose that a disinterested regard for the welfare of the country alone was the activating motive for these, or any other, works we have executed would indicate but a limited knowledge of their history.' Cautley, 'Canals of Irrigation', p. 80. Policy matters are dealt with in more detail in Chs. 5 and 6 below.

[23] *Papers on the Revenue Returns of the Canals of the North-Western Provinces, 1863–4* (Roorkee, 1865), p. 35. (Subsequent references to these annual reports will be abbreviated to *PRRC*.)

of the wish to justify the policy by making positive contributions to the annual revenue.

It was the combination of these commercial imperatives and the physical and climatic conditions in the region which impelled such a distribution policy. In contrast to conditions in the Punjab, where rainfall was lower, much of the agricultural activity in the Doab was fickle in its need for irrigation. Millets and pulses, for example, the *kharif* staples of the western districts, 'would derive no benefit from irrigation except in an unusually dry year'; a *rabi* staple, *gram*, in an ordinary year did not especially benefit either.[24] The best pickings, as far as the Canal Department was concerned, would be derived from the high-value crops which required substantial and repeated irrigation, and thus were stable in their demand and able to bear a reasonable water charge. Moreover, as far as possible, it was desirable to make use of the ample canal supply of the *kharif* season, when the rivers were high.[25] Naturally, such crops as would fulfil the financial requirements of the canal authorities were found mostly in areas of fertile soil – the very areas which best repaid investment in well facilities, since light (*bhur*) soil not only produced less impressive crops but absorbed the hard-won water virtually without trace wherever a supply from wells was feasible. Given the fact that, for economic reasons, the two elements most satisfactory to the canal's needs, fertile land and valuable crops, usually were associated with well irrigation, a clash was unavoidable.

Backward areas could hold out only more limited prospects as far as the Canal Department was concerned, and especially so in the short term. Complementary factors of production were vital to the extension and intensification of cultivation necessary for the full utilisation of an ample canal water supply. In the developed tracts, such as the carefully cultivated Jat villages of Jhinjhana *pargana* (Muzaffarnagar), the introduction of the canal at once produced a very high standard of prosperity. In villages 'sufficiently, if not fully protected by wells' the standard of cultivation promptly rose almost to the renowned level of the Jat villages.[26]

[24] *IIC*, pp. 183–4.
[25] These issues are dealt with more fully in Chs. 4 and 5 below.
[26] 'Report on the Permanent Settlement', p. 39, in Cadell, *Muzaffarnagar SR* (1882).

However, progress under canal irrigation in the hitherto precarious *parganas*, such as Purchhapar, was visibly held back by the relative shortage of complementary inputs, particularly labour, which moved only slowly into line with the new production potential. Purchhapar farmers were thus, for example, 'not in a position to make such inroads on the *bhoor* as were effected elsewhere', noted Cadell. 'The Jats and Tagas, notwithstanding the progress they have made, are still merely working on the edges of the sand waste, while the people of Khutowlee [Khatauli] have caused *bhoor* to disappear from estates in which it once formed a large area.'[27] Thus, although it was in Purchhapar that the effects of the canal in improving and stabilising agriculture were described in 1870 as the 'most marked' in Muzaffarnagar, this does not mean that the degree of improvement showed through in canal revenues. A similar amount of canal water supplied to nearby Khatauli, which previously had a substantial area under wells and which was more fully developed and more thickly populated, yielded more by way of revenue from the sale of canal water. Not only did the canal 'capture' much of the irrigation formerly done by wells (the 'irrigable area' from wells fell from 11,000 to 2,500 acres in the thirty years following the canal's introduction),[28] but the availability of complementary factors allowed a more intensive development of cash-cropping. The fact that Khatauli's farmers grew, by the 1880s well over 7,000 acres of sugarcane –

[27] Government of the North-Western Provinces, Revenue Department Proceedings (hereafter NWP Rev.), June 1870, (proceeding no.) 47. (In the IOR.)

[28] It should be noted that irrigation statistics appearing in the official reports take several forms. 'Irrigable area', alternatively 'wet' (as opposed to 'dry') area, refers to the area irrigated in the normal course of cultivation. Such fields might bear two irrigated crops in a year, or receive water less frequently within the rotational cycle. The essential aim of the settlement officers in calculating the 'wet' area was to exclude 'extraordinary' irrigation for purposes of revenue assessment. The broad rule applied in the settlements of the 1890s and early 1900s was that fields actually watered during one of three years (excluding 'abnormal' years) were recorded as 'wet'. Settlement officers also took the precaution of excluding from their calculations the year previous to settlement, when the cultivators and *zamindars* were apt to refrain from irrigation – often letting their *kachha* wells fall in – hoping to reduce the level of the new assessment. In the subsequent batch of settlements, the 'wet' area was derived from the more reliable indication of two watered crops out of five 'normal' years. 'Irrigated area' refers generally to the area (usually gross) actually *receiving* water in a particular season or year. Statistics in text from 'Report on the Ganges Canal Tract', p. 74, in Cadell, *Muzaffarnagar SR* (1882).

twice as large as the cane area in Purchhapar – was
undoubtedly significant in terms of Canal Department
revenue, since sugarcane was usually grown with canal water
and bore a comparatively heavy water charge.[29]

In Meerut, too, it was 'greatly in favour of the immediate
success of a canal that it should have entered a thickly
populated country', where the manure supply was sufficient to
'cope with the exhausting powers of a practically unlimited
water supply' and where there was 'sufficient labour available
for its utilisation';[30] while in Cawnpore's dry Sengar–Jumna
doab 'thinness of population' was the 'principal obstacle to
successful – i.e., *remunerative* introduction of canals'.[31] Moreover, canal engineers, even more than some settlement officers,
were disposed to regard the inhabitants of villages without
wells as 'an idle, thriftless lot', and thus, by implication, either
undeserving of canal water or, more pertinently, incapable of
making full use of it.[32]

It is easy to see the appeal of the richer tracts to
revenue-seeking canal engineers, confident of their ability to
displace wells. The net effects upon production were never
calculated, since only gross figures affected the direct canal
revenue. On the other side of the account were, of course, the
costs. Here, too, the margin between revenue and costs was
kept at its widest by this policy, since serving fertile developed
villages involved lower distribution outlays. A large proportion
of a village under irrigated crops meant compact irrigation and
low distribution costs per acre irrigated. In sandy areas, where
individual villages put down only a limited area to valuable
crops, distribution outlays were greater and water loss through
percolation and evaporation considerably higher, thus reducing the 'water duty'.[33] Thus it was 'generally acknowledged'

---

[29] Although *munji* (fine rice), which also bore a relatively heavy water rate, was more
widely grown in Purchhapar than in Khatauli (3,000 acres as against 1,250 in 1871),
this difference was more than covered by Khatauli's larger wheat area (greater by
4,350 acres, or 45%). Statistics from *ibid.* and from J.O. Miller, *Muzaffarnagar SR*
(1892), Appendix F-1.   [30] Buck, *Meerut SR* (1874), p. 9.
[31] F.N. Wright, *Cawnpore SR* (1878), p. 53. Emphasis added.
[32] E.g. NWP PWD(I), March 1873, 43, Note by Capt. J. Ross.
[33] The 'water duty' was defined as the crop area irrigated per cubic foot per second
(cu-sec) flowing continuously for a defined period of time. The 'duty' varied
considerably between canals: the relatively compact EJC having a significantly

that the system of distribution 'had for its object the least initial expenditure and . . . the prompt collection of water-rates for the largest possible area of more valuable crops'.[34]

The overall returns to the government, of course, consisted of more than direct returns from crop rates.[35] Although the gains in the form of direct charges for the irrigation of valuable crops were limited in the backward tracts, rising rent levels, reflecting larger and more certain returns in agriculture, would be partially acquired as revenue through assessment revisions at settlement, or, in the interim (from 1875), through the 'owner's rate'.[36] A long-run financial maximisation strategy on the part of the colonial authorities might well have looked primarily to increased 'indirect' returns (land revenue) in the early stages of canal development and, as complementary resources permitted the expansion of valuable water-demanding crops, so 'direct' returns in the form of water rates would grow. The canal authorities, of course, exhibited a high degree of time discount in this matter, and for this reason alone the

higher 'water duty' than the LGC, the irrigation of which was more widely spread (EJC, 111 acres *kharif*, 183 acres *rabi*; LGC, 44 acres *kharif*, 139 acres *rabi*). The comparisons are derived somewhat crudely from recorded discharge at the canal head; different results would be obtained where discharges at the head of distributaries are used. It should be noted that such comparisons ignore the relative importance of differences in crops, soils, the skill of the cultivators, and so on, while at the same time reflecting such factors. (For a detailed technical discussion, see Buckley, *Irrigation Works*, pp. 270–7.) It is sufficient here to note that the *value* of each cubic foot of water entering the canal (as reflected in the water rates paid by users) was, in 1884–5, Rs 885 on the EJC and Rs 375 on the LGC. Maintenance charges, like capital charges, were greater on the LGC (Rs 1.89 as against Rs 0.91 on the EJC). *Irrigation Revenue Report of the North-Western Provinces and Oudh, 1885–6* (Allahabad, 1887), accompanying Resolution, p. 4. (Subsequent references to these annual reports will be abbreviated to *IRR*.)

34 'Report on the Ganges Canal Tract', p. 100, in Cadell, *Muzaffarnagar SR* (1882).

35 The water rate was levied directly upon the irrigator on the basis of crop type and area, and did not vary according to the number of waterings given to the crop. For detail, see Ch. 5 below. Additional receipts (navigation charges, levies on water power supplied to mills, charges for filling ponds, etc.) made up only a small fraction of the total revenue (less than 2% in the 1880s). Receipts from plantations amounted to around 2% also. *IRR 1885–6*, accompanying Resolution, p. 5.

36 Arbitrarily set at a proportion (one-third) of the occupier's (crop) rate, the 'owner's rate' was intended to capture for the government at least part of the rise in the assessable value of land which occurred as a result of canal irrigation's having converted 'dry' land to 'wet' after the settlement had taken place. See Ch. 5 below for details.

quick-yielding strategy was appealing, particularly as 'indirect' revenues could only be realised through enhancements at fixed intervals. About the only immediate tangible benefit to land revenue brought about by the canal was the ability at least to collect a greater proportion of the *jama* in drought seasons. Moreover, there was another reason why the Canal Department should have opted for maximising direct returns. Through to the late 1860s the revenue administration, partly for reasons of antipathy towards the Canal Department, insisted that 'canal irrigation had little or no influence upon the revenue'.[37] The Muzaffarnagar Settlement Officer, for one, considered that insistence upon this view was 'at least partly' responsible for the distribution policies chosen: 'The Irrigation Department was prevented from looking to indirect revenue as a justification for a system which would have necessitated a larger outlay on distributaries and a greater annual expenditure on distribution.'[38] 'The good of the country', Cadell concluded, had not prevailed against 'departmental necessities', while the 'influence on land revenue was depreciated by the revenue authorities'.[39]

New distribution policies and problems of application

Although greater control over irrigation works was achieved by local governments as a result of Dalhousie's 1854 restructuring of the public works administration, and the substitution of the Military Board system by that of a Chief Engineer under each of the local governments, changes in the distribution policy away from the overriding short-term objective of securing immediate revenue from water rates only took place in the wake of the settlement revelations of the late 1860s. Effectively, these

---

[37] It transpired that this reference was to the canal's impact on the rent of land formerly irrigated by wells, rather than on land 'ordinarily dependent on the seasons . . . in which well irrigation is impracticable'. The critical point, however, was that the canals were not to be *credited* with any portion of the land revenue: 'The Board [of Revenue] do not . . . perceive that the rent paid for irrigated land can in any sense be regarded as containing in it any share of the water-rate payable to the canal.' NWP Rev., Aug. 1868, 2, C. Thornhill, Note on Connection between Water-rate and Enhancement of Revenue, 3 Nov. 1866.

[38] 'Report on the Ganges Canal Tract', p. 100, in Cadell, *Muzaffarnagar SR* (1882).

[39] *Ibid.*, p. 101.

investigations were the first specifically to examine the impact of the canals, and they not only brought to light the extent to which short-sighted distribution policies had produced undesirable ecological consequences, but made it clear that the land revenue potential of canal schemes had been neglected.[40]

It would have been very difficult to remove canal facilities from previously well-irrigated areas, even had the canal authorities desired to attempt this. 'The truth is', observed one settlement officer, 'that the canal officers dare not withdraw any existing irrigation, and consequently cannot abandon even grossly offending distributaries till the water they have hitherto given is provided from some other source';[41] indeed, attempts to do so on the Agra Canal had 'met with an awful howl'.[42] Such an act, moreover, would have necessitated outlay on a new assessment. However, with the government about to embark on a new phase of loan-financed irrigation expansion, the issue of distribution gained prominence. In the light of what had been revealed about past distribution, the Lieutenant-Governor argued for a 'general standard' to be applied to the new projects, and the guiding principle established was that well tracts which actually irrigated one-third of the cultivated area were considered 'sufficiently protected' and should thus not draw off canal supplies; where well irrigation fell short of this, canal water could be supplied up to the point at which *total* irrigation amounted to 42.5%.[43]

An inspection of the projections made for the LGC by Major W. Jeffreys – or those for the associated remodelling of what was to be the UGC by Captain (later Colonel) C.S. Moncrieff – illustrates clearly that this broad principle had been accepted.[44]

[40] See NWP&O Rev., Aug. 1879, 37, Note by E.C. Buck (Dir. of Agriculture).

[41] Wright, *Cawnpore SR* (1878), p. 55. An exception was made in some low-lying parts of Sikandra Rao (Aligarh), where canal irrigation was withdrawn from seventeen *reh*-affected villages around 1880.

[42] *Indian Irrigation Commission, 1901–03, Minutes of Evidence, United Provinces of Agra and Oudh* (Calcutta, 1903) (hereafter *IIC(E)*), p. 20, evidence of F. Bion (Executive Engineer (hereafter EE), EJC).

[43] NWP PWD(I), July 1871, 24; NWP PWD(I), Aug. 1874, 111. This policy is also discussed in Ch. 6 below.

[44] E.g. NWP PWD(I), July 1871, 21, Note by Maj. W. Jeffreys; NWP PWD(I), July 1871, 20, Note by Capt. C.S. Moncrieff. See also Moncrieff's report on remodelling the Ganges Canal, NWP PWD(I), Aug. 1874, 107. So zealously did Jeffreys appear

Making use of information supplied by revenue officers, where physical conditions allowed, the direction of the main channels and their distributaries were more carefully tailored to local conditions. The decision not to extend the LGC through north Fatehpur, and from there into Allahabad, was clearly a response to settlement officer opinion; the provision of the Bewar branch (LGC) supplying the *bhur* tract of northern Mainpuri, the Bhognipur branch serving the *ghar* tract in Etawah, and the Anupshahr and Deoband branches of the UGC were all part of this overall strategy. The details of the schemes also reveal this: the Fatehgarh branch of the LGC, running down Etah's Kali Nadi–Burhganga *doab*, was 'deliberately designed with a view to serving the weakest portions of the tract'.[45]

In practice the scheme had significant drawbacks. The fixed proportion of 42.5% took no account of local capacity actually to use the water,[46] while the provision of canal water sufficient to make up the total irrigated part of a village to the arbitrary percentage effectively ignored the subsequent effect the presence of the canal had upon the stability of the existing wells.[47] The system also led to many inadequately irrigated villages being denied canal water. This occurred where 'irrigable' areas calculated at settlement were particularly liberal, as was frequently the case before an even remotely reliable criterion of measurement was established. In Fatehpur, for instance, the 1878 'method of demarcation made it inevitable that many fields were included [as irrigable] which were not really in reach of water'.[48] In another case, a similarly

to pursue the newly formulated principle that the Chief Engineer, Col. W. Greathed, upbraided him for writing 'with all the ardour of a proselyte' and for exaggerating past errors. Even then, one of the first acts of Greathed's less conservative successor, Col. Henry Brownlow, was to take even Jeffreys's modifications a stage further. He altered the projected route of a 76-mile section along the high bank of Etawah's Rind Nadi on account of the fact that the tract was 'abundantly furnished' with 'inexhaustible' spring-fed *kachha* wells. See NWP PWD(I), Sept. 1878, 2. This adjustment was one of the first made as a result of the new policy of consulting with district officers in planning canal extensions. See the detailed case made against the projected line by Crosthwaite, *Etawah SR* (1875), pp. 18–19.

[45] H.O.W. Robarts, *Etah SR* (1905), p. 12.   [46] NWP PWD(I), Aug. 1876, 47.

[47] NWP PWD(I), Aug. 1876, 53.

[48] C.L. Alexander, *Fatehpur SR* (1915), p. 18.

vast overstatement by a settlement officer – that four-fifths of the tract due to receive canal water from the projected Mat branch was 'protected' by wells – led to the postponement, for over thirty years, of the much-needed scheme.[49]

The most intractable problem, however, was that it was difficult to be selective in withholding supplies to villages, or parts of villages, within tracts through which the canal ran. On grounds of increasing land revenue, 'good sanitation', and equal distribution of increasingly restricted canal water supplies, the policy of local government was that the supplanting of wells by canal water was to be avoided. But engineers would still have found it tempting to seek high canal revenue/low outlay solutions when determining local distribution patterns. The revenue authorities, even if now more inclined to concede that canals contributed 'indirectly' to land revenue, were steadfastly against actually allowing a 'paper credit' on the canal accounts reflecting that contribution, and therefore the motive for seeking to displace well irrigation still existed.

Thus although it was made a condition, when the Deoband branch was approved for introduction into Saharanpur and Muzaffarnagar in 1879, that 'no water would be supplied to those lands that were irrigated from wells', there continued to be a bias in distribution towards well areas and 'the condition was gradually overlooked' putting 'out of use . . . hundreds of wells'.[50] Even detailed and carefully applied schemes for 'debarring' well lands from canal irrigation, as attempted on the water-scarce Fatehpur branch, proved impractical. It was hard to exclude even entire villages: those villages barred under the scheme – like those whose inhabitants had originally refused to allow distributaries into their village – were, according to C.L. Alexander, now 'asking for it loudly'.[51] Where part of the village was already irrigated it proved particularly difficult to keep the canal off well land, since the cultivators often preferred to water this rather than the outlying

[49] See UP PWD(I), June 1903, 11, A.C. Evans, Report on Mat Branch Extension Project.

[50] *Royal Commission on Agriculture in India, Evidence*, vol. VII (London, 1927) p. 646, testimony of L.S. Sinha (UP Zamindars' Association).

[51] Alexander, *Fatehpur SR* (1915), p. 11.

har.[52] When drought struck, and the reach of the canal extended into fields the well could not service in such a season, 'rights' to irrigation from the canal were subsequently claimed by cultivators.[53]

Despite the policy adjustments, then, and the real gains on some branches serving dry tracts, the tendency for canals to replace wells continued. In Meerut, for example, a 50,000-acre increase in UGC irrigation between 1874 and 1901 was accompanied by 'an equal or even greater decrease in irrigation from wells'. Overall, while Meerut's canal irrigation moved from 280,000 to 380,000 acres, well irrigation fell by 71,000, leaving a net increase of less than 5%.[54] Part of this decline, no doubt, can be accounted for by a more realistic appraisal of well capabilities, but the overlapping continued, and the reduction in well irrigation in such villages was real.

Analysis of 'appropriateness' of the new technology

External effects on wells: perspective on losses

The intrusion of canal channels into well areas caused many *kachha* wells to fall in. In Muttra, the Settlement Officer's enquiries in 1878 revealed that 'all the *kucha* wells in villages through which the main canal passes, and from which 5,000 acres used to be watered', had been made useless, the destruction being pronounced where the rise in the water-table was greatest, and particularly where wells had been built along lines of drainage, now subject to stronger subsoil percolation.[55] In Bulandshahr, a general rise in the water level of six feet resulted in the 'wholesale destruction of *kachha* wells and their almost entire supercession by canal irrigation'.[56] Not every village was able to substitute canal water for their ruined wells.

[52] *Report of the Irrigation Conference, Simla 1904*, vol. II (Calcutta, 1905), p. 52; also Government of the United Provinces of Agra and Oudh, Revenue Department Proceedings (hereafter UP Rev.), July 1909, (proceeding no.) 89. (In the IOR.)

[53] Note from Figure 4 the way in which drought years were followed by a permanently enhanced area of irrigation.

[54] Gillan, *Meerut SR* (1901), pp. 1, 5.

[55] R.S. Whiteway, *Muttra SR* (1879), pp. 13–14.

[56] T. Stoker, *Bulandshahr SR* (1891), p. 47. (Unless stated otherwise, this and subsequent references are to vol. I.)

Occasionally, where it was necessary for the lines of the canal and the watershed ridge to diverge, villages in between could find their wells affected and yet, due to their level, be unable to draw alternative supplies from the canal.[57]

Such destruction, however, was by no means universal. Where subsoil contained more clay, temporary wells were less affected. In *pargana* Marehra (Etah), *kachha* wells in the vicinity of the canal appeared to last 'as long as ever',[58] and an efficient system of drainage, it seems, prevented a rise in the water level and thus the destruction of temporary wells in Aligarh.[59] Moreover, even though earthen wells were frequently affected, in most places masonry wells could still be built 'when required and with the same facility as before' and the supply in them was 'better than ever'.[60] Indeed, just as some villages could lose entirely their source of irrigation as a result of the canal, so it was possible for villages out of the canal's reach to have an improved water supply in their *pakka* wells, as was the case in parts of eastern Bulandshahr.[61]

The picture, then, regarding the effects of the canal on wells is mixed rather than one of sweeping destruction. It would clearly not be possible to gauge the extent of *kachha* well loss, since their numbers varied with the season. It is important, though, not to overstate the utility of such wells, particularly those supplied by percolation rather than springs.[62] Where the *mota* was absent, or lay too deep, a well shaft could be sunk to a little below the percolation level, whereupon water would filter through the porous material into the bottom of the well. A brushwood lining was usually required to impede the influx of sand and thus prolong the well's working life, which was seldom more than one season in any case. Its main problem, W. H. Moreland found, was its 'critical discharge': that its surface area determined the rate at which water could be drawn and that to exceed this rate was to cause an influx of sand which

---

[57] See e.g. Buck, 'Note on the Effect of the Canal', p. 225.
[58] S.O.B. Ridsdale, *Etah SR* (1874), p. 67.
[59] W.J. Burkitt, *Aligarh SR* (1903), p. 16.
[60] McConaghey, *Mainpuri SR* (1875), p. 40.
[61] Stoker, *Bulandshahr SR* (1891), p. 47.
[62] As F.N. Wright noted, 'The instances I have met with of a cultivator's well tapping the permanent spring have been rare.' NWP&O Rev., Oct. 1879, 11, F.N. Wright, Report on Well Irrigation in Pargana Ghatampur (Cawnpore).

eventually reached the percolation level. Since the 'cost and difficulty of construction' rose sharply with increased diameter, 'this simple well could never give a large yield'.[63] Such wells were 'universally dug' prior to the canal's intrusion into Etah's central *doab*, but their scanty supply was 'only sufficient for a single bullock run',[64] and even then, as in Aligarh, 'half a day's labour [often] would suffice to exhaust the supply'.[65] In Mainpuri's Bewar *pargana*, where also few *kachha* wells reached the spring, the supply was so deficient that 'bullocks could not be worked with profit'.[66] It was not uncommon for such wells to be worked by hand with the aid of a pulley (*dhenkli*) or a weighted lever (*rati*), irrigating 'often a mere fraction of an acre'.[67]

Although some spring-fed *kachha* wells might last for years, the majority succumbed in the rains, and in most areas an effective, enduring exploitation of subsoil water resources could only be ensured by the construction of a *pakka* well, one characterised by internal brickwork – dry or cemented – extending either part-way or entirely up the shaft length. Although in low-lying and submontane areas these could be of large diameter and thus tap the percolation supply, in most cases they were able to delve deeply enough to tap the *mota*. However, the cost of materials, and the frequent necessity for trial borings to ensure they tapped a strong spring, made *pakka* wells costly to construct, averaging Rs 200–300 for a two-lift well[68] – an outlay beyond the resources of most cultivators.

Previously, it is clear, local *zamindars* played an important role in the financing and construction of such wells, but by the 1870s this activity had clearly declined. According to Cawnpore's Settlement Officer, it was 'most noticeable how few wells the *zamindars* now build compared with the cultivators'.[69] In Bulandshahr, 'as a rule', the only *zamindars* who made wells themselves were self-cultivating members of *bhaiachara* and

[63] W.H. Moreland, 'Wells in the Gangetic Alluvium', *Agricultural Journal of India* (hereafter *AJI*), IV (1909), pp. 34–42.

[64] Ridsdale, *Etah SR* (1874), p. 5.   [65] *Aligarh DG* (1875), p. 381.

[66] McConaghey, *Mainpuri SR* (1875), p. 159.   [67] *IIC*, p. 201.

[68] Clibborn, *Papers Relating to the Construction of Wells*, p. 20. In addition, equipping costs would amount to around Rs 20 (exclusive of bullocks): bag (or *pur*), Rs 5–10; rope, Rs 8; wheel, Rs 2–3. *Ibid.*, p. 9.

[69] Wright, *Cawnpore SR* (1878), p. 6.

pattidari bodies; and on the 'rare occasions' when *zamindars* made wells for occupancy tenants they collected rents near to tenant-at-will levels.[70] Part of the problem, contemporaries recognised, stemmed from the purchase of land by the 'new zamindars', the traders and moneylenders 'disinclined to sink money in such improvements', 'whom', one settlement officer reflected, 'our laws have created'.[71] Certainly, the 'fine wells' constructed by Cawnpore's progressive landlord, Maxwell, had 'been allowed to fall into ruin through the apathy of the purchasing landlords'.[72]

While the *zamindars* had largely ceased to contribute their support to well-building, the operation of legislation relating to tenure was not conducive to encouraging construction by tenants. The Irrigation Commission was 'doubtful whether . . . the permanent tenants had received such full protection against enhancement of rent, in respect of the increased value imparted to their land or its produce owing to improvements made by them, as the landowners obtain from government against enhancement of the land revenue assessment on lands benefited by improvements which they have executed'.[73] While, on the one hand, the permanent tenant was legally protected from rent increases for periods of ten years, and, moreover, from any enhancement claimed on the basis of an increase in the productive powers of the land where such an increase was the result of the tenant's own investment (such as the sinking of a *kachha* well or the conversion from earthen to *pakka*), another provision enabled the landowner to claim a rent enhancement on the grounds that his rent was below that paid by nearby occupancy tenants for similar land enjoying similar advantages.[74]

Tenants with *pakka* wells were always liable to claim legal rights troublesome to the *zamindar*'s control of his estate. Thus, if there were landlords unprogressive in their attitude to well construction, there were others particularly obstructive in this

[70] Records of the Commissioner of Meerut, Meerut (hereafter MCR), Files of the Board of Revenue (hereafter BOR), VIII, 14(c), Ser. 2, D.C. Baillie, Assessment Report (hereafter AR) of Parganas Agauta, Siyana, and Shikarpur (Bulandshahr) 1889, p. 31.

[71] McConaghey, *Mainpuri SR* (1875), p. 9.

[72] Wright, *Cawnpore SR* (1878), p. 6.   [73] *IIC*, p. 57.   [74] *Ibid.*

regard. This was especially so in Bulandshahr, where the Settlement Officer noted that the proprietors 'not only failed to improve their property, but their policy had been directly and actively designed to prevent and obstruct improvements. It is almost universal practice for landlords to prevent their tenants from making masonry and half-masonry or, in extreme cases, earthen wells.'[75] One official had actually witnessed the refusal of a tenant's request made to the son of the Rajah of Mursan (Aligarh) to convert his well to *pakka*.[76] So intent were the Bulandshahr landlords upon gaining 'absolute control' over their tenants, and in preventing the growth of 'adverse rights of any sort', that they would submit to an abatement of rent in the case of earthen wells becoming impracticable rather than allow their conversion to *pakka*.[77] *Pargana* Dibhai *bania*–proprietor, Bhola Nath, actually dismantled a masonry well of his own when his tenants established claims to occupancy rights in the fields irrigated from it.[78] It is quite clear, also, that in their efforts to maintain control over their tenants, the Bulandshahr landlords had received the backing of the civil courts: suits of the Nawab of Chhitari, and even the lax Skinner estate, among others, had resulted in the issue of decrees closing wells constructed without permission.[79]

For a tenant to acquire a masonry well was a difficult enough problem without the opposition of his landlord. The financial impediment facing any cultivator was heightened for tenants by the non-transferability of occupancy rights (limiting his collateral), and by the rigidity of the rules relating to *takavi* advances.[80] This was also on age of increasing subdivision, when *khetbat* (scattered) partitions often reduced the area of land held in a compact block to below that which would justify

---

[75] Stoker, *Bulandshahr SR* (1891), p. 99. In cases where outright prohibition was not encountered, conditions were often insisted upon which neutralised the advantages to the tenant.

[76] *IIC(E)*, p. 257, evidence of E.A. Molony.

[77] Stoker, *Bulandshahr SR* (1891), p. 100. Proprietors felt, according to E.A. Phelps, 'that they lose not only in money but in prestige and power if occupancy rights accrue; for occupancy tenants not only pay less rent, but withhold extraneous contributions and sometimes even obedience and courtesy'. E.A. Phelps, *Bulandshahr SR* (1919), p. 10.

[78] Stoker, *Bulandshahr SR* (1891), p. 100.   [79] *Ibid.*

[80] NWP&O Rev., Oct. 1879, 11, F.N. Wright, Report on Well Irrigation in Pargana Ghatampur (Cawnpore).

a substantial well.[81] Such factors, together with the fact that landlords frequently regarded permanent wells built by their tenants as against their interests, meant that many cultivators had to be satisfied with *kachha* wells, which were precarious and 'liable to fail when most needed'.[82]

Well-users were dogged by a range of localised problems, from which those with *pakka* wells were by no means immune. Well systems were especially susceptible to variations in climatic cycles and movements in the water-table. In Ghatampur *pargana*, Cawnpore, a fall of fifteen feet in the water-table was recorded in 1877–8 due to the combined effects of deficient rainfall and increased irrigation from wells.[83] Many wells consequently suffered impaired irrigating capacities, as they did over a prolonged period during the dry cycle of the late 1890s.[84] More damaging long-term falls occurred in some tracts. A serious lowering of the table occurred in Muttra, where the level in Mahaban *tehsil* fell from 45 feet in the 1870s to around 65 when the Mat branch extension opened in 1904. By then, 'the cost of masonry wells was becoming prohibitive and the digging of earthen ones impossible'.[85] While the canal extension had the effect of restoring the water-table to the old level, parts of Agra, Aligarh, and Etawah continued to face this problem. The fall in western Aligarh was, by the early 1900s, 'affecting all wells alike'; earthen wells had become impracticable where the water was more than sixty feet down, and 'many of the masonry wells [had] gone dry'.[86]

[81] That this was a major problem for many peasants is confirmed in *ibid.*

[82] Stoker, *Bulandshahr SR* (1891), p. 100.

[83] NWP&O Rev., Oct. 1879, 11, F.N. Wright, Report on Well Irrigation in Pargana Ghatampur (Cawnpore).

[84] Gillan, *Meerut SR* (1901), p. 6; Alexander, *Fatehpur SR* (1915), p. 11; and S. Ahmed Ali, *Aligarh SR* (1943), p. 3.

[85] H.A. Lane, *Muttra SR* (1926), p. 6; *Royal Commission on Agriculture, Evidence*, VII, p. 379, testimony of Dr R. Mukherjee (Univ. of Lucknow).

[86] Burkitt, *Aligarh SR* (1903), p. 16; also UP PWD(I), June 1903, 11, A.C. Evans, Report on Mat Branch Extension Project. This is a case of the complex environmental impact of canal irrigation and related activities. While some localities benefited from improved well supplies as the canal raised water levels, attempts to reduce the incidence of waterlogging through drainage were sometimes disadvantageous to neighbouring well areas. According to Evans, the long-term fall in water-levels described in the text was mainly due to improvements made to the line of the Karwan Nadi and surface drainage installations in Aligarh. Other contributing factors were thought to be the dryness of the Jumna River (with two

A cycle of dry years at the end of the 1870s caused the establishment of the troublesome *baisuri* weed in the Central Doab. This was a particular problem in well villages, since the weed 'appeared to flourish in lands irrigated by wells'.[87] It was found in loam soil and necessitated repeated weeding to prevent it from crowding out the sown crops. The 'only method of dealing satisfactorily with this pest' was, settlement officers discovered, 'to drown it', and the only effective treatment proved to be canal irrigation.

The fall in the water-level in some localities could render well water brackish. This was not a new problem, for many wells in Mainpuri, Agra, and Muttra had long contained water 'brackish or oily and generally saturated with noxious salts'.[88] The water was not altogether useless for cultivation, but was invariably harmful to germination; sowing therefore depended upon the rains. Its harmful qualities, moreover, tended to be exaggerated by drought conditions, and could have lasting effects: 'Sometimes the autumn crop is a failure', wrote the Muttra Settlement Officer, 'because in the previous spring harvest the water of a particular well was used.'[89] As a result, the substitution of canals for wells in a large part of Muttra, as well as in other localities, had 'the same protective effect as would be given by the introduction of new irrigation in a tract which was before entirely dry'.[90]

The evidence points firmly to the conclusion that, although wells were not always destroyed by the canal, in many areas they displayed shortcomings which would have made their owners gladly turn to the canal. The disappearance of unprotected wells in the face of the canal was, as one settlement

canals drawing from it upstream), years of low rainfall, and a related increase in the use of wells. Clibborn noted, in fact, that parts of Aligarh had been 'collecting areas' for water, which had effectively kept up the level of water in neighbouring Muttra. Muttra's cultivators thus bore some of the costs of Aligarh's efficient drainage system. Clibborn, *Papers Relating to the Construction of Wells*, p. 7.

[87] H.R. Nevill, *District Gazetteers of the United Provinces of Agra and Oudh* vol. VI, *Aligarh* (Allahabad, 1909), pp. 17–18. (Subsequent references to this series will be abbreviated to the district name, followed by the letters *DG* and the date of publication. The latter will distinguish this series, published between 1903 and 1921, from that of E.T. Atkinson, which was published between 1874 and 1886. All volumes published at Allahabad.)

[88] W.J.E. Lupton, *Mainpuri SR* (1906), p. 3.

[89] Whiteway, *Muttra SR* (1879), p. 18.   [90] Lane, *Muttra SR* (1926), p. 6.

officer remarked, 'only what might be expected from their temporary and precarious nature'. The same might be said of brackish areas and the like; but, in fact, the officer went on, 'masonry and half-masonry wells also fall into disuse, though the rise in the spring level actually adds to their working power'.[91] The switch away from wells was wholesale, affecting all types of wells, even those without operational deficiencies. Cultivators preferred the canal. Thus, of Saharanpur, for example, it was reported that 'wherever canal water reaches a village well irrigation usually ceases', sometimes even where the canal water could not reach.[92]

There were exceptions to the trend, but they relate to fairly well-defined circumstances. Settlement Officer C.H.T. Crosthwaite could think of no instance where the Etawah cultivators did not take water, 'except in one or two cases where the fines inflicted for breach of canal rules . . . disgusted them, and in the case of market garden cultivation'.[93] Etawah's Kachhi smallholders, like the Kurmis in Bulandshahr,[94] were said to 'always prefer' wells for their garden cultivation, and wells 'could be found working in the *gauhan* lands of villages which could be irrigated flush from the canal'.[95] Similarly in Farrukhabad, where in the 'best and most highly manured fields' wells were 'kept up with some difficulty for the irrigation of opium and tobacco', while all the surrounding fields were 'resigned to the canal'.[96] Ahir cultivators, who were said to favour their *bara* cultivation, would take canal water where it was available 'without unreasonable labour', but where canal water had to be lifted on to the fields they often preferred wells, particularly where 'daily labourers' were absent.[97] Faced with this choice, numerous Ahir villages in eastern Bulandshahr preferred their *pakka* wells, which were kept in full supply by the effect of the canals.[98]

[91] Stoker, *Bulandshahr SR* (1891), p. 47.
[92] Porter, *Saharanpur SR* (1891), p. 13.
[93] Crosthwaite, *Etawah SR* (1875), p. 17.
[94] Stoker, *Bulandshahr SR* (1891), p. 47.
[95] Crosthwaite, *Etawah SR* (1875), p. 15.
[96] Buck, 'Note on the Effect of the Canal', p. 213.
[97] NWP PWD(I), March 1873, 43, Note by Capt. J. Ross.
[98] MCR, BOR, VIII, 14(c), Ser. 2, D.C. Baillie, AR Parganas Agauta, Siyana, and Shikarpur (Bulandshahr), 1889, pp. 10–11.

Operational comparisons between wells and canals

Clearly, wells had their shortcomings. But it is still to be explained why cultivators with effective well facilities chose to incur an increased cash outlay in exchange for a canal supply which was apt to be unreliable and which in many cases had to be lifted up to the level of the fields.[99]

On the face of it, a good well would seem to have served the typical holding adequately. Under the *lagor* system, where there was a single, moderately inclined ramp, up which the bullocks were backed after the leather bag (*pur*) had been emptied,[100] a pair of animals irrigated two to five *rabi* acres and up to 1.5 in the *kharif*. Using the more efficient *kili* system, with two (steeper) ramps, which allowed the use of more than one pair of bullocks (the animals being unpegged after each lift and walked up the ramp to a feeding trough), a single lift irrigated 5–7.5 *rabi* acres and 1–2.5 during *kharif*. With multiple lifts the norm on substantial wells, a *kili* well serving one or more families would irrigate an area ranging from an average of eight *rabi* acres in Muttra, where the water was fifty feet down (and where no *kharif* irrigation took place), to between eleven and thirteen *rabi* acres (and two to six *kharif* acres) in most Doab districts, where the water level was generally at 18–24 feet.

The problem with such wells was not so much that they tied up the cultivator's resources in the form of fixed capital, but that in order to carry out the irrigation they tied up productive factors in their operation. The extent to which they did this was very substantial indeed, for the time taken to give a single watering to one acre was considerable. One estimate suggests that with two pairs of bullocks it took a minimum of five to six days to water one acre with water at 25 feet, and Buck's measurements in *pargana* Tirwa revealed that with one pair of

[99] 'Lift' (*dal*) irrigation was most widespread in the Lower Doab. It constituted over half of the canal irrigation in Cawnpore, while in Meerut and Bulandshahr 80–90% of canal water was delivered to the fields by 'flow' (*tor*), also referred to as 'flush'.
[100] With powerful cattle, a load of 500 pounds could be lifted. For a description, see Clibborn, *Papers Relating to the Construction of Wells*, pp. 9–18. Data in this paragraph are taken from Clibborn's survey of 150 Doab wells. *Ibid.*, Appendix 1 (Table B).

bullocks, eight days was the norm.[101] In contrast, canal irrigation by lift could water an acre in a day, sometimes a little more,[102] and 'almost any quantity of land' could be watered by canal water given flow, although three acres per day was usual.[103] A single watering of one acre by well, therefore, on the basis of Buck's observations of three men to a well lift, amounted to 24 labour-days. The average of eight waterings necessary for cane would thus require the duty of 192 labour-days; the three waterings for indigo or wheat, 72.[104] With the canal, the actual operation required more labour for the duration – six men working two *beris* (baskets of split bamboo), with one or two guiding the water into the field – but since an acre could be covered in a day, eight labour-days sufficed for each watering. Thus, total irrigation labour requirements for an acre of cane was 64 labour-days (and wheat/indigo 24).

Clearly, such a vast difference in labour requirements, even where canal water had to be lifted, would seem to explain why the wholesale switch to canals took place. Yet the important point is how the peasant cultivator saw the new labour-saving technology; often, as Crosthwaite pointed out, 'to the average cultivator' the canal appeared 'an expensive business, more costly than his well'.[105] If there is doubt in the case of flow canal irrigation, in the case of lift the difference in cost clearly favoured the well, and hence the issue centred upon that sensitive area in peasant decision-making: the need to make payments to outsiders. The difference in outlays between the systems arose because of the need to hire so much of the labour required to lift canal water, while the drawn-out well irrigation could be manned mainly by family members. Of the 804 men working on canal irrigation in Buck's sample of Tirwa villages, 224 were owners of the field (or relatives), 72 were *jitas* (exchange labour), and 509 (or 63%) were hired labourers, on

101 F.N. Wright, *Memorandum on Agriculture in the District of Cawnpore* (Allahabad, 1877), p. 12; Buck, 'Note on the Effect of the Canal', p. 216, and discussion following *ibid.*, p. 228.

102 Buck, 'Note on the Effect of the Canal', p. 215.

103 *IRR 1873–4*, pp. cvi–cvii; J.F. Duthie and J.B. Fuller, *Field and Garden Crops of the North-Western Provinces and Oudh*, pt 1 (Roorkee, 1882), p. xviii; Wright, *Memorandum on Agriculture*, p. 13.

104 Buck, 'Note on the Effect of the Canal', p. 215.

105 Crosthwaite, *Etawah SR* (1875), p. 18.

two *annas* per day plus a quarter *seer* of *gram*.[106] Of every eight men working, therefore, five would be hired. This contrasts with well irrigation, where of the three operators only one was normally hired, and at a slightly lower daily rate.[107] Leaving aside the problematical issue of the costs of home labour and bullocks on the grounds that they had to be maintained in both cases, Buck calculated that the irrigation outlays for the two systems were Rs 6 (cane) and Rs 2.25 (wheat) with wells and respectively Rs 8.5 and Rs 3.5 with the canal; a difference, therefore, of Rs 2.5 (cane) and Rs 1.25 (wheat).

Where canal water required two lifts these margins in paid-out operating costs widened to Rs 8.5 and Rs 2.25.[108] At this point, even though two-lift canal irrigation was not significantly slower in watering an acre, entailed the same water rate, and still provided the substantial labour-saving over wells of more than 60 days for cane and 24 for wheat, some cultivators, when faced with such costs, preferred their wells. In Sookhee village, Buck observed that canal irrigation had only replaced wells when *rajbaha* adjustments made water available on one lift, as opposed to two or three previously; in Khundee village, wells were still used wherever two lifts were required.[109]

There were, however, operational advantages associated with the canal which explain why many villagers were prepared to use the canal even where labour payments associated with two (and sometimes three) lifts were involved.

[106] Figures from Buck, 'Note on the Effect of the Canal', p. 216.

[107] *Ibid.* Hire rates in canal villages tended to be slightly above those prevailing in well villages.

[108] The figure for canal outlay includes the water rate (Rs 3.5 for cane and Rs 1.5 for wheat). Hidden charges make these figures underestimates of the true costs, since there were a number of *haqs* and bribes related to the use of canal water, and the 'owner's rate' also had to be found by the irrigator (see Ch. 5 below). Against this, bullock costs were clearly higher in the case of well villages, since cattle working at wells required more substantial food than those lightly worked in canal villages, and could not be turned out to graze. In addition, due to the greater wear and tear they experienced, bullocks in well villages had to be replaced more often. See McConaghey's comments in discussion following Buck, 'Note on the Effect of the Canal', p. 229.

[109] Buck, 'Note on the Effect of the Canal', p. 218. Two-lift canal irrigation was nonetheless fairly common in Mainpuri at the time, which suggests that many cultivators were not deterred by the extra costs. See discussion following *ibid.*, pp. 230–1.

Two fundamental constraints upon production were dramatically affected by the substitution of one form of irrigation for another, and together they explain the cultivator's revealed preference for canal irrigation.

In the first place, even where good wells were concerned the irrigating capacity of wells was limited. As the Director of Agriculture remarked in the 1920s, 'The fact that even in the tracts most favourable for irrigation a larger area is not irrigated has only one interpretation – the cultivator has not the capacity to lift more than he does.'[110] The time taken to irrigate an acre placed severe limitations on the area which could be irrigated. Even if the wells could maintain the supply, bullocks could not work for more than twenty days a month 'without serious injury'.[111] In the dry months, therefore, with cane requiring water every fortnight, it was not likely that more than 1.5 acres could be irrigated by one pair of bullocks.[112] Not surprisingly, therefore, limited areas of one or two crops (mainly sugarcane and cotton) was the typical *kharif* pattern of well irrigation.[113] Wells, summed up Crosthwaite, 'require a large livestock and great labour . . . if a farmer has to work his well he cannot sow more sugar and wheat than he has time to irrigate'.[114]

The other major aspect of the innovation which made it such an advance on wells was that the factors tied up in the actual raising of water to the surface, labour and bullocks (leaving aside capital tied up in the facility itself), were the very factors which could otherwise be utilised in cultivating operations associated with expanding production. Well irrigation competed for resources which could only be used to raise water at the expense of the area under or the quality of cultivation. Both

---

[110] H.M. Leake, 'The Trend of Agricultural Development in the United Provinces', *AJI*, xviii (1923), p. 16.

[111] Clibborn, *Papers Relating to the Construction of Wells*, p. 18.

[112] Buck, 'Note on the Effect of the Canal', p. 222.

[113] In his sample of wells for Meerut, Clibborn found that the *kharif* irrigated area was half that of the *rabi*, and that while sugar accounted for 63% of the area (37% cotton), it took almost 90% of the water, since cotton received only one watering. In Cawnpore, two waterings for the cotton crop accounted for over 80% of *kharif* irrigation, and in Aligarh for 100%. Many wells simply did no *kharif* irrigation at all. Clibborn, *Papers Relating to the Construction of Wells*, Appendix 1.

[114] Crosthwaite, *Etawah SR* (1875), p. 18.

labour and bullock shortages occurred in the course of the
agricultural calendar, often their underemployment during one
period being an outcome of the production constraint caused
by their relative shortage at another. In some parts, labour was
the critical constraint holding back the *rabi* area,[115] while at the
same time the short supply of cattle was a major contributor to
inferior *kharif* production.[116]

As Buck demonstrated, for the outlay of a relatively small
sum, a peasant cultivator could on average release the labour of
two family members for a total of 56 days each by using lift
irrigation from the canal rather than working his well to irrigate
his acre of cane. In the case of wheat the number of days was 21
each. For cultivators with flow canal irrigation, the speed at
which irrigation could be accomplished, allied to the fact that
hired labour was not necessary, made the advantage even more
dramatic: as few as eight labour-days would suffice to water an
acre of cane to maturity. Moreover, the substantial labour-
saving effect was entirely internalised by the family enterprise.

On average, bullocks in well villages spent around 120 days
each year hauling water.[117] With the adoption of canal
irrigation they were free for redeployment. It was also possible
to have them properly rested and in good condition for peak
periods, such as the early *kharif*, when the animals – having
worked wells through the spring and assisted with harvesting –
were then under pressure to irrigate the hard-baked fields prior
to ploughing them: 'After treading out the corn, a day or two's
rest' was often 'absolutely necessary', and this the bullocks got
in the canal villages, while the land was being 'softened for the
plough by manual labour in canal villages'. Naturally, rested
animals accomplished the ploughing more efficiently.[118]

It is clear why, therefore, the canal had such a dramatic
effect even in well villages, since a vast supply of complementary resources was set loose, particularly where flow
irrigation was available. In Khatauli *pargana* (Muzaffarnagar),

---

[115] R.F. Mudie, *Agra SR* (1930), pp. 6–8.

[116] *Ibid.* 'The response of crop production to increments of bullock power is very high',
according to K. Singh, B. Goel, and V. Murty, 'Estimation of the Availability of
Bullock Power in Certain Tracts of India', *Agricultural Situation in India*, XXVI (1971),
p. 487.

[117] Buck, 'Note on the Effect of the Canal', p. 223.   [118] *Ibid.*, p. 219.

for instance, the bullock power released from the treadmill of water-raising on nearly 7,000 acres was 'sufficient for all the work connected with the increased cane cultivation [including processing] and for the ploughing of 10,000 acres besides'.[119] The saving in labour was unquestionably the reason behind the more largely increased area and the 'more generally maintained quality of the cane crop' in Khatauli than in neighbouring *parganas* formerly with fewer wells. Prior to the canal, many of Khatauli's dry estates yielded only inferior crops reared by absentee cultivators, who, 'however industrious, had plenty to do in their own (well irrigated) villages'.[120] When the canal arrived, 'the men and the cattle working the wells were set free, and at once began to improve the cultivation in their own and adjoining townships'.[121]

The reason why canal water was preferred to wells, even though, as critics have pointed out, some crop yields could be lower on canal-watered land, is also explained by the joint effect of increased irrigating capacity and released productive factors upon cropping patterns. The high unit costs of water obtained from wells ensured that the supply was carefully applied (via numerous *kiari* compartments)[122] to the limited area it would cover, and that such land was meticulously cultivated. This land would have priority over manure supply; it was put down to the high-value crops, and large amounts of labour were lavished upon it. Canal water was not paid for by volume, and its cheapness in terms of quantity encouraged the cultivator not only to dispense with the high degree of input intensity he had had to lavish upon his well-irrigated plot in order to obtain favourable returns on the expense – including opportunity cost – of lifting well water, but also to spread inputs over a larger area. An increased supply of cheap water meant that the cultivator could expect an acceptable rate of return on land which it was impossible to cultivate intensively when he

---

[119] 'Report on the Ganges Canal Tract', p. 71, in Cadell, *Muẓaffarnagar SR* (1882).

[120] Uttar Pradesh State Archives, Lucknow (hereafter UPA), Files of the BOR, 'Different Dis ricts', Muzaffarnagar, Box 8, File 35, A. Cadell, Rent-Rate Report (hereafter RRR) of Pergunnah Khataoli, 1872, pp. 35–6.

[121] *Ibid.*

[122] In well fields, an elaborate system of *kiaris*, or compartments made up of low earthen ridges, assisted the cultivator in distributing the water evenly over the field. See J.A. Voelcker, *Report on the Improvement of Indian Agriculture* (London, 1893), p. 74.

relied on his well. The cultivator therefore irrigated a larger area than before, though on average he cultivated it less intensively than he had the fields near his well. The sharp distinction, therefore, between *bara* and *har* was eroded, as, on balance, the trend was for the peasant to forfeit some of the output he could have expected from his well-watered fields, but to gain from the cultivation of better-quality crops over an additional area which was previously only infrequently irrigated, if at all. This caused the cultivator 'to extend the area of tolerably high cultivation',[123] and in this way to increase the total product of his holding. Hence, 'the whole tendency of Jat and Rawa cultivators' was to 'secure a large average produce, rather than the excessive fertility of a few fields'; indeed, the Jat in particular became well known for distributing manure and his best crops over his entire holding.[124]

Faced with a trade-off, the cultivators chose the option which, in some cases, meant that they achieved lower overall yields while increasing their income. Such a result was not the inevitable outcome of using canal water (although an unexpected degree of uncertainty of supply often had this effect). In many cases, the yield differential simply reflected the cultivator's adjustment to the new set of relative input scarcities. In villages where only a small area of canal irrigation was physically feasible, the especially high cultivation associated with wells was found on canal land.[125] The same applied

---

[123] *Muzaffarnagar DG* (1876), p. 545. Obviously, the scattered nature of holdings restricted the portion of any holding which could be irrigated from a well. The canal immediately permitted irrigation to be extended to the parts of the holding which were beyond the reach of the wells.

[124] UPA, BOR, 'Different Districts', Muzaffarnagar, Box 8, File 35, A. Cadell, RRR Pergunnah Khataoli, p. 55. This particular effect accounts for the outwardly surprising finding that the assessed rates in Muzaffarnagar had changed little in thirty years, despite the introduction during that period of the canal. According to Cadell, the average quality of each category of land had worsened. The explanation, in fact, was to be found in the *proportion* of total area in each category. As a result of the canal, a much larger area had come to be classified as 'irrigated', though much of it was not cultivated as intensively as the formerly small well-irrigated area. The larger 'irrigated' category had, of course, deprived 'dry loam' of much of its most productive land, while the extension of cultivation had, by moving onto the marginal land, reduced the average quality of *bhur*. This observation did not conflict with the fact that cultivation had increased overall. NWP&O Rev., June 1879, 74, Cadell to BOR, 8 June 1878.

[125] NWP&O Rev., June 1879, 74, Cadell to BOR, 8 June 1878.

where canal water had to be lifted and produced 'luxuriant' crops.[126] Any resource, in fact, available in relative abundance will not be used particularly efficiently. The farmer with canal irrigation, who could now cultivate 'thirty *bighas* with as much facility as the man dependent on well irrigation [could] his twenty *bighas*',[127] was not especially interested in maximising per acre yields, particularly with land as underutilised and water as freely available (and labour as scarce) as it was in most of east Muzaffarnagar's *parganas* in the 1860s. Where the balance of resources was different, as was the case in Khatauli *pargana*, where expanded cultivation was not a significant option in reorganised production, the differential in yields was much less pronounced, as Cadell indicated: 'It seems unquestionable that the more largely increased area and the more generally maintained quality of the cane crop in [Khatauli] may be attributed to the labour set free by the canal.'[128] Indeed, only as labour and bullock resources moved into line, and the supply of land became an increasing constraint on expanded production, would serious attempts have been made to improve yields.

Given this type of response to canal irrigation, it is not difficult to explain departures from the pattern of substitution. A particularly unreliable canal supply, an inordinate amount of interference by subordinate officials or other intermediaries, a scarcity of labour for hire to lift canal water – all of these factors could affect the cultivator's ability to gain from the new opportunities, and could make his well a better proposition. This was particularly the case where two lifts were involved, and the associated extra cash needs to be achieved through reallocation of resources grew larger. It was the case, too, where the cultivator, a Kurmi or a Kachhi perhaps, worked only a few acres. He not only relied on achieving outturns of garden-standard crops higher than the canal could be relied upon to produce (particularly given the limited voice he had in village water allocations), but had restricted scope for redeploying his freed resources on the rest of his holding.

---

126 'Settlement Report of the District of Muzaffarnagar', p. 59, in Cadell, *Muzaffarnagar SR* (1882).

127 *Meerut DG* (1875), p. 246.

128 'Report on the Ganges Canal Tract', p. 74, in Cadell, *Muzaffarnagar SR* (1882).

Complementarity of irrigation technologies

Although the early trends pointed to a dramatic demise of wells, their role changed to one often complementary to rather than in competition with canal irrigation. In Meerut canal *parganas* Chaprauli and Baraut, by the mid-1860s, wells had 'in great measure fallen into disuse' and well-sinking was 'almost entirely abandoned'.[129] One settlement officer recorded that 'for many years after the introduction of a new canal there was a cessation of well-making . . . even villages near the canal, but not getting canal water', seemed to 'lose heart and look on working a well as an intolerable toil'.[130]

The trends to the contrary quickly became evident, however. Already, in 1866, 'enterprising Jat proprietors' in Meerut could be found using their wells to 'guard against the uncertainties of canal supply';[131] cultivators right next to distributaries in Bulandshahr 'have sunk and are sinking numerous masonry wells', commented a canal engineer.[132] The trend gathered considerable force and continued throughout the period: in 1940, Meerut's Settlement Officer had all over encountered cases of cultivators lining their earthen wells, even where there was 'absolutely no necessity for wells'; more than 30% of upland wells constructed since 1901 were in canal circles.[133] Many canal villages in Muzaffarnagar were 'fully supplemented by *pakka* wells'.[134] This was particularly the case where villages were near the termination of a canal or distributary: in Cawnpore it was noticed that, even though the canal flowed right past their land, wealthier cultivators constructed masonry wells since in a dry year their water supply was frequently used upstream.[135]

That the motive was clearly to safeguard water supplies

---

[129] W.A. Forbes, Chaprauli Pargana Report, 1866, p. 8, in Buck, *Meerut SR* (1874).

[130] E.S. Liddiard, *Etawah SR* (1915), p. 14. It was common practice for cultivators from nearby well villages to rent some canal-watered land on which to grow their cash crops.

[131] Forbes, Chaprauli Pargana Report, p. 8, in Buck, *Meerut SR* (1874).

[132] NWP PWD(I), March 1873, 43, Note by Capt. J. Ross.

[133] C.H. Cooke, *Meerut SR* (1940), p. 10.

[134] NWP&O Rev., Jan. 1901, 47, AR Pargana Baghpat, 6 Oct. 1900.

[135] Voelcker, *Report on the Improvement of Indian Agriculture*, p. 70.

when the canal supply was inadequate in either timing or quantity is clearly indicated by the fact that a large number of new wells were 'rarely used'.[136] The process was partly facilitated by the wealth that accrued to cultivators in canal villages. The initial cessation of well-sinking noted in Etawah was, 'after a time', and 'partly due to the capital which the certainty of the canal has enabled [the cultivators] to accumulate', followed by a revival;[137] in *tehsil* Fatehabad (Agra) – at the tail of the system – canal cultivators were in a 'somewhat better economic position' than others and hence were 'able to afford the outlay on wells'.[138] The process was also facilitated by the reflection of prosperity in cattle quality, and in the buoyant water supply of wells in canal tracts.

The fundamental difference, however, related quite simply to the fact that deploying a well in this manner did not tie up resources of men and bullocks for long periods and, by and large, was a minimal interruption to cultivation as a whole. It coped easily with the uncertainty caused by a canal supply, the timing of which was governed more by administrative considerations than by agricultural needs. An independent source of supply permitted the timely application of the correct quantities of water, with significant benefits to yields. The cultivator did not have to jeopardise his outturn by taking too much water at a time in order to be sure there was sufficient moisture to last the uncertain period until the next watering. He used his well very much, in fact, as many of his descendants currently deploy tubewells.[139]

[136] Cooke, *Meerut SR* (1940), p. 10; see also Gillan, *Meerut SR* (1901), p. 7. In Bulandshahr, for instance, while 5,076 out of 5,914 earthen wells were recorded as having been used in 1901, this applied to only 10,530 out of the 22,252 *pakka* wells in the district. *Bulandshahr DG* (1903), pp. 41–2.

[137] Liddiard, *Etawah SR* (1915), p. 14.

[138] H.J. Frampton, *Rent-Rate Report of Pargana Fatehabad, District Agra* (Allahabad, 1931), p. 24.

[139] Recent studies in the UP of the comparative impact on yields, cropping patterns, and profits associated with different kinds of irrigation have shown that better performances on all three counts were achieved with private as against state tubewells. These results reflect mainly the more precise water applications which can be administered by farmers using their own tubewells. T. Moorti and J.W. Mellor, 'A Comparative Study of the Costs and Benefits of Irrigation from State and Private Tubewells in Uttar Pradesh', *Indian Journal of Agricultural Economics*, xxviii (1973), pp. 181–9.

Conclusion

Canal irrigation was, in the main, adopted by the Doab peasantry rather than imposed upon it. The 'satisficing' model of peasant behaviour is inadequate to explain the decision to adopt the new irrigation technology. Nor can the peasants' choice be fully understood by using the Schultzian peasant rationality framework which gained currency in the 1960s.[140] According to this formulation, the peasant was an efficient allocator of resources who maximised utility through applying factors until the marginal product of each corresponded to its market price. Within the terms of this neo-classical model, the canal represented a technological input which could disturb the 'equilibrium of traditional agriculture' through raising the marginal returns to investment.[141] This would prompt, according to Professor Schultz and his 'formalist' followers, a recognisably entrepreneurial response from the peasant.

While the notions contained in the Schultzian model, that the peasant was interested in increasing his 'income'[142] and that he weighed the marginal utility of each of his options, appear consistent with the findings of this chapter, the particular profit maximisation behaviour implicit in the model renders it of

---

[140] T.W. Schultz, *Transforming Traditional Agriculture* (New Haven, 1964).

[141] Professor Schultz's 'equilibrium of traditional agriculture' is characterised by static technology and by unchanging preferences and motives on the part of generations of cultivators. In the absence of technical advance, the marginal productivity of investment in additional agricultural factors declines to the point where there is no incentive to save for investment in them. The marginal rate of return on investment is known, having existed sufficiently long for an equilibrium to have become established between saving and investment and between the demand for and supply of agricultural factors as a source of income. With little incentive to add to the stock of traditional capital, due to its low returns, cultivation in the context of constancy of the 'state of arts' thus shows an inevitable tendency to stagnation due to the 'high price of permanent income streams'.

[142] There are, in fact, good reasons why the peasant should wish to increase production, whether or not the increment is destined for the marketplace, and whether or not he already produces a surplus. Dr T.G. Kessinger has described the nature of the local prestige systems and the way in which they set consumption standards within local society. Given the complex pattern of village relationships, a surplus of cash or grain for any family presented opportunities for political and social as well as economic gain. T.G. Kessinger, 'The Peasant Farmer in North India, 1848–1968', *Explorations in Economic History*, XII:3 (1975), pp. 328–9.

limited use in helping to interpret the adoption behaviour relating to the Doab canals. Schultz's critics have given much attention to his failure to take account of the risk factor in peasant decision-making, and the way in which insecurity modifies 'maximisation' behaviour.[143]

The typical peasant did not opt for the canal merely because it 'saved an infinity of toil' compared with working a well, nor simply because he could make more money by growing commercial crops. His decision is best understood in terms of the household producer–consumer unit; more in line, in fact, with a Chayanovian interpretation, which holds that it is misleading to perceive the peasant family as the functional equivalent of the western business enterprise (and the peasant *chef d'enterprise* as an 'entrepreneur') since the very structural characteristics of the peasant family farm rendered capitalist profit calculation an irrelevant basis for decision-making.[144] Thus, peasants have often shown themselves reluctant to save labour at the cost of adopting innovations which involve various extra costs and uncertainties. Labour for most families is, in one writer's words, more of an 'overhead' than a variable input cost.[145] Minimising imputed family labour costs – or, for that matter, those relating to farm capital, such as bullocks – is not likely to be the peasant's prime concern. However, he is likely to be adoptive when a technology relieves a production bottleneck which holds back the utilisation of his resources, and thus his outturn, provided that these advantages of adoption are not subjectively outweighed by the real costs and loss of independence associated with the technique. The canal was adopted because it released for productive redeployment those 'overhead' factors, labour and bullocks, which – in combination with canal water – permitted the peasant family to realise

---

[143] See in particular M. Lipton, 'The Theory of the Optimising Peasant', *Journal of Development Studies*, VI:3 (1968), pp. 327–51. Others have argued that the farmer exhibits 'proficiency' rather than 'efficiency' in that he accepts less than his maximum mean income in order to reduce his risk. M.G. Schluter and T.D. Mount, 'Some Management Objectives of the Peasant Farmer: An Analysis of Risk Aversion in the Choice of Cropping Pattern, Surat District, India', *Journal of Development Studies*, XII:3 (1976), pp. 246–61.

[144] A.V. Chayanov, *The Theory of Peasant Economy*, ed. D. Thorner, B. Kerblay, and R.E.F. Smith (Homewood, Ill., 1966).

[145] J.R. Millar, 'A Reformulation of A.V. Chayanov's Theory of the Peasant Economy', *Economic Development And Cultural Change*, XVIII:1 (1970), pp. 219–29.

more from its holding overall. Although the process involved a greater reliance upon the market, as will be subsequently confirmed, the very nature of the innovation improved rather than diminished the overall degree of security relating to the provision of family consumption needs. Where, of course, cultivators were able to combine well and canal irrigation, they could reap the benefits of both systems without incurring the disadvantages of either.

It is clear, therefore, that the overlapping of irrigation systems was far from a simple case of substituting equivalent forms of irrigation, one for the other. It should be noted, however, that there were dry areas, potentially irrigable, left without any irrigation source due to the chosen distribution policies. While it is not possible to establish whether the increment to production gained through the delivery of water into the well tracts would have been outweighed, or even rivalled, by the potential gains from supplying the drier, more backward estates, it is clear that the distribution biases had the effect of generally widening disparities between the estates making up the region.

4

Impact on agricultural production and organisation

While the peasant accepted the new technology for his own reasons, the implications were profound, for the effects of this new resource could not be confined to the crops which took water, but were felt across the whole range of peasant production and society. In a peasant system where no clear distinction was made between output produced for the marketplace and that destined for home consumption, farm investment, or savings, the disturbance to equilibrium at one point inevitably meant that effects repercussed upon staple food and fodder crops as much as upon those raised purely for the market. In a system the parts of which were aligned by mutual dependence, changing crops, techniques, and patterns of work implied changes in social and economic roles and relationships, and in ecological conditions, which reached beyond those villages receiving water.

The new technology implied a deepening involvement of the peasant with a money economy and a wider world. Increasing commercialisation did not mean, however, that the benefits would reach the bulk of the population, or that the changes would be ecologically sustainable in the long run. The relinquishment by degrees of self-sufficiency undoubtedly exposes parts of peasant society to the buffetings of often uncontrollable market forces, while commercial imperatives often give rise to localised environmental deterioration.

It is understandable, therefore, that both contemporary and latter-day critics of canal schemes should view with some scepticism the reality behind the commercial gains induced by the new technology. There is some consensus for the argument that the situation was running out of the peasant's control, partly as a result of the cultivator's own decisions relating to new production opportunities, and partly due to events over which he had no influence. In the first place, the lure of easy gains, so the argument goes, induced the peasant to ditch proven techniques and practices in favour of those which permitted higher but unsustainable returns. The peasant found not only that the canal 'disrupted [his] former pattern of work', but that his 'techniques were not adapted to deal with such sudden and radical change'. The canal, in conjunction with the other exogenous forces for commercialisation, evoked decisions from the cultivators which had undesirable consequences for soil fertility (through overcropping), per capita staple food supplies, and the quality and numbers of cattle stock (with the implications for manure supply and the quality of tillage).[1]

The problems, the argument proceeds, were not all of the peasant's making. He was often helpless against the tide of externalities which accompanied the canals. No single peasant, nor in the main any village community, could control or even influence some of the external forces which intruded into their decision-making processes. The most destructive of these were the physical manifestations of waterlogging and the increased salinity and malarial outbreaks associated with it. These took their toll of acreages and yields. Other forms of external effects arose simply because the private benefit resulting from the actions of individual peasants or communities adjusting to new opportunities exceeded the direct cost to them of that action. The costs were instead distributed (either widely or in a localised manner) through society. Hence individual decisions to extend cultivation produced community costs in terms of reduced grazing areas and a contraction in the supplies of draught stock from pastoralists.

The purpose of this chapter is to examine empirically the effects of canal irrigation to see if the outcome of its adoption was as unpropitious as has been suggested. It will examine the role of the canal in determining crop patterns and farming practices. This will provide the foundation for specific enquiries into the canal's effects on: work schedules and total employment; the maintenance and deployment of bullocks; the production of staple foods in relation to population; and the

---

[1] Whitcombe, *Agrarian Conditions*, pp. 70–91.

manner in which the canal affected crop yields. The empirical core of this section will consist of a statistical case study of the villages of Koil *pargana* (Aligarh), supplemented by evidence from the sugar-producing Upper Doab region and the less fertile tracts further south in order to maintain the necessary wider perspective.

The second section examines the main negative externalities confronting the peasant. It looks at the cause of the problems, the extent to which they might have been avoided, the actual degree to which they affected the tract's production, and the extent and effectiveness of corrective measures. Through careful sifting of the evidence it can be shown that the Doab's delicate ecological balance was indeed disturbed, and in some places seriously so, but that the most severe disruptions were frequently short-lived and, in terms of output – and the peasant's production system – easily exaggerated.

Comparative analysis of cultivation under different irrigation systems

Area of investigation: methodological considerations

Canal irrigation is, of course, only one influence on production decisions. Relative price changes, climatic variation, revenue policies, and communications developments would all impinge upon decision-making to an extent which would confuse the role of the canal in production adjustments made through time. Only broad trends can be gleaned from comparisons of production before and after the canal's introduction. A cross-sectional study clearly has the effect of freezing such influences, and this chapter's statistical base is drawn largely from the Koil *pargana* handbook compiled during the settlement survey of 1900–1 and containing agricultural statistics for each village unit.[2]

Situated in the Central Doab, the *pargana* includes the city of Aligarh itself. Twenty-two miles across, the area contained is 355 square miles, of which 130,000 acres were being cultivated annually around 1900. A *pargana*, of course, is primarily an

---

[2] Uttar Pradesh Board of Revenue Library, Lucknow (hereafter UPBORL), iv(c), Ser. 175/5, Handbook of Pargana Koil, Aligarh, 1900–01.

administrative unit and thus one of questionable value for economic or agrarian analysis. For the most part, however, Koil is especially suitable. In the first place, the canal and well tracts in the *pargana* are geographically distinct – the canal carves through the north-western parts of the *pargana* – thus permitting an examination of the contrasting agricultural systems under the respective irrigation technologies; inter-mixed villages would tend to generate very specific results, due to the integration of cropping patterns. This convenient feature is enhanced by the natural uniformity of agricultural conditions in the *pargana*: there were 'comparatively few villages below the general level, and not many above it'.[3] In part this uniformity stemmed from the even quality of the soil: by 1882 only 10% of the cultivated area was classed as 'unirrigated *bhur*', mainly localised in the south-west corner.[4] Uniformity was also found with regard to natural irrigation facilities. The *pargana* was noted for the facility with which earthen wells could be made: the *mota* existed 'in most villages', and even in the south-west *bhur* localities there was 'much good irrigation from wells'.[5] While neighbouring *tehsils* Hathras and Iglas were suffering a fall in their water-table, Koil farmers found water around seventeen feet down, and even less close to the canal.

For the purposes of this study, the salient point is that significant variations in crop patterns, population densities, and cattle numbers are unlikely to have arisen as a result of the relatively minor variations in natural agricultural conditions in the tract.[6] Moreover, distortions likely to result from the period of comparison containing abnormal rains or sequences of prices are reduced by the fact that five-year averages of cultivation statistics are available from the handbook. It is necessary also to mention that using the village as a unit of investigation may

---

[3] Smith, *Aligarh SR* (1882), p. 80. Also from a geographical standpoint it might be noted that in choosing a *pargana* situated on the border between the rich soils of the Upper Doab districts and the less advantaged districts further south, the extremes of conditions in the Doab have been avoided.

[4] The absence of insecure villages was reflected in the fact that the Settlement Officer found 'little or no *batai* (grain-rented) land'. *Ibid.*

[5] *Ibid.* The stability of *kachha* wells, and ease with which they could be constructed, was not impaired by the canal's presence in this *pargana*.

[6] This is confirmed in A. Aziz, 'Urban Gradients around Aligarh', *The Geographer*, xx:2 (1973), pp. 134–50.

have drawbacks which entail specific sampling precautions to avoid bias. There appears to be no systematic relationship between crop patterns and village size over most of the size range, but there does exist a danger of distortion from the inclusion of particularly small villages. Villages below around 125 acres show a high incidence of absentee cultivation, and many were uninhabited. The extra costs of cultivation this entailed – travelling to the hamlet, transporting equipment, watching and protecting crops – were especially likely to influence the nature of production in that unit. These villages have thus been excluded from the analysis, and a general preference has been shown for villages above 200 acres. The main criterion for the selection of sample villages was the level and form of irrigation. However, some villages, otherwise acceptable, had to be discarded where inaccuracies were plainly apparent in the statistical entries, or where the settlement officer's notes made it obvious that the village was in some way unrepresentative. The well villages of the south-western part, where more *bhur* occurred, were excluded from consideration.

Changes in cultivated area

It is very difficult to establish the quantitative impact of the canal upon cultivated area in the Doab. Even if reliable base figures could be obtained, the fact that it took time for cultivated areas to respond fully to the canal meant that the effects of price movements and transport developments inevitably intruded. Undoubtedly it gave a fillip to extended cultivation wherever there was suitable wasteland and sufficient population to break and work it. *Pargana* Chaprauli (Meerut) had been 'almost a desert at the British occupation', but the operation of the EJC quickly 'caused all the wasteland in the neighbourhood to be brought under cultivation'.[7] This was particularly the case wherever estates were held by Gujars and Rajputs, who were 'comparatively reformed' by what one official described as 'the most civilising agent at our disposal' – canal water given flush.[8] The Gujars especially responded to

[7] *Meerut DG* (1876), p. 240.   [8] *Muzaffarnagar DG* (1876), p. 708.

the canal by extending and improving their cultivation, relinquishing their former role as graziers of (usually stolen) cattle: 'The Gujars are creatures of circumstance. Give them a canal and teach them the profits of agriculture and they work their villages like Jats. Put them in a tract like Loni Khadir and they pay their revenue by stealing cattle and committing burglaries in Delhi.'[9]

The growth of cultivated areas was, however, far from dependent on intrusion into wastes. The opening of the Ganges Canal was undoubtedly the primary cause of a growth in cultivation in eastern Muzaffarnagar from 221,400 acres in 1841 to 264,000 in 1872. Only one-third of this 19% increase was accounted for by reduced 'cultivable' wasteland;[10] the largest contribution came from a reduction in 'recent fallow', which fell from 32,000 acres to 6,000. Net cultivated areas (n.c.a.) grew, therefore, to a substantial degree due to the canal's effect on temporarily idle land, particularly in closely populated tracts. This infrequently cultivated 'marginal' land was not necessarily intrinsically bad land: it might merely be poorly sited, and even uncultivated it would provide animal pasture. Given these factors, it would frequently have been the case that prior to the canal it was unable to compete for inputs during the periods when they were needed elsewhere. Such land was no longer marginal in the context of the operational holding when the canal enhanced the farmer's ability to manage his land, and when it released better (or better-placed) fields for the high-value crops.

The canal's impact upon net cultivated areas cannot be shown from the Koil statistics, since the necessary base statistics are not available. The important difference between net and gross cultivated areas can be shown, however.[11] *Dofasli,*

[9] Gillan, *Meerut SR* (1901), pp. 9–10.

[10] 'Cultivable' wasteland: land considered potentially arable which had never come under the plough, or had not been cultivated in the previous five years. It was the sparsely populated Bhuma Sambalhera *pargana* which saw the largest reduction in 'cultivable' wasteland; in Khatauli there was no significant reduction in this category. 'Report on the Ganges Canal Tract', p. 63, in Cadell, *Muzaffarnagar SR* (1882).

[11] 'Gross cultivated area' refers to the total annual area recorded as bearing crops, and fields bearing more than one crop during the year are thus counted twice in the total. 'Net cultivated area' includes only the actual area of land used, regardless of the number of crops taken from it in the course of a year.

land bearing two crops in the course of a year, increased markedly with the canal. The area varied with the quality of the soil (and with the area put down to sugarcane, which occupied the ground for two seasons), but even Mainpuri's sandy *pargana* Bewar – with around half its n.c.a. under canal irrigation – was double-cropping around 14% in the early 1900s.[12] Etah recorded one-fifth under *dofasli*, and the rich Upper Doab districts, even those growing sugar, at least one-quarter.[13]

Table 4.1 demonstrates the contrast in Koil's *dofasli* areas occurring between canal and well villages.[14] Figures from individual villages show clearly how *pargana*- or district-wide proportions can obscure individual differences. In the case of the best well villages, 20–22% were *dofasli*, while the fully canal-irrigated villages managed to double-crop 59% (allowing for sugarcane). Similar results are found in the sugar-growing districts: on five-year averages, 27% in Khatauli and 35% in Kandhla's canal circle were *dofasli*, with the addition in each case of 21% under all-year sugarcane.[15]

Cropping patterns

High-*dofasli* areas in the canal villages reflect their greater intensity of cultivation. As is shown by Table 4.1, more than four-fifths of the n.c.a. in canal villages was under *rabi* crops, almost half of which was the lucrative crop of pure wheat. Barley, sown alone or with *gram* and peas, accounted for the bulk of the remaining *rabi*, and generally followed the main *kharif* crops of indigo (one-fifth of n.c.a.), maize (a quarter), and cotton. In the well villages, a smaller share (63%) of the n.c.a. was put down to *rabi* crops, with pure wheat and barley less important – and wheat mixed more so – than in canal villages. With indigo non-existent, cotton was virtually the sole summer cash crop in well villages: while canal villages had one-third of the n.c.a. under non-food *kharif* crops, the well

---

[12] UPBORL, iv(c), Ser. 45, Handbook of Pargana Bewar, 1904.

[13] *Etah DG* (1911), p. 25; *Bulandshahr DG* (1903), p. 34.

[14] The Table compares the 'best' villages in each of these categories, as defined by the revenue assessment criteria.

[15] MCR, BOR, viii/ii, 14(c), H.A. Lane, RRR Pargana Kandhla (Muzaffarnagar), 1919, p. 9.

Table 4.1. *Comparison of crop patterns in sample canal and well villages,* pargana *Koil,* c. *1900 (in percentages of net cultivated area)*[a]

Crop	Canal villages[b]			Well villages[c]		
	Mean	Range Min. – Max.		Mean	Range Min. – Max.	
Rabi						
Wheat alone	35.4	21.7	46.8	20.7	10.9	30.8
Wheat mixed	8.2	4.0	17.6	14.2	3.5	27.8
Barley alone and mixed	29.3	19.0	47.6	19.6	8.8	27.6
Gram	3.5	0.5	7.5	4.2	1.2	11.7
Other crops	3.6	1.8	6.9	3.5	1.6	12.4
Total *rabi*	80.3	73.9	87.6	62.6	54.3	70.1
Kharif						
Juar alone	10.2	6.5	13.0	19.7	12.4	30.8
Bajra alone and mixed	1.7	0.6	6.7	5.0	0.8	12.0
Maize	26.0	22.2	34.0	10.4	4.6	15.4
Sugarcane	2.4	0.0	9.8	0.0	0.0	0.0
Indigo	19.1	12.6	34.8	0.0	0.0	0.0
Cotton	11.9	7.7	16.5	16.7	5.6	27.8
Rice	1.8	0.0	9.5	0.3	0.0	2.2
Other crops	3.5	1.1	6.8	6.2	0.8	11.5
Total *kharif*	76.6	64.6	91.1	59.2	51.4	72.2
Dofasli	56.8	45.1	70.2	21.8	11.5	42.3
Irrigation						
Canal-irrigated	85.3	68.6	98.7	0.0	0.0	0.0
Well-irrigated	8.7	0.0	23.6	90.5	78.5	100.0

(*Source:* Compiled from entries in UPBORL, iv(c), Ser. 175/5, Handbook of Pargana Koil, Aligarh, 1900–1.

[a] The villages whose statistics are included in this table comprise almost all of the first and second revenue assessment circles. Villages from these circles which are not included, seven in all, were excluded solely on the grounds that they did not fall comfortably into the canal/well groupings. Those villages included belonging to circle III were chosen according to irrigation criteria alone.
[b] Sample size: 13.
[c] Sample size: 16.

villages put effort into sowing twice the canal proportion of *juar* and *bajra*.

Although the cross-sectional nature of the comparison and the homogeneity of the tract minimises the effect of independent variables other than the area of canal irrigation in bringing about these observed contrasts in crop patterns, it is necessary to examine the significance of other potential influences which might have contributed to these contrasts.

Tenurial factors, proximity to markets, and caste type are the major ones. Caste type, as a potential influence upon the skills, crops, and methods of agriculturalists, can be for the moment dismissed in this instance, since the dominant castes in the highly cultivated canal villages included Chamars, Rajputs, and Brahmans[16] – none of them famous among district officers for the quality of its cultivation. Statistical analysis carried out respectively on samples of 74 and 112 villages indicated that the location cost and tenurial variables had no significant relationship with either specific crops or cropping in general.[17] The limited effect of the transport variable is explained by the relatively small distances involved (a maximum of twelve miles) and by the multitude of marketing outlets for cash crops. There were, for example, numerous indigo factories in the *tehsil* (33 in 1880);[18] cotton presses existed not only in the central city, but by 1900 in Harduaganj also;[19] moreover, cotton was exported not only by rail, but by canal barge from Barotha *ghat* to the Cawnpore mills. For wheat and other exports there was the choice of several railway stations in the *pargana*. The only indicators of tenurial differences are the legalistic ones of *sir* and *khudkasht*, plus occupancy and statutory tenancy. These are far from satisfactory in that they in no way reflect the diversity of tenurial relationships and ignore subletting, which was widespread. Though at the extremes they would usually reflect whether a village was in the hands of peasant proprietors or insecure tenants, there seems little to choose between the tenurial conditions prevailing in well as against canal villages and it appears unlikely that a significant relationship has been overlooked.[20]

---

16 *District Census Statistics: Aligarh* (Allahabad, 1895), pp. 45–52. This issue is returned to in Ch. 8 below.

17 For more details, see Appendix 2 below.

18 Smith, *Aligarh SR* (1882), p. 7.

19 *Report of the Board of Revenue on the Revenue Administration of the North-Western Provinces and Oudh, 1883–4* (Allahabad, 1885), p. 8. (Subsequent references to these annual reports will be abbreviated to *RAR*.)

20 It should be noted that any difference in tenurial structure which might exist between canal and well villages may be itself closely related to the canal presence, and thus 'tenure' would not constitute a truly independent variable. The role of tenure in its broader context is considered in Ch. 8 below.

Agricultural schedules and employment

Such adjustments to cropping patterns were facilitated not only by the fact that the canal released complementary factors from the wells, while also speeding up the irrigation process; through allowing greater freedom from weather-determined cropping it permitted greater flexibility in the timing of agricultural operations and meant that cultivation activity, and especially the deployment of labour and bullocks, could be spread more evenly through the year.

Usually, between the end of *rabi* harvesting operations in April and the onset of the rains in June, some ploughing might be accomplished in well villages in the wake of a shower, but work on the land was minimal and the heat and evaporation rates kept most wells idle (see Figure 1). Activity began in earnest with the coming of the rains, speed being of the essence as the farming community strove to put down its cotton and maize crops, followed by *juar* and then *bajra*. Wells might be used to make up in some degree for disturbances in the pattern of rainfall, but with his bullocks required for ploughing – and unable to do duty more than twenty days per month – the cultivator using wells was limited in his ability to operate independently of the rainfall pattern. Variability of rainfall was especially pronounced in June, and either inadequate or excessive falls reduced sowings.[21] Manured land bearing the quick-maturing maize could be prepared for a winter crop, usually barley and peas, but, in the main, other *kharif* crops occupied the ground too long to permit its use for *dofasli*: cotton, the chief summer cash crop, was ripe for picking from October to January; *bajra*, sown on poor soil, or following a *rabi* crop, was often not harvested until November; and nor was *juar* when grown for food, although if sown thickly as *chari* (fodder) and cut while green it could be off the ground almost as soon as maize.

Overlapping of crop seasons was thus a constraint upon

---

[21] The annual variability in June (166%) was significantly greater than that in July (27%) and August (42%). K.Z. Amani, 'Variability of Rainfall in Relation to Agriculture in the Central Ganga–Yamuna Doab', *The Geographer*, XIII (1966), pp. 35–47.

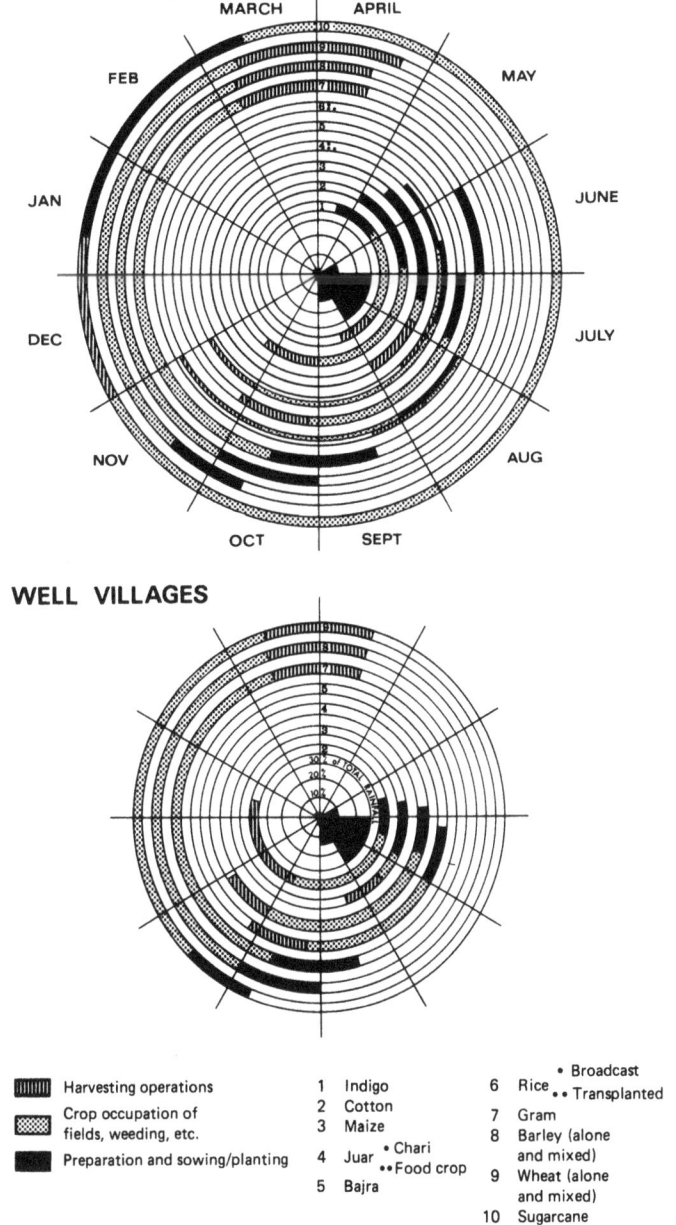

Figure 1. Seasonality of crops in relation to normal rainfall distribution. (Source: Crop data given in Table 4.1.)

double-cropping, but the practice was limited also by the cultivator's capacity to prepare and sow his fields with *rabi* crops in the time available to him. Just as *kharif* yields – particularly cotton, with its growing season in this area a little short for the plant's requirements – could suffer from delays in sowing, the outturn of *rabi* cereals benefited from timely sowings. *Gram*, for instance, when sown after mid-October, could suffer damage or even fail to germinate if the soil temperature was too low. A feature of agriculture in well tracts was the limitation on the cultivator's capacity – with the summer crops to gather in and make ready for the market – to prepare, sow, and sustain with well water during the cold season more than a certain proportion of his acreage. This phase of the calendar was also complicated by a widening of the rainfall variability, during September and October,[22] when the smaller quantities of rain were so important both to the maintenance of *kharif* crops and to the preparations for the *rabi*. A dry autumn, though a 'comparatively easy matter in the canal districts', occasioned 'great labour and delay in well districts'.[23]

In canal villages, where copious water could be turned onto the baked fields in May, and where the cattle would be comparatively fresh for ploughing, a substantial input of resources occurred before the rains. Indigo was the most striking example of this, being sown with the aid of canal water as soon as the *rabi* harvest was over. It was cut in August, after the *kharif* sowings were completed, leaving ample time to prepare the field for barley or *bejhra*. The cultivation of this dye crop was not the only pre-monsoon activity in the canal villages. Much of the cotton area would have been sown early with canal water, in order to prevent the plant's later growth from being checked by the onset of cold weather.[24] This practice gave the plant the opportunity to develop in the conditions most suited to it, and it enabled the best-quality *kapas* to be picked during September–November and the field to be

[22] October was the month with the greatest variability. *Ibid.*

[23] Duthie and Fuller, *Field and Garden Crops*, I, p. 2.

[24] This practice spread rapidly with the widespread switch to cotton from indigo during the late 1890s. W.H. Moreland, 'Conditions Determining the Area Sown with Cotton in the United Provinces', *AJI*, I (1906), pp. 37–43; see also W.H. Moreland, *The Agriculture of the United Provinces* (Allahabad, 1910), pp. 197–9.

cleared, irrigated if necessary, and sown with a barley mixed crop. The practice of sowing maize and *chari* prior to the rains with the aid of the canal also increased. In the latter case, on manured ground, it was not uncommon for two (and even three) crops of fodder to be raised during the *kharif*.[25] The 'pronounced increase' in the popularity of maize in the final decades of the century reflected not only its early contribution to foodstocks, but the fact that it facilitated the early preparation of the field for *rabi* sowings. Maize could, on manured land, be followed by wheat, though this valuable crop was normally grown in fields allowed a *kharif* fallow. Rice and sugarcane also intruded into the canal cultivator's schedules, though growing conditions were less suitable than those further north. High-quality rice was often found where canal water was available: either the transplanted *jarhan* variety[26] or, in some areas, *munji*, which, sown broadcast in May, was off the ground in time to allow a *rabi* crop. No crop was more affected by canal irrigation than sugarcane. Although in Koil soil conditions were largely unsuitable for the crop, with the immense amount of effort required to prepare the ground and irrigate the crop dramatically reduced by canal irrigation, some fields were put down to the crop even here.

That leisure preference was not at the root of the cultivator's decision to adopt the canal is affirmed by estimates of respective labour and bullock inputs per 100 acres of n.c.a., based on crop patterns found in Table 4.1. Figure 2 illustrates the comparison between well and canal villages of combined labour and bullock inputs in direct cultivation activity. The vertical axes represent 'imputed' costs of the units of labour and bullocks for each crop; that is to say, they utilise production cost estimates calculated on the basis of the average number of ploughings, weedings, etc., with bullocks and labour valued at market rates for each operation.[27] Ideally it would have been desirable to separate the two components, but the form of the estimates militates against this.[28] The total area shaded for each

[25] Moreland, *Agriculture of the UP*, pp. 178, 184–6.   [26] *Ibid.*, pp. 181–3.
[27] Duthie and Fuller, *Field and Garden Crops*, I, pp. 6–62 *passim*.
[28] For example, the hired ploughman with his bullocks made a consolidated charge for services. In addition, available data do not permit the calculation of a separate estimate of the number of days involved for men and cattle in the case of threshing operations or where sowing involved harrowing.

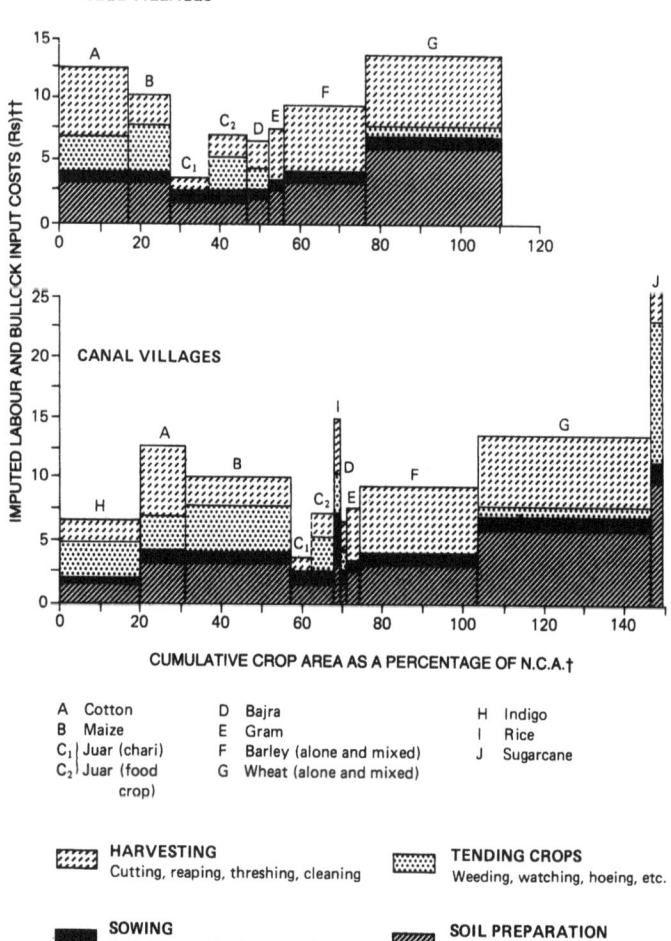

Figure 2. Comparison of labour–bullock inputs in direct cultivation activity (excluding irrigation), *pargana* Koil, Aligarh.* (Sources: Crop data given in Table 4.1; production input costs estimated from Duthie and Fuller, *Field and Garden Crops*, I, pp. 6–62 *passim*.)

* Inputs related to capital improvements, irrigation, crop-processing, transporting, and marketing *not* included.
† 'Other crops' not included because unspecified. They average 7.1% of n.c.a. in canal villages and 9.7% in well villages.
†† 'Imputed costs' explained in text.

group reflects, therefore, the total input per 100 acres of net field cultivation, excluding irrigation.[29]

As the illustration demonstrates, the adjustments to the cropping pattern involved an overall saving of labour only in the irrigation operations.[30] The increase in wheat area made large demands upon bullocks in particular, since it required thorough preparatory tillage – an average of eight ploughings – and the laborious threshing of a large output. The addition of indigo more than compensated for the smaller cotton area in canal villages, and the larger *rabi* coarse cereal area – at least partly switched from the coarse *kharif* millet area, which entailed little ploughing or care – increased the load on labour and bullocks. Maize, though requiring relatively little from draught animals, was especially labour-using, since it was a sensitive crop requiring careful sowing, weeding, and tending, including the earthing up of its shallow roots.[31]

The employment of men and bullocks was clearly more continuous throughout the year in canal villages, and the amount of work put into the land was itself raised. The peaks in activity were still there, but they had a less limiting effect upon the utilisation of resources due to the crop management flexibility imparted by the canal, replacing to some extent the pattern of hurried responses to weather conditions. Moreover, the constraining peaks under the canal were more usually peaks of labour demand rather than of bullock power in Koil-type canal villages.

[29] Not captured in this diagram is the lifting of water, to which well villages would have devoted labour far in excess of that needed in canal villages to lift and/or direct water into the fields. On the other hand, though, the labour and bullock days devoted to processing, transport, and marketing activity, which would clearly have been more pronounced in the commercialised canal villages, is also not included.

[30] It might be argued that the use of the same estimates for both groups ignores any operational differences between well and canal villages, such as ploughing. The crop most affected by this difference would be sugarcane, though that presents no problem in this instance, since the sugar is confined to Koil's canal villages. To the charge that copious supplies of water could be substituted in part for better tillage, it might be noted that severe time pressures upon well farmers could produce the same effect. Whereas cotton sown with canal water could be put into carefully prepared soil, the speed with which cotton was sown with the rains in well villages often meant that only two hurried ploughings were possible. *Indian Cotton Committee, Report* (Calcutta, 1919), p. 48.

[31] Moreland, *Agriculture of the UP*, pp. 178–81.

There is little evidence here, then, to support the claim that 'released from the lengthy rhythms of the well lifts the cultivators and their bullock teams had no alternative employment to offset this disruption in their work pattern and were condemned to periodic idleness'.[32] The pattern revealed for Koil – the broad concentration upon indigo–wheat *dofasli*[33] – was repeated throughout the Central and Lower Doab; and over and above the increased need for hired field labour occasioned by the greater intensity of cultivation, the demand for labour to lift canal water became pronounced in the lower canal divisions, especially in Cawnpore and Farrukhabad, where half the canal irrigated area was by 'lift'. In Farrukhabad, Buck observed, 'there is always such a demand for labour during the open week that a labourer can rarely be obtained for less than two annas'.[34] The low-caste labouring Chamars in this area were said to be 'hardly as a mass . . . removed one degree from starvation, unless where demand for labour, as for lifting canal water' put them in 'comparatively easy circumstances'.[35] In Cawnpore they were most thickly gathered in the highly irrigated Bilhaur *tehsil*, where they were 'extensively employed by Brahman and Rajput cultivators' and could 'support themselves adequately by working two days a week'.[36]

To the north of Koil lay a notable variant to the pattern established in the lower parts of the Doab: here were to be found the great sugar-growing tracts, strewn with neat blocks of the cash crop *par excellence*. Much of the effort which in Koil was devoted to indigo, *dofasli*, and the valuable wheat fields was lavished first and foremost upon the cultivation and processing of sugarcane, to which 'everything else is made subservient'.[37] It was regarded in Muzaffarnagar 'above all as the rent-paying crop', and wherever the tenant paid a fixed rent rather than a crop rate he put 'under cane as much land as the available supply of manure and due regard for the rotation of crops' allowed.[38] There were numerous villages which could boast 20–35% of their area under cane, reflecting the fact that the

[32] Whitcombe, *Agrarian Conditions*, p. 10.
[33] Irrigated cotton largely replaced the outmoded indigo dye crop around 1900.
[34] Buck, 'Note on the Effect of the Canal', p. 216.
[35] *Cawnpore DG* (1909), p. 104.   [36] *Ibid.*   [37] *IRR 1872–3*, p. 17A.
[38] Miller, *Muzaffarnagar SR* (1892), p. 48.

canal had truly revolutionised sugar production there. It permitted a relaxation in the full year's preceding fallow, previously universal; it almost halved the laborious field preparations designed to retain maximum moisture; it freed bullocks and men from unceasing irrigation by wells; it made available to the soil sufficient moisture to justify the adoption of higher-yielding canes, replacing the thin *daulhu* strain previously 'all but universal'; and it encouraged the early adoption of several developments of cane-crushing technology which increased extraction rates and improved the quality of the juice, while also reducing labour and bullock inputs.[39] If bullocks were largely idle during January–March in many parts of the Middle and Lower Doab, this was far from the case in cane villages, since the harvesting, processing, and transport of the crop took place then, followed by the sowing of the new crop in fields ploughed ten to fourteen times. Little indigo was grown in the Upper Doab, but in many localities the ample water supply made viable the cultivation of the valuable *munji* rice, the sowing of which followed closely the busy *rabi* harvest.[40]

In cultivation alone, therefore, there was 'continuous employment for the labouring population', and 'peculiarly profitable employment at short intervals'.[41] Not surprisingly, a significant amount of local migration took place, much of it focussing upon the fully irrigated estates of 'strong cultivating communities', which could 'hold out the prospect of employment, but not land, to newcomers'.[42] In eastern Muzaffarnagar, *khadir* labourers moved into the upland canal villages, and the more 'thickly populated districts' of Meerut contributed noticeably to the increase in the population of Jansath and Khatauli which followed the canal's opening.[43] Despite the fact

[39] See Ch. 8 below.

[40] So profitable was this crop in areas where cheap canal water was available that it in some places 'retarded the extension of cane cultivation'. UPA, BOR, 'Different Districts', Muzaffarnagar, Box 8, File 35, A. Cadell, RRR Pergunnah Moozuffur-nuggur, 1872, paras. 7 and 32.

[41] 'Report on the Ganges Canal Tract', p. 18, in Cadell, *Muzaffarnagar SR* (1882). This is shown clearly in Figure 2, where sugar cultivation is shown to involve almost double the inputs of labour and bullocks required for wheat. In addition, it required a disproportionate amount of resources in processing operations. See Ch. 8 below.

[42] *Muzaffarnagar DG* (1876), p. 715.

[43] 'Report on the Ganges Canal Tract', p. 78, in Cadell, *Muzaffarnagar SR* (1882).

that, in the words of one official, the 'largeness of profit that rewards industry therein' tended to 'attract population from without', the 'difficulty and expense of procuring labourers' continued to be a 'frequent subject of complaint amongst well-to-do cultivators'. Such was 'the call for labour' in canal areas that the Saharanpur Settlement Officer had actually 'met with instances of a high-caste proprietor – even a Rajpoot, associating with a chumar in partnership with himself in his seer holding, on condition that the latter should enjoy an aliquot portion of the profits'.[44]

The vast expansion in demand for hired labour may well reflect to some degree a reduced labour input on the part of those cultivators doing well. As one official observed, 'Where the people are well off, sharers and tenants alike prefer to take their ease and leave the hard work to servants.'[45] From the evidence presented here, however, it is clear that the continuing and intense demand for labour was mainly a reflection of the canal's effect upon crop patterns. Indeed, such was the demand for labour that it gave rise to certain social changes. It was, for instance, pointed out in a 1922 report that the Chamar caste had unmistakably 'imbibed new ideas', and that its members were 'endeavouring to raise themselves in the social scale' by refusing to handle carcasses and 'discouraging their women from going out and working with their hands'.[46]

Cattle stocks

One result of the introduction of canal irrigation is said to have been a reduction in the cattle stocks of villages formerly working wells. Indeed, it has been argued that the canal helped precipitate a reduced cattle supply as pastoralists such as the Gujars were attracted away from hereditary pursuits by the enhanced returns to cultivation.[47] In addition, the canal not only appeared to reduce the need for cattle but played a major role – through its effect upon grazing resources – in increasing

---

[44] Le Poer Wynne. *Saharanpur SR* (1871), pp. 138–9; see also Miller, *Muzaffarnagar SR* (1892), p. 23.

[45] MCR, BOR, viii/ii, 14(c), Ser. 5, J.O. Miller, AR Tahsil Muzaffarnagar (Muzaffarnagar), 11 April 1891, p. 25.

[46] *RAR 1921–2*, p. 30.   [47] Whitcombe, *Agrarian Conditions*, p. 85.

rearing, purchase, and maintenance costs. Together, it is argued, these forces caused a rationalisation of cattle resources by cultivators. Contemporary commentators not only identified this trend but linked it with the broad critique of the canal's impact on husbandry standards: the reduced manure supply consequent upon a decline in cattle numbers was matched by a deterioration in stock quality, and both contributed to slovenly cultivation and deteriorating soil quality.

There are indeed numerous comments by contemporaries to support the notion that canal irrigation caused reduced cattle numbers. William Crooke noted in 1881 that the canal 'enabled the cultivators to dispense with a large number of their plough cattle' and that, in addition, 'excessive use' of canal irrigation had 'tended to lower the class of cattle used in agricultural work'.[48] Crosthwaite and McConaghey claimed the same for Etawah and Mainpuri respectively.[49] Even the local population in Aligarh were inclined to this view:

All village authorities agree, patwarees, zemindars and cultivators alike, that where canal irrigation is extensively employed and is of long standing plough cattle have largely diminished. Only just enough are retained in the village to get through the ploughing and what homeland irrigation there may be, and the rest are gradually sold off, and the proceeds invested in other ways.[50]

Rationalisation of cattle resources was unlikely to have been as uniformly straightforward as these observations imply, and in point of fact the qualitative evidence does admit of exceptions. Crosthwaite pointed out that such reductions occurred in 'purely agricultural villages', but also noted: 'In some villages there are large numbers of cattle kept for carts, and cultivators ply a carrying trade in the intervals of agriculture. In such places the number of cattle is not affected by the canal.'[51] In

[48] 'The number of cattle now maintained is, in comparison with the area under cultivation, inadequate.' NWP&O Rev., April 1883, 63, W. Crooke, Note on Fuel and Fodder Reserves in Central Ganges–Jumna Doab, 22 Nov. 1881; *Etah DG* (1884), Appendix 1, p. 8.

[49] Crosthwaite, *Etawah SR* (1875), p. 16.

[50] UPA, BOR, Canal Shelf, Box 2, File 8, Memorandum by W.H. Smith, 11 May 1874, p. 3. Smith's successor, S. Ahmed Ali, reiterated this view seventy years later. Ahmed Ali, *Aligarh SR* (1943), p. 3.

[51] Crosthwaite, *Etawah SR* (1875), p. 16.

formerly dry villages (as opposed to those using wells prior to the canal) the cattle numbers not only increased, but were needed more 'constantly in the village, and thus not sent miles away for pasturage'.[52] Certainly, wherever sugarcane was extensively grown, substantial cattle resources were needed to process the crop.

Empirical evidence, however, suggests that observers inaccurately gauged the effect upon cattle stocks – even in villages which did not fall into the above 'exceptional' classifications. Buck's 'systematic enquiries' into *pargana* Tirwa had turned up only one village in which cattle had been 'materially reduced in number since the advent of the canal'.[53] Koil *pargana* data add strong support to this view. The *pargana* is not a sugar-producing one, and although some of the canal villages possessed a substantial number of carts, there is no relation within the canal sample between carts and recorded numbers of draught animals. A comparison of stock numbers per acre for highly irrigated canal and well villages (Table 4.2) produces a ratio in favour of canal villages: one bull, bullock, or bull-buffalo per 3.3 acres of n.c.a. as against 3.8 acres for the well tract.

It is possible that distinct changes in cattle management had set in during the 25 years between Smith, McConaghey, and Crosthwaite's time and 1900 (although this would still not account for Buck's Tirwa findings); it is possible also that the observed decline in cattle numbers had something to do with the tendency – 'among Jat cultivators especially' – for some canal villages outside the sugar areas to sell off cattle for that part of the year during which revenue officers made their rounds.[54] Whatever the explanation of the difference between available empirical evidence and the impressions of district officers, the canal's effects upon cattle numbers proved less dramatic than observers feared in the 1870s. Cultivators, it seems, responded to the growing shortage of grazing facilities

---

[52] UPA, BOR, 'Different Districts', Muzaffarnagar, Box 8, File 35, A. Cadell, RRR Bhoomah Sumbulherah, 1872, para. 19.

[53] Buck, 'Note on the Effect of the Canal', p. 218.

[54] The practice was for the bullocks to be sold when *rabi* ploughing was completed and for fresh cattle to be bought in time for *kharif* ploughings. UPA, BOR, Canal Shelf, Box 2, File 8, Memorandum by W.H. Smith, 11 May 1874, p. 3.

Table 4.2. *Comparison of cattle stock numbers in canal and well villages, pargana Koil, c. 1900*[a]

Village type (and number in sample)	(1) Cows, cow-buffaloes	(2) Bulls, bullocks, bull-buffaloes	(3) Calves, heifers	(4) Total livestock[b]	(5) Net cultivated area (in acres)	(6) Number of cultivated acres per mature male animal $[(5) \div (2)]$	(7) Number of cultivated acres per mature animal $[(5)/(1)+(2)]$
Canal (13)	2,260	2,953	1,821	7,034	9,965	3.3	1.91
Well (21)	2,121	2,671	1,967	6,759	10,056	3.8	2.05

[a] Villages were selected from assessment circles I–III on the basis of their irrigation characteristics and subject to their respective stock entries being complete.

[b] Sheep and goats excluded.

Source: Compiled from village entries in UPBORI, IV(C) Ser. 175/5, Handbook of Pargana Koil, Aligarh, 1900–1.

by planting fodder crops and stall-feeding their cattle. The Doab was in any case particularly unsuitable for pasture, producing mainly a coarse grass (*panni*) which was 'injurious to cattle'[55] and which invariably failed during drought. 'Zamindars and tenants', reported the Bulandshahr Settlement Officer, 'vastly prefer arable land to pasture and waste, since it is far more profitable . . . cultivation will feed far more cattle than jungle will'.[56]

Canal villages especially had the capacity to grow *chari*, and fodder crops covered one-fifth of Meerut division's *kharif* area in 1906.[57] So widespread did the practice become in the Doab – Aligarh cattle by 1940 were reported to be 'mostly stall fed'[58] – that when drought scorched the pasture of wastes, the effect was to kill off the old cattle which crowded there rather than to 'seriously diminish the number of useful cattle'.[59] While the quality of working animals appears to have been improved by stall-feeding,[60] it was not, it appears, at the expense of the number of 'useful' cattle kept: plough animals amounted to less than half the cattle kept in both samples of villages and, as with plough animals, the cows and cow-buffaloes kept by canal villages exceeded those in well villages (by 7%).[61]

Commercial crops and food staples

The extent to which canal-fostered commercialisation of agriculture affects the supply of staple foods is clearly an important question, particularly because of the implications for the poorer sections of the community. In the Indian context this issue has been examined through comparisons of food and non-food crop areas, and although this provides a somewhat crude indication of trends, the results have not generally

---

[55] *Meerut DG* (1876), pp. 202–3. *Usar* land was of no use for grazing.
[56] UP Rev., Jan. 1919, 13, E.A. Phelps to Meerut Commr., 17 Dec. 1917. According to Moreland, where large millets or spring cereals were grown, an acre would often yield more fodder as a secondary product alone than would pasture.
[57] *Ibid.*   [58] Ahmed Ali, *Aligarh SR* (1943), p. 3.
[59] NWP&O Rev., Oct. 1885, 15.
[60] Moreland, *Agriculture of the UP*, p. 118. In addition to being fed *bhusa*, chopped *juar*, and *bajra* stalks, stall-fed cattle also received a 'ration of gram or some other cheap foodgrain, or of cotton seed, or of oilcake'. *Ibid.* Cattle quality is returned to in Ch. 8 below.   [61] Figures taken from Table 4.2.

supported the contention that increased cash crop production jeopardised food supplies. Professor Harnetty, examining the 1861–70 increase in cotton production, noted that the increases took place as part of an overall expansion in cultivation and did not normally interfere with other crops.[62] More recently, Dr Michelle McAlpin has employed acreage figures for the later part of the last century to show that for five cotton-growing provinces (not including the UP) there was no apparent shift away from food to cotton.[63]

The sample statistics from Koil yield similar results. In the highly irrigated canal villages almost 70% of the n.c.a. bore a cash crop,[64] as opposed to 37% in well villages. Far from squeezing out the usual food crops, though, the crop patterns were adjusted to take full account of food needs. In the crop comparisons contained in Table 4.3, the area under what might be described as 'consumption' food crops is greater (by around 7%) than that in the well villages. In canal villages, during the *kharif*, low-quality slow-maturing millets were pushed out by maize, which replenished foodstocks by early September; and in the *rabi* much larger areas of barley – either alone or with *gram* or peas – were grown as the staple for the poorer villagers.[65]

This seemingly clear comparison is complicated, however, by doubts over the specification of '*juar*', which appears to cover both *juar* grown as a millet and allowed to mature for human consumption, and *juar* seed sown thickly as *chari* to be cut while green and used as animal fodder. The introduction of the canal caused *chari* production to increase substantially, but it is not clear that such an increase occurred in well villages, which, theoretically, could have more pasturage. It is probable that

[62] P. Harnetty, 'Cotton Exports and Indian Agriculture, 1861–70', *Economic History Review*, Series 2, XXIV:3 (1971), pp. 414–29.

[63] M.B. McAlpin, 'Railroads, Prices and Peasant Rationality: India 1860–1900', *Journal of Economic History*, XXXIV:3 (1974), pp. 662–84.

[64] Viewing wheat, for the moment, as a crop destined for the market.

[65] The results are even more pronounced for canal villages with a slightly smaller proportion of their n.c.a. under canal irrigation. A sample of nine such villages drawn from assessment circle III (selected on the basis of irrigation levels, subject to their being more than 200 acres in size and to the handbook entries being complete), with an average of 73% of their n.c.a. recorded as under canal irrigation, showed an advantage in 'Total food crops' of 11.5% over the well sample.

well villages had need of specially cultivated fodder, particularly as their *rabi* cereal area – and hence their *bhusa* (straw) supply – was less than that in canal villages. It is also likely that half (11% n.c.a.) of the double-cropped area (22% n.c.a.) not under maize (10%) was under the early-maturing *chari*, since none of the other *kharif* crops (cotton, *bajra*, and *juar* as millet) was harvested in time to allow a following *rabi*.[66] On this reckoning, at least half the well village '*juar*' crop was fodder. Since this likely *chari* proportion is much the same as the total '*juar*' area in the canal villages, doubts over the specification of '*juar*' would not reduce the percentage differences in favour of canal villages shown in the right-hand column of Table 4.3.

In addition, treating wheat solely as a commercial crop may bias the results against the canal villages, since it was a feature of the Meerut division in general during this period that a larger proportion of the population came to consume wheat.[67] Naturally, the extent to which the poorer village inhabitants ate wheat varied according to its price, but the increase in the intake of wheat among the better-off villagers undoubtedly reduced the demand for the poorer grains. Wheat production was 65–75% greater in canal villages.[68] It would be reasonable, too, to point out that the food crops which canal cultivators switched towards were, if more costly to rear, also higher-yielding. *Bajra* outturns were rarely more than 5½–7 *maunds*, and *juar* 8–10, but maize and barley generally yielded 10–16 *maunds* (see Table 4.4).[69]

The discussion of food supplies is of course incomplete without taking account of the denominator – the number of

[66] See Table 4.1. Note in this connection that the area under 'Other crops' refers mainly to minor crops grown mixed with a main crop. The respective areas were calculated according to the proportion of the mixture.

[67] See T.W. Holderness, *Memorandum on Food Production in the North-Western Provinces and Oudh* (Allahabad, 1891), p. 11.

[68] There is always, in such comparisons, the danger that specialisation will distort the pattern. Where irrigation is localised, cultivators in a position to do so will move in to canal villages (through renting or purchase) and grow their cash crops there, while the reverse will be the case for canal villagers, who will move some of their food crops out to nearby non-canal villages with dry fields available for rent. In this particular discussion, any such trends would tend to produce a result which *understates* the food situation of the canal villages relative to well villages.

[69] Similar trends can be identified in Muzaffarnagar's Ganges Canal tract. See Stone, 'Canal Irrigation and Agrarian Change', pp. 94–101.

Table 4.3. *Cultivation of food and non-food crops: comparison of canal and well villages,* pargana *Koil,* c. *1900 (in percentages of net cultivated area)*

Crop category	(A) Canal villages	(B) Well villages	Difference [(A)–(B)]
Commercial crops (wheat, indigo, cotton, sugarcane)	68.8	37.4	+31.4
Rabi food crops (excluding wheat)	41.0	38.0	+ 3.0
Kharif food crops (including 'juar')	39.7	35.4	+ 4.3
Total food crops (excluding wheat and sugarcane)	80.7	73.4	+ 7.3

Source: Compiled from statistics in Table 4.1.

Table 4.4 *Estimates of outturn of principal food crops in the western UP (in* maunds *per acre)*

Crop	Irrigated	Unirrigated	Dry fodder
Sugarcane	35	25[a]	?
Rice	16[b]	12[c]	?
Maize	14	10	?
Juar		8–10	45–60[d]
Bajra		5½–7	30
Wheat	15	10[e]	⎫
Barley	16	11	⎪ approx. 1½ times
Wheat–barley (*gojai*)	13–15	10	⎬ grain weight
Gram	12	8	⎪
Barley–gram (*bejhra*)	14	9	⎭

[a] Average of estimates from 'Report on the Ganges Canal Tract', p. 19, in Cadell, *Muzaffarnagar SR* (1882).
[b] *Munji.*
[c] *Dhan.*
[d] When sown as *chari* (entirely for fodder), the yield was 280–300 *maunds* of green fodder (equivalent to 100 *maunds* of dry fodder).
[e] Average for unirrigated land put at 10 *maunds* in Meerut and 7 *maunds* in Agra.
Sources: Moreland, *Agriculture of the UP*; Duthie and Fuller, *Field and Garden Crops,* 1, pp. 6–62 *passim.*

mouths to be fed. Is there a population density difference between the two groups? A comparison was made for 47 villages drawn from assessment circles I–III: 22 canal villages and 25 well villages.[70] The average population of the canal

[70] Again, villages were selected on the basis of whether or not they fell clearly into either the canal-irrigated or the well-irrigated category. In effect this meant that if a

sample was 1.4 persons per acre of net cultivation (0.95 per acre of gross cultivation). The corresponding figure for well areas is lower, at 1.07 (0.88 per acre of gross cultivation). In terms of net cultivated areas the population difference amounts to 23.5% and 7.3% in terms of gross cultivation. When these figures are placed alongside those of Table 4.3 and considered in the light of the points made above, it seems wholly reasonable to conclude that the respective systems gravitated to their own equilibrium. Prospects of commercial gain did not outweigh the principle Moreland noted: namely, that the 'first object of the cultivator is sufficient food to carry his family and his labourers over to the next harvest, and perhaps a little beyond'.[71]

Comparative crop yields

How did the systems differ with respect to crop yields? In the absence of quantitative data, this question can only be answered through piecing together the qualitative strands of available evidence. Localised salinity and waterlogging are likely causes of canal-related yield deterioration, and will be considered in more detail below. This section is concerned with the causes other than external ones: those which stemmed from the decisions and actions of the cultivators themselves. It has already been noted that, according to critics, reduced cattle numbers contracted the supply of manure needed to counteract the drain on fertility caused in turn by a cultivating community which, armed with the labour-saving canal facilities, had an apparently reckless urge to overload the land with crops, and compounded the problem by overwatering the fields. It has also been noted that, with respect to certain valuable crops – including sugarcane – canal cultivators often opted for lower

village had more than 55% of its n.c.a. recorded as under canal irrigation, it was classed as a 'canal' village; all of the villages in the 'well' sample had above 78% of their n.c.a. recorded as under well irrigation. Population figures are for 1891. Jalali has been omitted from the canal sample on the grounds of its disproportionate size (9,000 inhabitants), its position on a trade route, and its flourishing bi-weekly market, which may have attracted grain supplies from outside. Its inclusion would have exaggerated the population density in the canal sample (which already contains some smaller market centres) relative to that of the well villages.

[71] Moreland, 'Conditions Determining the Area Sown with Cotton', p. 39.

per acre yields than were realised with wells, achieving higher overall returns through a larger proportion of the holding bearing irrigated crops.[72] This section looks at the yield impact over a broader range of crops.

Leaving aside 'garden crops' (including sugarcane), E.C. Buck was adamant that 'canal-water produces finer crops on good land than well-water: natives universally admit this. The large supply of water gives more plant food to the crop.'[73] In places where irrigation was reported to have 'outrun the manure supply' a 'certain degree of exhaustion' was to be observed, especially in villages for the most part dry prior to the canal.[74] 'Overcropping' – or the depletion of soil nutrients through inadequate manuring and fallowing given the frequency and quality of crops reared – appears to have taken place in unmanured land. The initial impact of canal water was to produce good wheat crops on intermediate *manjha* fields formerly bearing inferior crops. After a couple of seasons, though, the yields began to decline, falling steadily over four or five years.[75] 'Does the unmanured canal land produce such good crops as before in the rains', asked Buck, 'or even such good rabi crops in a year when there are, as often happens, two or three good rainfalls after Christmas? I think not.'[76]

It quite naturally took cultivators some time to adapt, via the 'learning curve', to the new form of irrigation. A period of experimentation lasting quite a few seasons was necessary before they could gravitate towards the farming practices most appropriate to the use of canal water. Once it was clear that canal water alone could not sustain in unmanured fields the initially high yields, rotational and other practices were adjusted in line with overall production objectives. The argument that the actual manure supply diminished in canal

---

[72] See Ch. 3 above.   [73] *IRR 1873–4*, p. cx.
[74] GI Home, Revenue, and Agriculture Departments, *Selections from the Records of the Government of India*, no. CLX of 1879, *The Wheat Production and Trade of India* (Calcutta, 1883), p. 66, W.C. Benett (Dir. of Agriculture NWP&O) to Sec. NWP&O Govt., 20 Nov. 1881.
[75] *Ibid.*; see also Buck's comments in *IRR 1873–4*, p. cx.
[76] *IRR 1873–4* p. cx. Crosthwaite disagreed with Buck: to his mind, deterioration occurred 'not so much with regard to the average yield of crops irrigated by other means as to the extraordinary large produce of the first years of canal irrigation'. Crosthwaite, *Etawah SR* (1875), p. 16.

villages during this period is clearly questionable in the light of the data presented on livestock numbers. Indeed, district officials reported that, even in the highly cultivated district of Meerut, as far as manure was concerned, there was 'usually as much as is required'[77] in the villages. Focussing on the total supply of manure may be misleading, however, in a discussion of its application to outlying canal-watered fields. It may be more relevant to consider the problem of manuring fields in terms of the costs of transporting and spreading available supplies, a process which often involved the hiring of *gadhwalas*.[78] In this connection, several observers pointed out that Jat cultivators, on obtaining canal water, had taken to manuring in turn their outlying fields. It appears that only when it was possible to irrigate these fields were the returns sufficient to justify the manuring costs. Similarly, it was reported by the Settlement Officer for Mainpuri in 1905 that the manured *manjha* area in Shikohabad *tehsil* had extended with increased canal irrigation, and that the fertility of such land was frequently equal to the *bara* or *gauhan* fields lying close to the village site.[79]

Not only agricultural practices but the nature of the crops themselves had potentially an important bearing on soil fertility levels and yields. Many canal villages had large areas under indigo. Since, when grown with canal water, the growing season of this crop was such as to permit a following *rabi* crop, it appeared unavoidable to many observers in the 1860s and 70s that severe fertility loss would ensue due to overcropping. In the event, the doom-laden prognoses appear to have been unfounded. 'It may be doubted', George Watt observed from the vantage point of the 1890s, 'whether the effects are practically as bad as they first appear.' His explanation as to why the sterility predicted in canal tracts due to indigo had 'not occurred in any material degree' was that the long tap roots prevented the plant from trenching on the nutrients of

---

[77] *Meerut DG* (1876), p. 225; see also Crosthwaite, *Etawah SR* (1875), p. 103.

[78] *United Provinces Provincial Banking Enquiry Committee, 1929–31, Evidence* (Allahabad, 1931) (hereafter *UPPBEC(E)*), vol. II, p. 258.

[79] 'Report on the Ganges Canal Tract', p. 57, in Cadell, *Muzaffarnagar SR* (1882); W.J. Lupton, *Rent-Rate Report of Pargana and Tahsil Shikohabad, Mainpuri District* (Allahabad, 1905), p. 11.

following crops, and the practice of ploughing in the decaying roots and stalks imparted nitrogen to the soil.[80]

The general impression district officers gained on their travels through canal and well tracts was not one of deterioration. In the few cases where crops were noticeably poor in Aligarh's canal villages, Settlement Officer W.H. Smith noted that these estates were invariably 'rack-rented and overworked'.[81] In 'the great mass of cases', Smith was 'unable to perceive any difference between the quality of crops in well and canal irrigated tracts'.[82] Certainly the evidence of rent movements does not suggest deterioration: competition rents in the well circles of Koil increased by 54–72% over the period 1880–1900, compared with percentage increases of 139, 120, and 91 in canal circles.[83] Other commentators fall in behind the general view that 'where the farming is good, canal crops are quite as good as well crops'[84] and that the prevalent system of rotation of crops was 'adequate to the restoration of the necessary properties of the land'.[85] For some crops, such as cotton, yields could in the run of seasons be even higher, due to the more timely sowings.[86] If unmanured canal land did not yield quite as well when the rains were favourable, it is clear that crops grown on such land were less subject to fluctuation; while in a dry year, Buck was in little doubt, 'a canal village presents a richer appearance than a well village'.[87]

Overall, with stability of outturn of importance to peasant

[80] G. Watt, *Pamphlet on Indigo* (Calcutta, n.d.), p. 25. *Gram* and peas also played an important part in fixing nitrogen in the soil. See A. Howard, *Crop Production in India: A Critical Survey of its Problems* (Oxford, 1924), pp. 25–6.

[81] NWP Rev., May 1877, 69.

[82] UPA, BOR, Canal Shelf, Box 2, File 8, Memorandum by W.H. Smith, 11 May 1874, p. 6. In the parts of Mainpuri where the canal had been in full operation for years, Settlement Officer McConaghey was adamant that the crops were 'as good as in other pergunnahs with similar soils where wells alone are used'. See discussion following Buck, 'Note on the Effect of the Canal', pp. 227–31.

[83] Burkitt, *Aligarh SR* (1903), p. 15.

[84] Crosthwaite, *Etawah SR* (1875), p. 16.   [85] *Etah DG* (1911), p. 26.

[86] Although in good years there was little to choose between the cotton yields from well and canal land, over a number of years the outturn average would favour cotton sown early with canal water since it matured before the cold weather set in. According to Moreland the average was 20 lbs. per acre difference (150–170 lbs.); according to Hailey, the average yield on canal-irrigated land was 230 lbs. *Indian Cotton Committee, Minutes of Evidence*, vol. 1 (Calcutta, 1920), p. 5.

[87] *IRR 1873–4*, p. cx.

farmers, it is not surprising to find their decisions consistent with a strategy of achieving a moderate and consistent return, which maximises outturn over a period of time, rather than either a less frequent, but higher, yield, or a highly variable outcome under dry cultivation. The farmer was still able to do this even though the water supply was unreliable and he had deliberately to trade off yield against security by 'overwatering' when water was available in order that his crops would be reasonably sure to survive the uncertain interval to the next watering.

The operation of wells so affected the resources available within the family or village as to impose constraints upon 'overcropping'. But there was no automatic mechanism by which well irrigation allowed villages to make optimal use of their land resources. In many villages well irrigation entailed constraints upon the cultivation of land, and a good deal of 'undercropping' occurred, even in villages fully irrigated by wells. Indeed constraints were such that 'underwatering' was likely to have been as frequent a yield-depressant as 'overwatering' in canal villages. But comparisons of yields per acre of specific soil types are, arguably, of less importance to a peasant than output reliability and the produce of a holding overall. In part, this is what the rent and population differentials reflect.

Negative externalities and production

Floods and waterlogging

Early design faults and a niggardliness over funding were the main contributors to the waterlogging problems (and the related ones of *reh* and malaria) which ebbed and flowed from the 1860s onwards, and only gradually receded. The initial government-constructed distributary system consisted of a trunk line running parallel with the main canal on either side and fed by numerous short channels which made up a herringbone pattern.[88] Since these distributaries did not follow the watersheds, they interfered considerably with natural

[88] NWP PWD(I), Aug. 1874, 69, Note by Col. W. Greathed.

drainage lines.[89] Canal engineers, moreover, showed little concern for the need to allow waterway under the raised irrigation channels, compounding the problems arising from similar omissions by railway engineers[90] and by the cultivator–builders of channels who made it 'the rule rather than the exception to align their channels so as to cross and interfere with . . . natural lines' of drainage.[91]

A typical example of the drainage problems caused by the canal comes from *pargana* Tirwa (Farrukhabad), where, prior to the canal, floodwaters annually drained from Lakha *jhil* to Bahosi *jhil* and from there to the River Pandu. The construction of the Hussarun *rajbaha* blocked the flow between the *jhils*, while between the Bahosi *jhil* and the river lay not only the Bahosi *rajbaha* but the main canal itself. At first, cultivators around Lakha *jhil* were able to utilise the build-up of floodwaters to extend their rice cultivation, but the benefit was short-lived: as the *jhil*'s water level rose, 170 *bighas* were thrown out of cultivation. Bahosi cultivators, also, had temporarily benefited from a full *jhil*, but heavy rainfall and flooding problems around the Lakha *jhil* eventually persuaded the canal officer to cut through the Hussarun *rajbaha* to relieve the build-up of floodwaters, swamping in the process the Bahosi lands and causing the loss of 233 *bighas* of cultivation.[92]

Such losses were largely borne by the cultivators. The government was 'slow in giving up revenue on swamped lands', and actual remissions covered only a small proportion of land so damaged.[93] Within the rural areas, of course, some

[89] Buckley, *Irrigation Works*, p. 287. Cawnpore division appears to have been particularly badly affected in this regard. In the Ganges Canal Inspectorate Report for this zone, Capt. C.S. Moncrieff noted that many distributaries had been 'aligned on an utterly vicious system'; constructed at a time 'when we all knew nothing about rajbuhas', the result was 'a series of peculiarly bad lines which have been left almost unchanged, and that in a most difficult country, where rice patches are interspersed between oosur and swamp, and where drainage is very complicated'. NWP PWD(I), May 1874, 23, 27 Jan. 1874.

[90] E.g. *Muzaffarnagar DG* (1903), p. 45.

[91] NWP PWD(I), June 1873, 96, Note by EE, Cawnpore Division.

[92] NWP Rev., June 1873, 77, H.F. Evans, Report on Canal's Effect on Natural Drainage, Pergunnah Tirwa Thuttia, Farrukhabad District. See also NWP Rev., June 1873, 73–6 and 78–95.

[93] NWP PWD(I), Aug. 1874, 100, Note by Capt. J. Ross; see also NWP PWD(I), Oct. 1875, 59, H.B. Webster (Meerut Collr.) to Meerut Commr., 28 July 1875. See Ch. 6 below for discussion of remissions procedures.

cultivators would be hit harder than others, and in those villages where landlords were keen to extinguish *maurusi* (occupancy) rights held by their tenants, *zamindars* were said to 'rather rejoice in a year of calamity from floods'.[94]

For several reasons, however, the actual production loss can be prone to overstatement. This is illustrated by the case of *khadir* land, which was widely damaged as the water lavished upon the upland fields percolated down into the valley, where it exacerbated a natural tendency to saturation in the soil.[95] Over several thousands of acres *khadir* cultivation fell off and population dwindled as the fertility of the land declined. That this was related to the effect of the canal is beyond dispute. Set in its wider context, however, the net effect was comparatively small. Prior to the canal's introduction, cultivation in upland areas adjacent to the river valleys was usually insecure, and parts of the *khadir* were, for some villages, the only areas in which 'produce was tolerably assured'.[96] It was widespread practice for cultivators from upland villages to descend into the comparatively unhealthy *khadir* to grow rice, sugarcane, and wheat in the moist conditions there. When the canal transformed production conditions on the upland it was a perfectly rational redistribution of resources which lay behind the switch in production away from the *khadir*. As Alan Cadell put it, 'The very prosperity of the [Muzaffarnagar] upland is the hardest blow which the *khadir* could have received.'[97] But while the *khadir* became less important for its agricultural produce, its importance as a grazing ground and fuel source was enhanced by the contraction of upland waste, and by the increased demand for bullocks. Through releasing resources to upland cultivation the *khadir* contributed to the progress of the canal villages at the expense of its own cultivation.

[94] NWP PWD(I), Aug. 1874, 100, Note by Capt. J. Ross.

[95] For more detail on the canal's effect on the *khadir*, with specific reference to the Ganges Valley in eastern Muzaffarnagar, see Stone, 'Canal Irrigation and Agrarian Change', pp. 104–6.

[96] *Muzaffarnagar DG* (1876), pp. 478.

[97] UPA, BOR, 'Different Districts', Muzaffarnagar, Box 8, File 35, A. Cadell, RRR Bhoomah Sumbulherah, 1872, para. 19. Landlords were reported to have 'difficulty in keeping their cultivators together', particularly the 'migratory class' of *khadir* Chauhans, who were 'constantly deserting villages' to which they felt 'none of the attachments, which bind an upland cultivator to his ancestral fields'.

Even on the upland plains, though, the extent of the damage to production from waterlogging was especially difficult to gauge. However, where settlement operations coincided with periods of flooding, a rough measure can be obtained of the actual loss in cultivation. In badly affected Bulandshahr, at the end of the abnormally wet cycle in the 1880s, the area put out of cultivation by floods and waterlogging came to little more than 1% of the districts's n.c.a.[98] However, there is no way to discover the extent of that 'large, but undefined and undefinable [area] so saturated or soured that it ceased to yield full crops'.[99]

The effects of waterlogging upon agriculture varied considerably, and were not always negative: rice, for instance, which took its oxygen from the water, was little affected by poorly aerated low-lying soils; saturated fields commonly provided forage and water for village herds; fields effectively waterlogged in the *kharif* could, by November, be merely moist, and allow rabi sowings without the need for irrigation;[100] and where waterlogging was of comparatively short duration, the paying cane crop usually did not suffer serious injury.[101] Such reasons help explain why settlement officers found that the 'test of competition rents' was 'against the view that there has been any serious deterioration'; it remained the case, they insisted, that a full supply of canal flow irrigation was the most valuable quality land could have, and that it commanded a high rent 'even where there is a danger of saturation'.[102]

It is clear that the canal authorities anticipated that a drainage system would have to be installed eventually,[103] but right through the 1860s, when the effect of canal irrigation upon the spring level was becoming ever more apparent, only 'minor expedients' were employed to meet directly the problems of a

---

[98] The damage, affecting 155 villages, was mainly confined to the western part of the district. 5,656 acres were recorded as 'under water'; 4,527 acres as insufficiently dry to permit *rabi* sowings. Stoker, *Bulandshahr SR* (1891), pp. 9–10.

[99] *Ibid.*   [100] *IRR 1873–4*, pp. 1A–2A.   [101] *Ibid.*, p. 20A.

[102] Miller, *Muzaffarnagar SR* (1892), pp. 9, 53. Such reasons may also help explain the fact that even some of the most badly affected villages around Jansath town (those which, in fact, canal officers seriously considered depriving of canal water due to drainage defects) showed substantial increases in population and numbers of houses at the 1881 census. NWP&O Rev., June 1890, 20.

[103] *PRRC 1862–3*, p. 25.

wholly inadequate drainage system.[104] A Meerut conference held in the wake of a serious outbreak of fever in the Upper Doab during 1869–70 concluded that the 'increased moisture of the country since the opening of the Ganges Canal' was in some way linked with the fever, and that the drainage of irrigated lands had 'not hitherto received the attention it required';[105] the 'absorbing power of the soil' was becoming seriously impaired.

By the early 1870s, only 560 miles of drainage cuts existed in the entire Doab. This situation reflected the dire shortage of funds made available for such works, which right through to 1874 relied upon Imperial funds.[106] The year 1874, however, was one of 'unparalleled disasters', with such heavy and prolonged flooding and such 'unusual sickness and mortality' in Meerut division as to prompt a personal field inspection by the Lieutenant-Governor. At Meerut it was decided that a list of 28 drainage projects were 'urgently required', and that they should be constructed before the beginning of the 1875 rains. Under the new financing regulations arising out of its 1870 commitment to do 'all in its power to obviate the evil', the NWP Government was able to make available provincial funds for these works, and designated Rs 8½ *lakhs* of the estimated cost of Rs 11 *lakhs*. A more comprehensive scheme, costing around Rs 45 *lakhs*, and designed to restore the Upper Doab to a 'thoroughly satisfactory sanitary and agricultural condition', was approved the following year. Almost immediately, however, shortages of available funds forced the planned cuts to be reduced in width 'so that for the present the volume of drainage carried off per square mile of catchment will be less

[104] Indirectly some improvement was effected through the distribution remodelling schemes described in Ch. 6 below.

[105] NWP PWD(I), Aug. 1875, 22, Resolution by Lt.-Governor on Drainage Operations, 6 April 1874.

[106] Replying to criticisms made of engineers for failing to correct drainage problems, one senior canal official pointed out to the NWP Government in 1872 that the previous Chief Engineer, Col. W. Greathed, had actively discouraged schemes which did not promise a direct return to outlay. The government made it clear that Greathed had merely been following official policy: namely, that unless schemes did promise to be directly remunerative, removal of surface water could only be undertaken out of irrigation funds if its presence were caused directly by the canal itself. It did concede, however, that 'want of funds' often caused delays. NWP PWD(I), Aug. 1872, 292 and 297.

than is desirable'.[107] Within another two months, the 'state of finances' had compelled the postponement of all parts of the scheme which had not actually commenced.[108]

The drainage programme only began in earnest during the 1880s, when a series of wet seasons created serious waterlogging problems and prompted 'an outcry against canals'.[109] Guided by the extensive water level monitoring scheme established in the early 1880s,[110] engineers had, by 1900, expended around Rs 43 *lakhs* on drainage installation in the Doab. The aggregate length of drainage channels stretched 3,327 miles, more than a third of the total length of canals and distributaries.[111] The weight of evidence suggests that these measures were effective in reducing flooding and waterlogging problems: 'entirely satisfactory' was the verdict of one senior canal official.[112] This was possible because, as the results of investigations into the precise cause of waterlogging showed, even in the most heavily irrigated areas rainfall had a much greater effect on the level of subsoil water than did canal irrigation.[113] The canal did not prevent a marked fall in the spring level in dry years, as ten years of observations on the Agra Canal made clear.[114] What this meant, therefore, was that, provided that excess surface water could be conveyed away through a system of surface drainage, the operation of the canal did not imply an inevitable water-table imbalance.

Installing surface drainage was itself far from straightforward. In the lower divisions, where drainage lines had a fairly deep section and where the country consisted of a series of narrow *doabs*, drainage cuts needed only to be short ones, and were very effective. In the Upper Doab, long deep cuts were needed to span the wide *doabs*, while the more defined river lines

[107] NWP Rev., June 1877, 82, Sec. NWP PWD(I) to GI PWD, 21 July 1876.

[108] NWP Rev., June 1877, 40.

[109] C.H. Hutton, 'Rainfall, Irrigation, and Subsoil Water-level of the Gangetic Plain in the United Provinces', *AJI*, XIII (1918), p. 201.

[110] *IRR 1885-6*, p. 146. Levels were recorded using non-irrigation wells at two- to three-mile intervals along a series of east–west lines crossing the Doab.

[111] *IIC*, p. 185.

[112] Hutton, 'Rainfall, Irrigation, and Subsoil Water-level', p. 201.

[113] *IRR 1884-5*, p. 21, Report by Chief Engineer (hereafter CE).

[114] *IRR 1892-3*, p. 105B; see also NWP&O Rev., March 1889, 45, Note by CE on EJC, 15 Feb. 1884.

of the Central Doab posed the problem of being relatively shallow and of impeding the flow of drainage lines during high floods.[115] In some areas, engineers found that there was not sufficient outfall for drainage lines to utilise. The depression running for 68 miles along the western side of the Agra Canal posed particular problems in this regard,[116] and the east Muzaffarnagar depression in the vicinity of Jauli Jansath proved so resistant to attempts at drainage that it eventually led to the Jansath distributary being abandoned.[117] Some *khadir* lands, too – most notably the Solani–Ganges *khadir* in north-eastern Muzaffarnagar – clearly called for more resources than the authorities felt they were worth, and in consequence the deterioration was not halted. *Khadir* land elsewhere, however, was responsive to the measures: improvements along the Choiya Nadi allowed good *rabi* crops to be reared even during the wet 1884–5 season;[118] and in Etah's Burhganga tract, drains had done 'excellent service' in removing swamp and permitting timely winter sowings.[119]

In many areas, it might be argued, the canal engineers were rather too successful in draining the land. Complaints were numerous from cultivators who, in exchange for having their field-to-field communications interrupted by drainage lines (and often their canal supply diminished as part of the drainage package), found that the measures adversely affected the *jhils* and the saturated fields which before 'always gave water and forage of some kind'. Drainage often rendered such land 'dry and bare for nine months in the year', or else the land would be taken up by someone for cultivation so that it was no longer 'common' for grazing purposes.[120] So effective, in fact, were the measures centring upon the Kali Nadi (which was itself

[115] NWP PWD(I), Aug. 1875, 42, R.E. Forrest (EE), Note on Drainage Cuts.

[116] *IRR 1892–3*, p. 107B. Part of the area prone to waterlogging along the Mat branch (UGC) also proved resistant to drainage measures designed to relieve a chain of depressions below the watershed ridge in Sikandarabad and Khurja *parganas* (Bulandshahr).

[117] *Muzaffarnagar DG* (1903), p. 45. The expenditure of Rs 42,267 on drainage had halted neither the rise in the spring level nor an associated increase in the incidence of malaria. With average death rates during 1881–9 put at 42 per thousand, irrigation was stopped altogether for 1,000 acres around Jansath from the early 1890s. NWP&O Rev., June 1890, 35.

[118] *IRR 1884–5*, p. 11.   [119] *IRR 1892–3*, p. 6A.

[120] *IRR 1880–1*, p. 16A, Note by H. Marsh (EE); *IRR 1883–4*, p. 13.

deepened and straightened) that not only was the Kali–Kharon *doab* freed from floods and supersaturation, but the efficient surface run-off caused the water-table to fall thirty miles away in Muttra.[121] Indeed, towards the end of the period (after two prolonged cycles of dry weather), when the attention of some officials was directed towards the future irrigation possibilities of tubewells and mechanical lifting power, one revenue officer particularly well informed on such matters argued that there existed in the Doab a 'less steep gradient in the subsoil water-level than twenty years ago and a permanent tendency for the water-level to fall'. He considered it a matter for 'serious consideration' whether the 'policy of surface drainage has not been carried too far'.[122]

Salinity

The spread of saline deposits, or *reh*, over the land's surface in the north India plains is closely connected with an enhanced water-table level. When the table rises to within five feet of the surface, the 'capillary fringe' of the water film extends upwards upon reaching the high-temperature zone in the top layers of soil. Poorly aerated, compact, and stiff soils are particularly prone to this capillary process, which leaves dissolved salts on or near the surface as the water evaporates.[123] Such 'efflorescence', according to its alkaline intensity, might leave crops virtually unharmed, or might put the field (or patches of it) entirely out of production.[124]

The Doab had always contained extensive areas of salt-infected land, and the accumulating evidence relating to the

---

[121] See Ch. 2 above for details. The Muttra area along the Kharon Nadi (and to a lesser extent in parts of eastern Aligarh) was apparently the only one in which the fall in the water level was so pronounced as seriously to impair well irrigation in non-canal tracts. For a description of the Kali Nadi improvement scheme, see *Bulandshahr DG* (1903), p. 10.

[122] E.A. Molony, 'Rainfall, Irrigation, and the Subsoil Water Reservoirs of the Gangetic Plain in the United Provinces', *AJI*, xii (1917), pp. 84–9.

[123] These deposits can be either snow-white encrustations, made up largely of sulphate and sodium chloride, or of a brownish-black colour, where sodium carbonate appears in addition. See Howard, *Crop Production in India*, pp. 43–4.

[124] According to Howard's Punjab experiments, it was 'only as the proportion reaches a certain level that [the salts] first interfere with and finally prevent growth'. *Ibid.*, p. 43.

cause of the phenomenon, set alongside the manifest effects of the canal upon the spring level, was consistent with contemporary visions of the north-western plains being converted into a 'howling wilderness'.[125] In the 1860s and 70s, with the rime 'spreading like a disease',[126] officials were understandably beginning to fear the monster of their own creation. To what extent this alarm was based on a fear of the *process* then known to be at work – the effects of which were considered practically irreversible – rather than of the actual damage taking place is worth consideration. Gauging the extent of *reh* damage caused by the canals cannot be done with any accuracy, but a crude assessment is possible from the scanty evidence available.

Available statistical information on *reh* is marginally more useful than that on waterlogging because *reh* not only was less variable in relation to the season but was also highly visible. That being so, too much should still not be expected, since even by 1972 Uttar Pradesh had no 'reliable assessment of land affected by salinity or alkalinity'.[127] Moreover, there is still the problem of gauging its impact in reducing yields rather than preventing cultivation. *Reh* was fairly rare in the Upper Doab. In Meerut, where there were no *usar* plains, the 'comparative absence' of *reh* was 'marked';[128] as in Muzaffarnagar and Saharanpur, such salinity as occurred was mainly confined to the *khadirs*.[129]

It was in the Central and Lower Doab, where there were extensive *usar* plains, that *reh* mainly occurred. There, much of the canal-related *reh* which did appear was confined to these *usar* expanses. In Etah, for instance, around 1870, 'large quantities' of *reh* were to be found adjacent to the canal in the rich southern *dumat* tract, but 'little appreciable damage . . . would appear to have been done to cultivation from this cause',

---

[125] Whitcombe, *Agrarian Conditions*, p. 11, quoting a remark made by the Superintendent of the Geological Survey in 1877. See also *ibid.*, pp. 77–9.

[126] R.G. Currie, *Bulandshahr SR* (1877), p. 42.

[127] *Report of the Irrigation Commission*, I, p. 310.

[128] *Meerut DG* (1876), p. 202.

[129] In Muzaffarnagar, in 1901–2, a total of 8,272 acres were recorded as '*reh*-infected'. Most of this was in the *khadir* and did not affect cultivated land at all. *Muzaffarnagar DG* (1903), p. 16.

since it was mainly confined to the *usar*.[130] McConaghey's 1870s settlement of Mainpuri revealed the same pattern: *reh* had spread but was confined 'practically wholly to usar'.[131] 'Injury from *reh* is much less frequent than is sometimes asserted', insisted Crosthwaite, Etawah's Settlement Officer, adding:

If one village is injured from reh, the people talk of it, and raise a clamour about it, till others think the whole pargana is ruined. When settlement is in progress the cry of reh is raised from every canal village. I know of only two or three villages where, owing to the stoppage of drainage, reh has accumulated so as to affect the cultivated land. And in these, although the outcry is great, the land is cultivated and the same rent paid as heretofore.[132]

These are by no means the comments of men indifferent to the problem. Crosthwaite and McConaghey were both prominent in arguing for canal rate enhancements to encourage the more careful use of canal water, and any evidence of extensive *reh* formation, linked closely at the time with 'over-watering', would clearly have provided backing for their argument.[133]

Figures relating to the subsequent settlement period – when rainfall was heavy and drainage needs only slowly being attended to – give support to the notion that the outcry over *reh* was disproportionate to the actual damage it caused. In Bulandshahr, despite the waterlogging, comparisons using maps and fieldbooks of the previous settlement did not suggest to the Settlement Officer the material extension of *usar* plains, and while in the western parts some new *reh* had appeared, the area 'actually sterilized or permanently thrown out of cultivation by it' since the previous settlement was 'absolutely inconsiderable'.[134] Indeed, wherever figures are available, the areas involved are relatively small. This can be demonstrated by the case of Shikohabad *tehsil*, which was located near to the canal in Mainpuri, one of the Doab's most salt-affected districts. During the thirty years up to 1900, the Settlement

[130] Ridsdale, *Etah SR* (1874), p. 5. Ridsdale knew 'only of two instances where several fields of good land have been rendered altogether unculturable by the recent appearance of *reh* and saline qualities in the soil'.

[131] Quoted in W.J. Lupton, *Rent-Rate Report of Karhal Tahsil, Mainpuri District* (Allahabad, 1904), p. 5.

[132] Crosthwaite, *Etawah SR* (1875), p. 17.

[133] This debate is examined in Ch. 5 below.

[134] Stoker, *Bulandshahr SR* (1891), pp. 10–11.

Officer discovered, *reh* had spread along the canal channels and 'infected' 44 villages. However, the actual area of the fields affected in these villages amounted to 337 acres (in a *tehsil* of 125,000 acres), and in many cases the damage was only serious in patches.[135] Again, in Karhal *tehsil*, with *reh* apparently 'virulent' in land lying between the two canal branches, a fall of 650 acres of cultivation in 24 villages occurred in the thirty years to 1904, with 'some further areas' in those villages suffering yield reductions. Seventy-eight villages in this canal tract were not affected, however, and increased their cultivation by 1,250 acres. Overall, in a tract of 40,000 acres, there was a net gain in cultivation of 600 acres.[136]

Experiments aimed at reclaiming alkaline fields, whether bordering on the bizarre or merely ingenious, made it clear that reclamation was more costly than land purchase price. The most straightforward corrective action involved draining the field, lowering the water-table, and leaching down the salts (either by rainwater or by irrigation). Many areas recovered quite naturally as drainage measures combined with dry cycles to lower the water-table. The Kali *khadir*, for example, several thousands of acres of which became *reh*-infected when the river was used as a canal escape, was in many parts reported to be 'completely recovered' after the river had been deepened and straightened. By the 1900s the formerly *reh*-afflicted areas contained 'large areas of continuous and stable cultivation, some of it rich and luxuriant'.[137]

Malaria

Villages in the canal districts regularly experienced prostrating bouts of malarial fever, and periodically fell prey to serious epidemics which claimed many lives. The role of 'fever' as a cause of death is spectacularly prominent in the annual statistics put out by the Sanitary Department. They show that during the 1880s, for example, it was common for more than four-fifths of UP deaths to be caused by fever. In the canal

[135] Lupton, *RRR Pargana and Tahsil Shikohabad, Mainpuri*, p. 8.
[136] Lupton *RRR Karhal Tahsil, Mainpuri*, p. 5.
[137] P. Harrison, BOR Review, p. 3, in Phelps, *Bulandshahr SR* (1919). See also *Bulandshahr DG* (1903), pp. 16–17.

districts this proportion was higher: even the comparatively healthy year 1891 saw 95% of the deaths in Meerut and 93% of those in Mainpuri and Bulandshahr attributed to fever.[138] In 'good' years, the crude death rates (c.d.r) due to fever hovered around 20–25 per thousand. Frequently they were in the 30s, and rose in years of epidemic to over 40. Locally high fever death rates were recorded for 1885, 1890, and 1894, with bad epidemics in 1908 (when the Doab districts experienced a c.d.r. of between 50 and 70) and 1922. Perhaps the greatest epidemic, though, occurred in 1879, when fever deaths amounted to 114 per thousand in both Aligarh and Bulandshahr (reaching 83 and 80 for the respective areas during the September– November period, compared with an average of 19 per thousand during 1874–8), and 82, 78, and 67 respectively in Meerut, Etah, and Muttra.[139] In such years, crops were in some villages known to rot in the fields 'for want of hands to harvest them', since 'money could not buy labour worthy of the name'.[140] Such was the demand for fuel for cremations in 1879, observed the Sanitary Commissioner, that in 'some places, fuel . . . had become so scarce as to hinder the proper cooking of food', and the bodies of the fever victims littered the countryside, in rivers and stagnant ponds, in the bottom of dry wells and pits, and in the grass patches skirting the canal and railtracks.[141]

The mortality was normally at its greatest in the towns. Deoband recorded a c.d.r. of 123 per thousand in 1884;[142] Shamli one of 162 in 1885, when apparently none of its inhabitants escaped the epidemic and virtually no infants

[138] *Report of the Sanitary Commissioner for the North-Western Provinces and Oudh, 1891* (Allahabad, 1892), p. 9. (Subsequent references to these annual reports will be abbreviated to RSC.)

[139] Fever deaths per thousand for September, October, and November were: Bulandshahr, 16, 38, 26; Aligarh, 22, 37, 24; Meerut, 15, 23, 14. Figures given in NWP&O PWD(I), July 1882, 5, Army Sanitary Commission Memorandum on Sanitary Conditions in NWP.

[140] Autumn farming activity could be hampered by fever. *Rabi* sowings were sometimes delayed and even prevented by severe local attacks. See UP PWD(I), Feb. 1909, 3; and UP PWD(I), March 1909, 3. 'As I write', recorded the Meerut Commissioner in 1870, 'there are wide expanses of Khureef fields, long since ripe, and which are dropping their seed, which are still standing because the cultivators cannot cut them.' NWP Rev., June 1871, 6.

[141] *RSC 1879*, pp. 56–7.   [142] *RSC 1884*, p. 35.

survived.[143] Infants and young children were always prominent among the mortality victims of fever, and in many villages in Muttra 75% of the infants died during the 1908 attack.[144] In the town of Saharanpur, more than half the 1908 malaria deaths were of children under five.[145] Unlike famine, malaria struck at all classes, although it was observed that 'mortality, without doubt, seemed greatest among the poor'.[146]

The horrific statistical impression given by mortality returns is unreliable, however. Deaths from malaria, as represented by 'fever' returns, were grossly exaggerated, particularly for normal years. The collection of statistics on births and deaths relied fundamentally upon the reporting agent, the *chaukidar*. He may have been, as the Sanitary Commissioner pointed out in 1904, 'probably the cheapest registering agency in the world', but at Rs 3 per month, the government could hardly expect accurate diagnoses of the causes of those deaths the *chaukidar* registered.[147] The medically ignorant registrar was expected to record the causes of any deaths from information provided by the *chaukidar* on his monthly visit to the circle registrar at the police station. The *chaukidar* was usually able to identify roughly deaths from smallpox and cholera, but could not distinguish between the various types of fever:[148] 'The chaukidar persists in calling everything *bokhar* to which he cannot obviously give another name', complained the Sanitary Commissioner as late as 1921.[149]

Mortality figures of a more reliable – if restricted – nature show a much lower incidence of death from fever. Thirty-one per cent of the deaths among European troops in 1888 were put down to this cause; for native troops the figure was 16% and for prisoners, 9%.[150] From his observations and experience,

---

[143] *RSC 1885*, p. 75.

[144] Government of the United Provinces of Agra and Oudh, Sanitation Department Proceedings (hereafter UP San.), May 1909, (proceeding no.) 15, Maj. J. Chaytor White, Report on 1908 Outbreak of Malarial Fever in UP. (In the IOR.)

[145] UP San., June 1910, 34, Maj. J.C. Robertson, Report on Enquiry into Prevalence of Malaria in Saharanpur City.

[146] *RSC 1879*, p. 59.

[147] *RSC 1904*, p. 5. For a description of the way statistics were collected, see *RSC 1877*, pp. 27–8; *RSC 1885*, p. 10; and H.C. Cutcliffe, *A Sanitary Report on Certain Districts in the Meerut Division* (Allahabad, 1868), pp. 19–20.

[148] *RSC 1877*, p. 28.   [149] *RSC 1921*, p. 11.   [150] *RSC 1888*, p. 49.

particularly of children's diseases, the Sanitary Commissioner asserted that of the more than one million recorded UP fever deaths in the current year (81% of the total), 'probably not more than one-half was due to any recognised form of fever, but to other causes . . . Death from the malarial type of fever is comparatively quite rare, and the average mortality from such a fever may not in ordinary circumstances exceed twenty-five in the hundred from all causes . . . The enormous fever death rate has no actual existence.'[151]

Support for this view can be found in the seasonality of the mortality figures. Malaria is essentially a post-monsoon phenomenon, the transmission of which comes to a sharp halt as the temperatures fall with the onset of winter. A closer examination of the 1890 figures reveals that the excessive mortality from 'fever' actually took place in the spring months; well below half the Meerut fever deaths actually took place in the second half of the year.[152] Sample surveys aimed at determining more accurately the causes of death confirmed this general picture: in normal years a large proportion of fever deaths were more likely to be the outcome of diseases such as dysentery, pneumonia, influenza, diphtheria, and smallpox. A 1921 sample survey, involving over 6,000 cases, reduced the figure for the proportion of deaths due to malaria from 58% to 10.6%.[153]

While it is certain that malaria was to some extent an indirect cause of death – those debilitated by fever were undoubtedly more susceptible to death from the respiratory diseases of winter[154] – the typical level of malaria mortality was much less dramatic than the c.d.r. figures suggest, although the degree of statistical exaggeration is proportionately less significant in times of epidemic outbreaks. Certainly the impact of malaria upon the mortality rates did not prevent even the most affected districts in the Upper Doab from increasing their population during the 1872–1911 period relative to the increase in the UP as a whole (15% as against 12%).[155] However, people may not have died from the disease as frequently as the official statistics suggest, but they certainly suffered its symptoms. A

---

[151] *Ibid.*, pp. 49–51.   [152] Figures from *RSC 1890.*   [153] *RSC 1921*, p. 11.
[154] *RSC 1923*, p. 13; see also Review at end of Report.
[155] *Census of India, 1921*, vol. xvi, *United Provinces of Agra and Oudh* (Allahabad, 1923), pt 1.

benign and protracted form of malaria was endemic over much of the western parts of the provinces. 'Very few people', observed the Sanitary Commissioner in 1908, 'pass through a normal year without fever, but if they are well nourished they throw off an attack from which they are otherwise likely to succumb.'[156] So, if adult fatalities were rare, everyone was still liable to prostrating attacks of fever, causing them to take, on average, a week off work, followed by perhaps a month at half-efficiency.[157]

The problems did not end there, since male sufferers frequently experienced the side-effect of impotence. In carrying out his investigations for his *Sanitary Report on Certain Districts in the Meerut Division*, H.C. Cutcliffe found that 'sometimes the men of a village would come to me in collected numbers, and unhesitatingly assure me that they were all impotent and beg me to cure them'. The problem was apparently 'extraordinarily frequent in the most malarious tracts', confirming an 1864 report on Bhynswal village by the Collector of Muzaffarnagar, who wrote in reference to its waterlogged and unhealthy state: 'The Jats say they cannot now get girls to marry their sons as people declare the residents have become impotent from disease.' Cutcliffe was in no doubt that very few children were being born to the women of these localities.[158]

That the canal system was linked in some measure and by some means to this phenomenon was established relatively early. While there are instances of malaria outbreaks prior to the widespread introduction of canals – there was apparently one in 1817 – the Upper Doab was 'as a matter of popular opinion . . . in former times considered a very healthy land to which people from the cities of Delhi, Agra and Lucknow resorted for a change of air after illness'.[159] Indeed, Meerut – which had become so unhealthy that its abandonment as a cantonment was discussed in the early 1870s – was reputedly 'once the healthiest station in the provinces'.[160] There does

---

[156] *RSC 1908*, p. 12.

[157] *Chambers Encyclopaedia*, vol. viii (London, 1973), p. 856.

[158] Cutcliffe, *Sanitary Report*, pp. 41–2; NWP PWD(I), March 1872, 53, Cutcliffe to Meerut Commr., 9 Nov. 1864.

[159] *RSC 1879*, p. 23.

[160] J.D. Graham, *A Report on an Enquiry into the Prevalence of Malaria in Meerut, 1911* (Allahabad, 1913), Appendix iv, Report by Maj. Thomason, 3 Dec. 1874.

appear, then, to be some suggestion of a long-term change in the incidence of the disease, and evidence from the UP, alongside that from the Punjab, Madras, and Sind, suggests that the environmental implications associated with the construction and operation of extensive canal systems played an important part in this development. Indeed, in this respect, the 'epidemic increase' in northern India has been described as 'man-made' by the late Professor G. Macdonald of the WHO Expert Committee on Malaria.[161]

Whereas the real scientific breakthroughs relating to human malaria parasites and their transmission were not made until the 1890s,[162] some of the earliest work in India on epidemics is considered by medical scientists to be remarkable for its exact attribution of the causes, and indeed for the discovery made by army surgeon T.E. Dempster in 1847 that the spleen rate in children was a reliable guide to the incidence of malaria. Dempster was on the committee appointed by the Governor-General in 1845 (which also included the military engineers W.E. Baker and Henry Yule) to enquire into the cause of the 'unhealthiness' affecting the line of country along the Western Jumna Canal, and particularly that afflicting the Karnal cantonment.[163] The committee was also to make a judgement as to whether the proposed Ganges Canal would have similar effects on the health of the local population. In their investigations the men visited 300 villages, and built up, with the aid of oral testimony and observation, a picture of the 1843 epidemic. They were able to show that although a large part of the NWP experienced a fever outbreak, it was more prevalent and severe in the canal-irrigated areas than elsewhere. 'Excess

[161] L.J. Bruce-Chwatt and V.J. Glanville (eds.), *Dynamics of Tropical Disease by G. Macdonald* (London, 1973), p. 155.

[162] It was in 1894 that Patrick Manson began to argue that the disease was passed from man to man by the mosquito. The poet–scientist Ronald Ross, working in Sikandarabad, discovered that the disease was transmitted exclusively by the female mosquito of the genus Anopheles in 1897.

[163] W.E. Baker, T.E. Dempster, and H. Yule, *Report of a committee assembled to report on the causes of the unhealthiness which has existed at Kurnaul, and other portions of the country along the line of the Delhi canal, and also whether any injurious effect on the health of the people of the Dooab is, or is not, likely to be produced by the contemplated Ganges Canal, with appendices on malaria by Surgeon T.E. Dempster* (Calcutta, 1847); reprinted in *Records of the Malaria Survey of India*, 1:2 (1930), pp. 1–68, and in Cautley, *Ganges Canal Works*, III, Appendix B. Note: 'Delhi canal' refers to the Western Jumna Canal.

of moisture' they identified as the common link between the badly afflicted areas, and they identified a sharp contrast on the EJC between the light soil, efficient drainage, and *rajbaha* irrigation in the north and south divisions, and the heavy soil, obstructed drainage, and imperfect distribution network in the central division, where the fever had been as bad as that on the WJC. The planned Ganges Canal, they concluded, should not be abandoned: those (upper) sections of its route they had seen consisted of light soils, and as long as the work was designed so as to avoid drainage problems, and irrigation prohibited in environmentally unsuitable areas and in the proximity of towns and cantonments, then the health of inhabitants would be largely unaffected.

This 1847 report, as a commonsense delineation of the causes derived from empirical sampling, was hardly improved upon until the early decades of this century, when studies into the mechanisms involved in malaria epidemics, making use of the scientific developments in the field, were pioneered by S.R. Christophers.[164] H.C. Cutcliffe's investigation into an 1867 outbreak of fever in Meerut division – in which he had the task of determining whether there was any 'ground for attributing its appearance or decrease to the action of the canals, or to the general disregard of sanitary principles' – involved visiting 140 villages, but added little to the findings of the earlier study.[165] Swampy conditions were broadly correlated with fever; the canal was often accompanied by high water-tables, interrupted drainage, and standing surface water, particularly in clayey soils where rice was grown. The canal was thus linked with fever because it was associated with the conditions in which fever seemed to thrive. The fact remained that the disease was to be found in swampy conditions away from the canal; moreover, there were many efficiently drained canal villages where 'fever was unknown'.[166] It was impossible to be more precise about the culpability of the canal in this matter as long as little was known about the transmission of the disease.

---

[164] See, for example, S.R. Christophers, 'Epidemic Malaria of the Punjab, with a Note on a Method of Predicting Epidemic Years', *Paludism*, II (1911), pp. 17–26.

[165] Cutcliffe, *Sanitary Report*. A summary of the content and findings of the early investigations made into malaria is contained in *Meerut DG* (1875), pp. 125–9.

[166] *Meerut DG* (1875), p. 127.

Numerous suggestions were advanced, including 'defilement of drinking water' and the 'great diurnal range of temperature' on the plains. According to the available statistics, the case against the canal in this matter was mixed. Furthermore, it was possible, through attending more carefully to drainage needs in the design and layout of new systems, to minimise the creation of conditions associated with fever. Hence the NWP Government, heavily committed to a programme of canal expansion, sharply rebuked the Sanitary Commissioner, Dr C. Planck, for oversimplifying matters when he pointed out in 1879 that canal irrigation extension would certainly be followed by an increase in fever.[167]

It is possible to explain the way in which the canal probably affected the incidence of malaria through analysing the nature of the disease in this region and assessing historical evidence in the light of modern knowledge on the circumstances of transmission and the epidemiology of the disease. The western third of the UP is an area of unstable malaria, characterised by a lack of continuity of transition. This is because the malaria-carrying anopheles vector, *A. culifacies*, is a comparatively inefficient vector, being zoophilic (i.e. preferring animal to human blood) and relatively short-lived. The climatic circumstances of the western UP zone are also a factor in limiting the transmission of the disease. Below a certain temperature, the cycle of extrinsic development of the disease within the mosquito is infinitely retarded.[168] This, plus the fact that high temperatures with low levels of humidity shorten the life expectation of the mosquito, means that the anopheles cannot maintain transmission during the dry season. The lack of continuity of transmission has an important effect upon the development of immunity within the population: immunity is simply not well developed in conditions where malaria is a seasonal phenomenon. Maximum immunity (associated with stable malaria) is achieved where regular transmission of the disease takes place and so stimulates in the infected person the proliferation of antibodies in the lymphoid-macrophage system of the spleen, liver, and bone marrow.[169] In areas of irregular

[167] *RSC 1880*, 'Orders in Government'.
[168] 19°C in the case of *P. falciparum*, 15°C for *P. vivax*,
[169] L.J. Bruce-Chwatt, *Essential Malariology* (London, 1980), pp. 58–64.

transmission, the population can only build up a partial immunity. In the case of the most common form of malaria in the Doab, *Plasmodium vivax* ('benign tertian' malaria), the pattern of relapses associated with it stimulates a renewal of the antigenic response as successive merozoites are released into the blood.[170] This means that an affected person has some degree of immunity for perhaps a number of years, and although he is prone to reinfection and attacks of fever during the autumnal outbreaks – particularly if heavily infected – most adults have sufficient tolerance to shake off the attacks. Because young children and infants – with the exception of those who may inherit a degree of short-lived immunity from the mother – do not have this protection, it is understandable why mortality among them is very high.

Partial immunity to one species of malaria does not carry over to others, however, and *P. falciparum*, which was often as prevalent as *P. vivax* in the later part of the malaria season, causes a particularly lethal form of the disease. This 'malignant tertian' malaria was essentially responsible for the epidemics in this region, and – because it is characterised by dangerous complications – was the main cause of mortality during epidemics.[171] One reason for this is that this form of malaria, in areas of seasonal malaria, is less susceptible to the development of immunity because the pattern of relapses observed in *P. vivax* does not occur.[172] Immunity is lost quickly, usually well within the year.

This combination of circumstances made the western areas of the UP particularly vulnerable because the disease here was

[170] Merozoites are developed and multiplied in the liver from the sporozoites injected by the bite through a process known as pre-erythrocytic schizogony. Some of these merozoites re-enter the liver cells and initiate successive exo-erythrocytic schizonts, which accounts for the pattern of relapses. See *ibid.*, pp. 13–22.

[171] The complications associated with *P. falciparum* – identifiable from the descriptions contained in the contemporary reports – account for the great variety of symptoms which baffled those investigating the cause of the outbreaks. In cases of this type of malaria there is a tendency for infected cells to agglutinate in the capillaries of internal organs, producing symptoms referable to the site of agglutination. Blackwater fever and cerebral malaria are just two of the common and serious complications. See *Manson's Tropical Diseases*, 17th edn (London, 1972), pp. 55–63.

[172] In *P. falciparum* there are no exo-erythrocytic phases, and the antigenic stimulus lasts only as long as the recrudescences occur and tends to die down after several months. *Ibid.*, p. 54.

so volatile. Fluctuations in the incidence of malaria are likely, in all areas of unstable malaria, to be marked. The typical pattern is for endemic levels to be very sensitive to minor changes in the environment, with interruptions of transmission common, at times prolonged, and frequently succeeded by epidemics, the severity of which is aggravated by the size of the non-immune population which has grown up in the meantime.

As far as the transmission of the disease is concerned, it is important to note that, as a result of the mosquito's zoophilism, its high mortality, and its long extrinsic cycle, the probability that the insect will pass on infection once it has swallowed gametocytes is very small in areas of unstable malaria. The number of bites required for transmission is high, and the disease is thus perpetuated only where mosquitoes are very numerous. Anophelism without malaria is quite possible; certainly different levels of endemicity will exist within a region, depending on local conditions. However, environmental changes of quite small degree are potentially of great importance to the incidence of the disease, and it is quite feasible that such changes can induce not only short-term but long-term changes in the character of the disease in a region. A climatic shift, or simply a series of seasons suitable to the breeding needs and longevity of the anopheline mosquito, will sometimes give rise to major regional epidemics.[173] At the local level, the transmission of the disease sufficient to produce either exaggerated seasonal epidemics or exacerbations of endemicity tends to result from changes in breeding conditions; equally, it could be the result of a usually unimportant vector anopheline being converted into an important one by a change in its biting habit. The *A. culifacies* species has been found to feed on humans to a greatly varying extent – from 2% to 80%[174] – and this variation has a great potential effect on the incidence of the disease.

[173] Studies of major epidemics in Sri Lanka, Sind, and the Punjab all show the general interplay of interrupted transmission and varying immunity. Bruce-Chwatt and Glanville, *Dynamics of Tropical Disease*, p. 132. For a major epidemic to develop it is necessary for transmission at an enhanced rate to continue for at least three times the combined incubation period of malaria in the mosquito and in man to the time of the appearance of infective gametocytes. This amounts to around sixty days for *P. vivax* and ninety days for *P. falciparum*. *Ibid.*, p. 153.

[174] *Ibid.*, p. 152.

Canal irrigation undoubtedly caused a numerical increase in anophelines through its effect upon breeding conditions in the Doab. *A. culifacies* breeds mainly in pools,[175] and the combination of excess irrigation, interrupted drainage, and high subsoil water levels tended to make pooling more common and is known to be a frequent cause of high epidemicity through simply increasing the number of anophelines.[176] The existence of the canal also undoubtedly led in many areas to greater humidity, which would in certain circumstances increase the life expectancy of the mosquito and thus the mean number of infective bites it could carry out.

Historical evidence is consistent with this picture of larger numbers of anopheline mosquitoes resulting from the construction and operation of the canals, and of more widespread and intensified seasonal attacks of the disease. Cutcliffe found that the sickness due to fever was 'general and intense' along the tracts bordering the Ganges Canal right down to Meerut, and all the way down to Baraut on the EJC.[177] Often the initial attacks in newly irrigated areas were of particular ferocity, reflecting the fact that the communal immunity levels were low. Thus one Sanitary Department official noted how, towards the end of the 1878 rainy season, the

remarkable malarious fever peculiar to the northern portions of the Doab appeared to cross over the Jumna near to Delhi and follow the course of the new system of canals starting from Okhla. And, as is usually the case when a disease occurs for the first time in a locality, in a very fatal type, the canal will probably be blamed for nearly, if not all, of it . . . In my opinion . . . the system of irrigation as at present practised must be charged with very much of the fever and mortality, where extensive irrigation is introduced for the first time.[178]

The bulk of the July–December fever deaths in Agra and Muttra (66,000, compared to a normal 13–16,000) were in the trans-Jumna parts of the districts particularly near to the new Agra Canal. A fever map for west Muttra for 1878 shows at least eight villages along the canal with a death rate due to fever of over 100, with Pisawa recording 187.[179] The timing of the

---

[175] *Ibid.*, p. 161.   [176] *Ibid.*, p. 155.

[177] Cutcliffe, *Sanitary Report*, p. 38.

[178] *RSC 1878*, Appendix by Surgeon-Maj. R. Pringle (Deputy Sanitary Commr.), p. 4A.

[179] *RSC 1878*, p. 63.

outbreak and the descriptions of the symptoms indicate that *P. falciparum* was prominent in the outbreak.

Available evidence does indicate, however, that under normal circumstances the presence of canal irrigation was likely to be insufficient in itself to bring about high levels of transmission. Historical data show, quite clearly, that the epidemic outbreaks were closely associated with climatic conditions, with the precise geographical pattern being influenced by local circumstances. Research into 100 badly affected villages lying between Dasna and Aligarh, conducted in 1885, revealed that while villages with water-tables more than six feet below the surface consistently returned a fever mortality figure of 36 per thousand, those villages where the water was nearer than six feet suffered a c.d.r. of 56. Over time, the general level of fever mortality was closely related to the autumn rainfall.[180] Excessive rainfall accompanied the serious fever outbreak with its fulminant areas in 1879. Relatively low fever rates in the Doab in 1883 were followed by heavy rains and humid conditions in 1884 (when September's rainfall was ten inches higher than usual) and 1885, resulting in death rates from fever in the 30s and 40s per thousand. In the epidemic year of 1908, rainfall was normal on an annual basis, but nearly all of it fell in about two months. Furthermore, the high death rates many districts recorded during such years were not confined to the canal districts. Excessive rainfall in Rohilkhand tended to have the same effects, and on published statistics there is often little to choose between these contiguous tracts.[181] Rohilkhand was a poorly drained division, and tended to experience fever bouts which drove the overall figure for fever deaths into the 40s, which was often higher than the level recorded for Meerut, particularly, it seems, after the installation of extensive drainage in the latter division.

Combining historical evidence with present-day knowledge, therefore, it seems reasonable to accept the view of Dr C. Planck, expressed in the 1870s after an investigation of fever in Muzaffarnagar, that 'one may justly arrive at the opinion that ague has not been introduced as a new thing into the canal

[180] Note that the years given for high mortality from fever coincide with the above-average rainfall shown in Figure 4.
[181] This was the case, for example, during the quinquennium of 1893–8.

irrigated country, but that its area and period of prevalence and its intensity of attack have very greatly increased since irrigation from the canals was introduced'.[182] There is clearly much room for more and closer research on this issue. Detailed analysis using local and village statistics might reveal more about the mechanisms of the disease in this area, since district totals mask local variations. It is quite clear, in this connection, that the Doab towns were particularly vulnerable to the disease. The year 1887 was not untypical in this regard: of the twenty towns with the highest fever death rates, nineteen of them were in the Doab districts. Kosi, through which the Agra Canal flowed, had an average mortality rate of fifty per thousand over 22 years,[183] during which time the town's population fell from 12,500 (before the canal) to 7,000 in 1911. The reason for the higher c.d.r. from fever in the towns was possibly linked to the fact that the high proportion of people to animals in the town caused a change in the biting habit of the mosquitoes; certainly there is evidence for this in the case of Delhi.[184] Even in the towns, however, investigations in 1909 revealed the close relationship between fever and the rainfall curve.

Many towns were particularly exposed to fever attacks on account of their location. Sites of villages and townships had been established at a time when water was less in abundance, and many were located in depressions, where tanks could be constructed. With drainage lines frequently obstructed by canal lines and railway embankments, many settlements became completely surrounded by water in the rains.[185] The Muzaffarnagar market town of Shamli, for instance – once 'a pleasant city in a green valley' – had by the 1880s become 'a noisome fever hole', where the drainage of thirty square miles flowed 'past, under and through it', with the result that the town had reverted to a 'crumbling ruin, where the people are sterile and die out'.[186]

[182] *Meerut DG* (1875), p. 127.

[183] Death rates are from 'all causes'. J.D. Graham, *Report on an Enquiry into the Prevalence of Malaria in Kosi (District Muttra)* (Allahabad, 1910), pp. 3, 8.

[184] G. Covell and J. Singh, 'Anti-malaria Operations in Delhi, Part IV', *Journal of the Malaria Institute of India*, v (1943), pp. 87–106.

[185] *Muzaffarnagar DG* (1903), p. 48.

[186] NWP&O Rev., March 1889, 8, C.G. Palmer (EE, EJC) to Superintending Engineer (hereafter SE), EJC, 25 Nov. 1885. Shamli's population fell from 9,728 (1865) to

The extensive corrective measures of drainage and improved water control on the canals undoubtedly improved matters generally, but additional measures had to be adopted in particularly difficult cases where the numbers at risk were high. Thus – some sixty years after such measures were first recommended by the 1847 committee – prohibition of irrigation, and of rice and sugarcane cultivation, was adopted around badly affected towns, notably Saharanpur, Kosi, and Nagina. These policies, combined with other measures by the health authorities to limit local breeding sites – altogether costing Rs 4 *lakhs* – led to significant reductions of spleen incidence in all the cases, with Saharanpur registering a reduction from 84 to seven.[187] Overall, however, the canal areas continued to be, in the words of one witness to the Agricultural Commission, 'usually associated with a high malarial index'.[188]

Conclusion

This chapter has examined in some detail the direct production effect of the canal, and has argued that the peasant response to the new production possibilities was generally consistent not merely with the achievement of improved incomes but with the maintenance of long-term viability and balance within the farming sector. Against these gains, however, have to be set the losses, and these relate – as with most irrigation schemes – to

6,800 (1881). While part of the fall may have been due to its reduced importance as a market town historically based on an overland trade route to the Punjab, the rise in the spring level from forty to seven feet, along with the high fever incidence, was clearly an important factor. The death rate (all causes) for the last half of 1885 was put at 162 per thousand. *RSC 1885*, pp. 72–5.

187 *Royal Commission on Agriculture, Evidence*, vii, pp. 192–3, 198. Such measures, of course, worked against some local economic interests, and in 1921 – apparently without consulting the health officials – the UP Government (itself not without an economic interest in the matter) permitted the Saharanpur farmers to resume irrigation within the boundaries of the prohibited area. In the event the relaxation of restrictions endured, though eventually on the understanding that the restrictions could be reintroduced if the malarial index showed a serious tendency to rise. Splenic enlargement is part of the process of acquiring resistance to the disease, and is the mark of a partially immune person with some remaining infection. The percentage of children affected (which is what the figure refers to) is the principal yardstick for measuring the intensity of endemic malaria.

188 *Ibid.*, p. 198.

the canal's environmental impact. For the first time in the history of the Doab, man was able to affect the physical conditions there through his large-scale meddling with key environmental processes. However ingenious the engineers, their capacity to influence the environment was – particularly during the first half-century of canal development – ahead of their understanding of the forces involved. Moreover, as problems manifested themselves, and corrective measures came to be within the ken of the engineers, restrictions on funding delayed and often limited the effectiveness of ameliorative programmes. Nevertheless, environmental catastrophe did not overtake the Doab in the wake of canal construction, though such an outcome seemed likely to many observers in the 1870s, when the problems seemed to be on the increase. In addition, therefore, to taking into account the spatial extent and significance of the problems of salinity and waterlogging when attempting to assess the quantitative impact of these externalities – and allowing also for the *positive* externalities – it is clearly important to take into consideration the time dimension to these problems. Overall, the canals might be said to have been accompanied by periods of unnecessary disruption within specific localities, the incidence of which, in some areas at least, declined after the 1880s. Arguably, it is the effect of the canal upon the incidence of malaria which was the most significant of the externalities: not only because of the discomfort suffered by the afflicted villagers, but due to the fact that the disease must have held down internal population growth within a prosperous community and thus assisted in the perpetuation of labour scarcity within the canal tracts.

5
Pricing

The terms on which irrigation water is made available to cultivators is an important determinant of the way in which the water is used. Decisions on the form, level, and structure of charges inevitably affect the economic, social, and agricultural impact of canal schemes. In the case of the Doab canals, it is clear that the freedom of irrigation officials to pursue their objectives through the pricing mechanism was very much constrained. More precisely, it was constrained by factors other than those relating to the system's technological characteristics, the specific nature of the areas and communities it served, and the multiple (and often conflicting) aims of the department itself. There were broader aspects to the functioning of the canals, to do with Imperial administration, that effectively impinged upon pricing strategies. Indirect returns were sought in terms of political stability and social welfare, and no matter how strong the practical case for specific pricing adjustments according to the priorities of those operating the service, broader political and administrative considerations exerted their influence. As a result, the pricing system conditioned the impact of canal irrigation in a manner best described as benign; the potential for it to be used in a more formative way was not exploited. This chapter examines the factors determining the character of the pricing system, and then proceeds to discuss the impact of the adopted price regime upon agrarian change in the region.

Pricing policies

Entanglement of canal and land revenues
One of the striking features of UP canal irrigation during the

period under review is the low level of charges made for canal water. In relation both to the cost of irrigation from wells and to the value of commercial crops raised, canal water – particularly that given 'flush' – was widely acknowledged to be underpriced.[1] Moreover, as a proportion of the value of average gross product, it became cheaper over time. While the price of wheat, for instance, roughly tripled between the mid-1860s and 1920, the water rate for wheat less than doubled, the flow rate per acre increasing from Rs 2.25 to Rs 4 and the lift rate rising even less, from Rs 1.5 to 2.[2] *Gur* prices increased less spectacularly over the period – being influenced more by local markets than by international ones – but the crop value was generally high and significant yield improvements took place on canal land, so that an acre in the 1910s often realised in excess of Rs 150. Despite this, the canal charge rose only slightly, from Rs 5 in 1865 to Rs 7.5 in 1920, and the charge for lift irrigation actually fell on the Upper Doab canals from Rs 3.33 to Rs 3.

The low and, in real terms, falling level of charges by no means made the schemes unprofitable. It has already been pointed out that the Doab canals did not involve the installation of storage capacity, and the earlier projects operated with inexpensive headworks arrangements.[3] Even so, the cheap water policy needs to be explained, particularly in the light of the fact that the pressures to increase price levels were substantial. The most consistent pressure for increased charges came from the Government of India, and was partly motivated by a desire to improve the returns from 'productive' works in order to help finance the loss-making 'protective' schemes in the insecure regions. But there was support for the idea of raising charges from revenue officers as well, who argued that a generally higher and more flexible rate system

[1] On the basis of imputed cost estimates, the Director of Agriculture H.R.C. Hailey calculated that 'a fair average' difference between well and canal irrigation was Rs 15 per acre (excluding the costs of well construction). UP Rev., July 1909, 88.

[2] In real terms, on the basis of average wheat yield from canal-irrigated fields, the cultivator in the mid-1860s would require 1.5 *maunds* of his acre's fifteen-*maund* crop to cover the water rate, while in the late 1910s two-thirds of a *maund* sufficed. This result is not dissimilar to the trend in Ludhiana. See *Indian Cotton Committee, Evidence*, VI, Appendix 23.

[3] Profitability is discussed in Ch. 2 above.

would allow the encouragement of improved agricultural practices and reduce the damaging effects of the wasteful use of water, and from those irrigation officers who wished to see their efforts better represented in strictly accounting terms. Why, then, given the monopoly powers possessed by the authorities in the matter of canal pricing, were the rates purposely kept down?

The key to understanding canal pricing policy is contained in a statement made by Professor Eric Stokes. 'At the heart of the Indian administration', he wrote, 'lay the land revenue system. The one was to react with the other with results of deepest importance for Indian society.'[4] Quite simply, the canal revenues became entangled with the land revenue, and from that point there was no chance that pricing could be pursued independently by the canal authorities.

The influence of the land revenue system over the canal administration's freedom to manipulate prices had its origins in the early years of the operation of the EJC. This canal, like others constructed prior to the 1860s, was built out of current revenues and classified along with ordinary expenditure to be charged against the annual Indian revenue. As with the Western Jumna Canal, the authorities 'looked alone to the general improvement of the country as the source from which they should derive a return adequate to the outlay'. Initially low direct water charges encouraged cultivators to incur the considerable costs of bringing water to their fields and was part of an admitted policy of inducing cultivators 'to become dependent on [the canal] instead of their wells'.[5] Indeed, about the only reason for charging a water rate at all was 'not so much to form a productive source of revenue . . . as to give . . . an efficient control over its expenditure'.[6]

---

[4] E.T. Stokes, *The English Utilitarians and India* (Oxford, 1959), p. 81.

[5] R. Baird Smith, *Revenue Reports of the Ganges Canal, 1855–6* (Roorkee, 1856), p. 19. As with any irrigation scheme of such magnitude, the sizeable component of indivisible costs in total outlay necessitates that a large part of total capacity has to be taken up before capital costs can be covered by receipts. Low direct charges for water are commonly used in the initial stages to build up demand.

[6] *PP* 1878–9, IX, *Report from the Select Committee, Minutes of Evidence*, p. 94, Col. J. Crofton quoting Col. J. Colvin. Note: 'direct charges', 'occupier's rate', and 'water rate' are all equivalent terms used to refer to the charge levied upon the actual user of canal water.

In the longer term, direct charges were clearly intended to rise. On the newly opened Ganges Canal, the Director's policy was to have

> low rates and large areas of irrigation ... I consider it of great importance to the Revenue to encourage in all reasonable degree, the surface spread of irrigation, for when once its advantages have been realised, the cultivators will cling to them tenaciously and in course of time Government will find that the water has so risen in value as fairly to authorise an increase in the rates of payment for it.[7]

Even the rates initially set 'purposely low' on the Ganges Canal were, along with those on the EJC, 'purposely reduced' in 1863 with the aim of encouraging the irrigation of commercial crops.[8]

The deliberate policy of setting low water rates involved leaving a considerable proportion of the enhanced value of outturn with the cultivator. Inevitably this was reflected in increased rent levels and thus in the assets upon which revenue assessment was based. Thus, while the pending resettlements of the Upper Doab districts were expected to result in a considerable enhancement of the land revenue demand, much of this increase, it was clear, was due to the spread of cheap facilities of canal irrigation. One of the problems, however, of realising part of the returns to canal irrigation through the revenue system was that under the 'half-assets' rule on which revenue assessments were based, landlords automatically realised half the increase in rental assets resulting from public expenditure on canals. Moreover, they realised the whole of any subsequent increases in rental assets for the duration of the period of settlement. With canal irrigation projected to expand considerably, this was in itself unsatisfactory to many officials, but the problem was heightened at this time by the imminent switch from temporary (thirty-year) to permanent settlements. Under the proposed system, a large proportion of the profits arising from canal expansion would be surrendered to the landlords in perpetuity.

As a response to this predicament, a group of canal and revenue officers, prominent among whom was A.O. Hume,

---

[7] Baird Smith, *Revenue Reports of the Ganges Canal*, p. 14.

[8] Movements in water rates over the period are shown in Figure 3.

proposed a change in the system of revenue assessment. The system they advocated involved the exclusion of the influence of the canal from the process of assessment, and the basing of the water rate on more or less commercial principles so that a 'permissible' level of the profits due to canal water could be collected directly.[9]

The argument in essence was founded upon the same kind of fundamental principles that had provided the basis for the revenue systems; indeed the analysis developed along the same lines as the debate over the doctrine of rent and the land tax. To those in favour of direct pricing, the surrender to *zamindars* of a share in the returns to state investment was unacceptable. To them, water, like land, was considered a natural monopoly, the payment for which was strictly rent. By this formulation, if the owner of one factor, land, was not the owner of another factor, water, then it did not follow that the rent recoverable by the landowner should increase. Where, they argued, the land-owner had dug a well on his land, he was entitled – as owner of both elements of production – to a rent enhancement which, although practically merged in rent, was strictly divisible into the old rent plus the return to the improvement. By such an argument the state, as exclusive owner of canal water, was bound to ascertain and appropriate the full water rent, only reducing it to encourage the cultivator to use the water or where there was excess supply.[10] The scheme, they insisted, could be implemented by assessing as irrigated only land irrigable from wells and classing the rest as 'unirrigated' for revenue purposes. Water rent levels could be determined by experimentation on a regional basis.

Leaving aside the theoretical justification for the new scheme, there were, its advocates claimed, sound practical arguments in its favour. For one thing, it would permit the pursuit of a more flexible policy towards distribution: adjustments could take place without endangering land revenue; and failure of canal supply (or the need to withdraw it for health reasons) could be accommodated without the complications of revenue remissions. The existing canal

---

[9] A basic statement of this argument was published anonymously: 'Canal Rent vs. Land Revenue', *Calcutta Review*, XLIX (1869), pp. 1–36.

[10] *Ibid.*, pp. 15–16.

irrigation distribution pattern – effectively locked into place by the structure of land revenue payments – could then be adjusted with a view to achieving specific aims, among them realising a higher return.[11] Higher charges would encourage careful distribution of water and prevent the neglect of wells and other irrigation sources. It was also, it was claimed, self-evidently more effective than either temporary or permanent settlements in coping with the extension of cultivation after settlement, since at least a full water rent was obtained if canal water was used on newly broken land.

When, in January 1865, revenue officials gathered at a special Agra conference to discuss the problems posed for the system of land revenue assessment by the spread of canal irrigation, it certainly looked as if the new system would be taken up. It undoubtedly had the backing of the Governor-General, who had stated earlier that theoretically it would be best to assess land according to its 'natural productive powers and capabilities without regard to artificial irrigation' and to leave the water rate to be fixed according to the 'additional fertilising properties' of canal water.[12] In fact, the Government of India was known to be considering the idea of grafting what was effectively the new system onto the old through the use of two scales of water rate: one for villages already assessed at a higher rate as a result of the canal, and a higher one for those where the land revenue had not yet been adjusted to take account of the canal. Recognition would thus be given to cultivators whose rents effectively incorporated an element of water rate. Not only would such a system allow permanent settlement to proceed without involving serious loss of future revenue as canal irrigation increased, but it would permit water rates to be raised in line with the central government's wishes.[13]

In the event, despite the broad support for a change in the system, the existing arrangements emerged intact from the Agra conference. This outcome was mainly due to the firm

[11] The authors of the *Calcutta Review* article were generally in favour of redistribution (so as to get 'absolutely [the] highest return for the whole of the water'), though they added: 'Politically, and we even venture to think commercially, a modified distribution of the water supply along the whole line of the canal seems imperative.' *Ibid.*, pp. 16–17.

[12] *Ibid.*, p. 2, citing GI PWD Resolution, 15 Aug. 1864.

[13] For details, see *ibid.*, p. 3.

opposition to the proposal on the part of the Lieutenant-Governor, Sir William Muir, supported by another senior official, Auckland Colvin. Muir declared that he was 'satisfied with the present system', which was the 'matured result of the experience of many years' and was well understood by both officials and populace. He objected to the proposal not only on the basis of envisaged practical problems – among them that of the assessment becoming more 'a conjectural process of guessing what might have taken place without the canal' – but also because, while he accepted that the present diversion of profits from state works to *zamindars* was 'unfair to the public', from the NWP Government's point of view 'it would be neither just nor politic to withdraw from the land-owners the whole of their share of the profits in the shape of increased rents'.[14] Nor did Muir accept that the alternative proposal would realise the sum relinquished as a component of the revenue demand, arguing that there were limited possibilities for increasing direct water rates. The commitment to uniform rates over geographical areas, the desirability of using up the supply, the lapse in well arrangements caused by the canal (along with the physical destruction in some areas), and the canal's role in times of drought – all imposed a constraining influence on pushing up rates significantly. A margin would still remain, and the *zamindar* would no longer be bound to release even half.[15] Muir's preference for the existing system meant that the returns to canal irrigation would remain entangled with the land revenue. The state was thus going to have to continue to share the benefits of the canal with the *zamindars*.

In the light of this decision it was clearly going to be difficult to proceed with arrangements for permanent settlement in an area of rapidly expanding canal systems. Permanent settlement, even under the best circumstances, involved a 'great financial sacrifice' on the part of the government, and settlement operations in these north-western districts during the 1860s were making it clear just how large this sacrifice would be. Cadell's report on *pargana* Kandhla (Muzaffarnagar)

[14] For a summary of Muir's position, see NWP&O Rev., May 1888, 10.
[15] Uttar Pradesh State Regional Archives, Allahabad (hereafter UPA(R)), Files of Meerut Commr., Dept. IV, File 9/1869, Ser. 17, 'Irrigation Works in India', Note by W. Muir, 10 May 1865, p. 5.

showed that the recently introduced canal irrigation was far from achieving its potential, and that even where estates had quickly improved their cultivation standards, 'so much consideration was shown at settlement, in order to avoid the evil effects of too sudden a rise, that the demand was raised so cautiously that even now there is a large .margin for enhancement'.[16] The same situation applied in the case of Bulandshahr: 'Whatever the loss from underassessment may now be', the Meerut Commissioner wrote to the Board of Revenue, 'I am quite certain that the loss twenty years hence would be enormous . . . The district is in a transition state, and this is not the time for permanent settlement.'[17] Faced with such a prospect, the Secretary of State duly decreed that permanent settlement should be withheld in areas likely to have canal irrigation within twenty years, or where the extension of present systems was likely to cause an increase in existing assets of 20% or more.[18]

Even the system of temporary settlement was ill-designed to cope with conditions of rapid change, however. The prospective revenue loss was undoubtedly reduced by the decision not to opt for permanent settlement, but the problem did not disappear, since in the case of irrigation expansion after settlement, half of the resulting increase in assets was relinquished for up to thirty years. To get round this problem, Muir himself, in 1865, proposed that a system of quinquennial reassessments be adopted for villages where canal irrigation since settlement had increased by 20%. He suggested that the 'wet' rates determined at the previous settlement should be applied to such land. Although this proposal did not meet with the approval of the Secretary of State – whose main objection was that it constituted an interference with contract[19] – the idea

---

[16] 'Report on the Permanent Settlement', p. 6, in Cadell, *Muzaffarnagar SR* (1882). The judicious (revenue) rate enhancement was apparently accentuated in terms of its effect upon revenue assessment by the settlement officer's practice of classing as 'wet' for revenue purposes only the land actually irrigated in the measurement year, thus recording as 'dry' some of the land which was 'irrigable' and which received water in the course of a rotational cycle. Such errors were common prior to the establishment of more systematic procedures.

[17] Quoted in C.A. Daniell, *Bulandshahr SR* (1869), p. 12.

[18] NWP Rev., June 1867, 32, Sec. of State to GI, 23 March 1867.

[19] Other objections were that it was overly mechanistic, entailed a virtually permanent settlement staff, and was likely to lose as much direct water revenue by discouraging

of some form of interim arrangement to stem the loss of revenue was revived by Calcutta following a report favourable to the notion by the Inspector-General of Irrigation, and following the gleaning of estimates as to the size of the losses which would otherwise be incurred.[20]

The scheme which resulted, and which was incorporated in the 1873 North India Canal and Drainage Act,[21] involved the levying of what was called an 'owner's rate'. This was a levy made on proprietors in all estates into which the canal had been introduced for the first time since the previous settlement. The levy amounted to a proportion (one-third) of the (occupier's) water rate. In point of fact, the levy – which was determined by the Lieutenant-Governor in consultation with revenue and irrigation officials – was inherently contradictory in terms of fundamental principles: it was strictly part of rent, yet it varied according to the water rate, which was an element of production cost. Moreover, although the device did produce a levy which bore some relation to the profits derived from irrigation, and to the land quality – since it was based on a scale of charges for canal water which were broadly calibrated according to crop value – it was irritatingly uneven in the way it fell upon the cultivating community.[22] The most obvious manifestation of this was the way in which the new system took no account of villages which had *increased* their irrigation, being confined – for ease of administration – to villages newly in receipt of canal water. From a financial viewpoint, it was not even satisfactory in the villages where it could be applied, since

irrigation as it realised through land revenue increases. See NWP&O Rev., May 1888, 10.

[20] Figures from Bulandshahr showed that, as a result of the expansion of irrigation since settlement, the average revenue enhancement per acre recorded as irrigated in 1870–1 was only around one-quarter of that in Saharanpur's EJC tract, where the irrigation was well established at settlement. The Bulandshahr settlement had taken place when only 25% of the irrigated area in 1870–1 was actually able to receive canal water. The enhancements per acre were fourteen *annas* in Saharanpur as against four *annas* in Bulandshahr. *IRR 1871–2*, p. 23.

[21] Refe red to hereafter as the 1873 Canal Act.

[22] Proprietors whose tenant–cultivators used flow irrigation (which bore a higher water rate), double-cropped their irrigated fields, or lived in areas where the water rates were higher than elsewhere automatically made a larger owner's rate payment than those cultivators who used lift irrigation, single-cropped their canal fields, or lived on the systems with lower rates.

it was found that the mechanistic levy fell short by about a quarter of the sum which 'might fairly be claimed' under the 'half-assets' rule.[23] Nevertheless, the *ad hoc* device was regarded as the best available, and was adopted in the Punjab as well as the UP. The best alternative, a charge based on the quality of the land receiving irrigation, would have been expensive to administer. For all its contradictions and defects, the owner's rate was cheap and simple to operate.

The adoption of the new levy effectively submerged the debate over the form the charge for water should take, particularly in relation to the land revenue system. It surfaced again in the 1880s only to sink permanently from view shortly afterwards. The basic view of the then Lieutenant-Governor, Sir Auckland Colvin, had not altered since 1865, when he had given Muir his support in this matter: no system had been devised for assessing the profits derived from canal irrigation, as distinguished from profits due to all ordinary causes, which did not involve either serious administrative difficulties or a failure to reach more than a comparatively small share of the profits liable for assessment.[24]

Entanglement constraints upon pricing adjustments

Once it was clear that the overall revenue system was not going to be adjusted so as to separate from land revenue the effect of canals, the energies of those with an interest in canal pricing policies were devoted to bringing about the adjustment of the direct rates within the constraints imposed by the revenue system. Their focus, in essence, was now more specific, and had switched away from the *zamindar* to settle upon the actual cultivator.

Of the three distinct sources of pressure for higher user charges which can be distinguished, the Government of India was the most persistent. Its reaction to the 1863 reduction of water rates marks the beginning of its long campaign to have the rates increased. The Governor-General called for 'readjustment and general increase' of the NWP canal charges and insisted that 'such rates as are shown to be fairly chargeable

---

[23] NWP&O Rev., May 1888, 10.   [24] *Ibid.*

having regard to the just claims of Government to a suitable return to the capital sunk . . . and to the known capabilities of the land . . . should be adopted at once, and generally increased by degrees as the occasion allows'.[25] The government's concern to 'obtain for the general taxpayer, by whom the capital is provided . . . the full value of canal water'[26] was behind its resourcefulness in seeking ways of pushing up canal charges. It even attempted, in 1869, to have inserted in the North India Canal and Drainage Bill a clause permitting the levy on irrigable land not actually taking water during a particular season. The precise amount of the levy was to be fixed so as to give a 7% net return on capital. When this proposal was rejected by the London Office,[27] it devoted its efforts to enhancing water rates to levels based on the principle of 'what the market will bear'.

Although their reasons differed markedly from the Government of India's undisguised concern to improve financial returns, many revenue officers, too, in the early 1870s argued the case for enhanced direct water charges within the existing pricing framework. There was broad agreement that existing rates were wasteful of water to the detriment of distribution efficiency, standards of husbandry, and land revenue; and that the rate structure should be adjusted to encourage the more careful application of water (perhaps even by encouraging lift irrigation), and should attempt to exclude the watering of poor soils since the water could be more efficiently used on better soils elsewhere.[28] Their arguments were given greater force by the conviction that careless use of flush canal water was compounding the environmental problems so obvious to observers in the 1870s. Not every officer, however, went as far

---

[25] *PP* 1865, xxxix, *Resolution by the Government of India Public Works Department Relative to the Canals of the North-Western Provinces*, p. 543.

[26] Quoted in *Report of the Indian Taxation Enquiry Committee, 1924–5* (Madras, 1926), p. 109.

[27] The proposals and objections to them are contained in *PP* 1870, LIII, *Report on Irrigation Works in India*, pp. 96–8, Sec. of State to GI, 11 Jan. 1870. The Secretary of State agreed in principle with obtaining a full return from users, but felt it iniquitous that the cultivators should bear the ultimate burden of cost overruns, inaccurate demand projections, and so on.

[28] UPA, BOR, Canal Shelf, Box 2, File 8, Memorandum by W.H. Smith, 11 May 1874, pp. 6–7.

as Edward Buck, who insisted that the only way to avoid 'over-irrigation and consequent deterioration of the soil . . . in the wake of the canal' was to raise the rates to 'a level which could only be paid on manured lands'.[29] Nor was there much support for those few revenue officers who wished to turn the clock back and use price enhancements to promote a return to using wells. The Saharanpur Settlement Officer, for example, was pleased – if probably mistaken – at signs in *parganas* Nakur and Gangoh that 'the recent [1865] enhancement had happily driven the people to use some of their wells again'.[30]

Canal officers, too, frequently looked to rate enhancements to check apparent trends they considered undesirable. In the early 1870s increases were proposed for the sugarcane water rate, 'partly with a view to checking the increase of this crop, which has spread rather beyond the cultivator's powers of tillage'.[31] A decade later such an increase was still being sought. The spread of cane cultivation, encouraged by more efficient processing methods, appeared to be interfering with the irrigation of grain crops. An engineer on the EJC, arguing for a doubling of the cane rate, described the problem:

No matter what the period of the year a full demand comes on, some 35,000 acres of sugarcane have to be watered to the prejudice of the other crops . . . In 1878 I watched the effects of the sugar demand on the Eastern Jumna Canal. I found it interfered considerably with the rabi crops, the early and late maize, and the rice.[32]

At the same time, other officers were concerned that the low existing rate on indigo was likely to induce its expansion in the northern districts. The Superintending Engineer on the EJC, for one, noting the way in which indigo cultivation could be 'forced upon the tenants to the detriment of sugar-cane', reflected Chief Engineer J.G. Forbes' concern that, whether or

---

[29] NWP PWD(I), Sept. 1873, 92. See also UP PWD(I), Feb. 1908, 1, E. C. Buck, Note on Fertilization of Sub-Himalayan Lands by Alluvial Deposit; and *IRR 1873–4*, p. cix.

[30] Le Poer Wynne, *Saharanpur SR* (1871), pp. 98–9. So advantageous were the canals in comparison with wells that even in Cawnpore the cultivators were 'ready to pay three times the schedule rate provided they get canal water in preference to their wells'. (*Report of the Irrigation Conference*, II, *Discussions*, p. 52, Note by N.F. McLeod.) It was left to non-price 'debarring' strategies (discussed in Ch. 3 above) to attempt to preserve well irrigation.

[31] *IRR 1871–2*, pp. iv–v.   [32] *IRR 1880–1*, p. 2A.

not it was desirable for indigo to spread in areas where canal water had not yet been devoted to other crops, it was not desirable that excessively cheap canal water for indigo should oust the established pattern of cropping on the EJC:

It is to be hoped that the attempt of the manufacturers to introduce this dye into the upper parts of the canal may not lead to any further increase, since, broadly speaking, it can only do so by elbowing out other crops which bring a better return to Government, and by occupying land that might be utilised in increasing the food produce of the country.[33]

Despite, therefore, the 1865 enhancement of rates which resulted from the Government of India's intervention, it is clear that the rates were still widely considered to be too low, and higher rates than those generally prevailing were indeed fixed for the Agra Canal and the LGC when they opened in the 1870s. The adjustment, however, was a relatively minor one. Lift rates were kept at the same level as on older canals, with flow rates being twice rather than 1½ times the lift rates.[34] This somewhat unimaginative change was felt to be justified in the light of the clear evidence that the difference between lift and flow rates was too narrow, given the differential in labour costs.

Higher rates, of course, could be instituted on new schemes with a minimum of complications. On established canals there was the question of whether increased direct charges for water would entail reductions in rent (and, therefore, in land revenue) due to the fact that the two had become entangled. As it turned out, the whole thrust of the Government of India's attempt to have the rates raised involved the search for ways of bringing about increases which could not be opposed by the local government on the grounds that they were unfair to cultivators whose revenue assessments had been affected by the canal. The quest for a 'legitimate' rate enhancement involved a prolonged and intricate chess-like game played between the authorities in Calcutta and those concerned with the day-to-day administration of the province and its irrigation.

[33] *IRR 1881–2*, pp. 9, 29, 10C.

[34] For more details on adjustments to occupier's rates since 1845, see Ansari, *Economics and Irrigation Rates*; and *Report of the Irrigation Rates Committee, United Provinces* (Lucknow, 1939), pp. 12–24. (A copy of the latter is available at the Central Design Directorate, Uttar Pradesh Irrigation Dept., Lucknow.)

One important round in this interesting contest took place in 1879. The Government of India elicited an enquiry in the NWP into the possibility of moving the water prices on the older canals into line with those on the new canals. It reasoned that unless it could be shown that the flow–lift distinction constituted a recognisable element in the determination of land rents, there was no reason why the flow rates on the older canals should not be raised. The investigating officers, W. Irvine and E.B. Alexander, could not, as it turned out, detect a systematic link between the mode of irrigation and rent levels. Each settlement officer in the Central Doab had, noted Irvine, 'paid no regard to the character of irrigation in his assessment of irrigated lands'. 'Little doubt can be entertained', he concluded, 'that up to the present, rents are paid for irrigated land without regard to the source or comparative cost of bringing the water on the land.'[35] Here, then, was a case where the existing system of assessment would outwardly seem to permit an opportunity for direct charges for flow irrigation to be enhanced: a demonstrable cost advantage to specific cultivators not apparent in the determined rent levels.[36]

There seemed no logical reason why the increase should not be approved. None, that is, until the Chief Engineer, Colonel (later General) Henry Brownlow, voiced his reservations, the principal among which was his concern for specific groups likely to be adversely affected by such an increase. He agreed that there would be villages (and groups within every village) willing to take water at enhanced rates, but that there were many others, particularly on the EJC, where the rents had long previously come to reflect the availability of cheap water, and where tenants could not afford higher canal rates.[37] On the older canals, only for those tracts which had been recently

---

[35] NWP&O Rev., Nov. 1879, 20; for the entire discussion, see NWP&O Rev., Nov. 1879, 3–24.

[36] As was pointed out in Ch. 3 above, Buck's estimate was that in the case of sugarcane, a water rate differential between lift and flow of Rs 1 effectively represented a difference of forty to fifty man-days in the irrigation of one acre of cane to maturity.

[37] Brownlow pointed out to the BOR that in some Meerut *parganas*, cultivators were frequently found who paid the high rent of Rs 16 for their sugarcane fields, and statutory tenants who paid the high average (all-round) rate of Rs 12 per acre. In Bulandshahr, he noted, there were many Thakur and Brahman *maurusi* tenants 'in great difficulties'. NWP&O Rev., March 1880, 23.

incorporated into the canal system – and thus where rent levels had not had time to absorb the gains – could Brownlow endorse the proposed enhancement.

The Lieutenant-Governor was quick to back the Chief Engineer. To lower the rents would clearly have involved going back on the earlier decision to leave the revenue system unchanged. It would have been unpopular with *zamindars* and 'totally at variance with the whole policy of Government'. Without such adjustments, the tenants – even those in what was the richest part of the Doab – would be set at risk: 'There was a strong consensus of opinion', it was stated in the *Report of the Board of Revenue on the Revenue Administration* for 1880–1, that such a water rate increase 'would be prejudicial to the interests of the right-of-occupancy tenant and would tend in great measure to reduce him to the level of his rack-rented brethren'.[38] With the issue set in these terms – and against a background of official concern for the protection of hereditary land rights – the Board of Revenue was quick to move into line with the stance of the NWP Government. 'This virtually ends the discussion', they informed the latter in 1881. 'Nothing but the most absolute and pressing necessity could justify the Government in even running the risk of destroying the occupancy tenures of these provinces in this way.'[39]

Given the circumstances, the only way the local government considered it could push up water rates was at resettlement, when it could take advantage of the 'invariable pause' in rent increases as landowners attempted to gain a lower assessment.[40] Thus it was in 1892 that the rates on the older canals were moved into line with those of the newer systems (see Figure 3). But the moves and the countermoves did not stop here. When the Indian Government pointed out that there was no relation in theory between the owner's and occupier's rates, and thus no reason why increases in the latter could not be made, the local government replied that a *de facto* link existed and that the owner's rate was generally levied from the cultivator, and that to increase the occupier's rate would increase the cultivator's burden.[41] When the Board of Revenue argued that the occupier's rate was treated by landowner and

[38] *RAR 1880–1*, p. 26.　　　　[39] NWP&O Rev., Sept. 1881, 56.
[40] NWP&O Rev., Aug. 1888, 48.　　[41] *Ibid.*

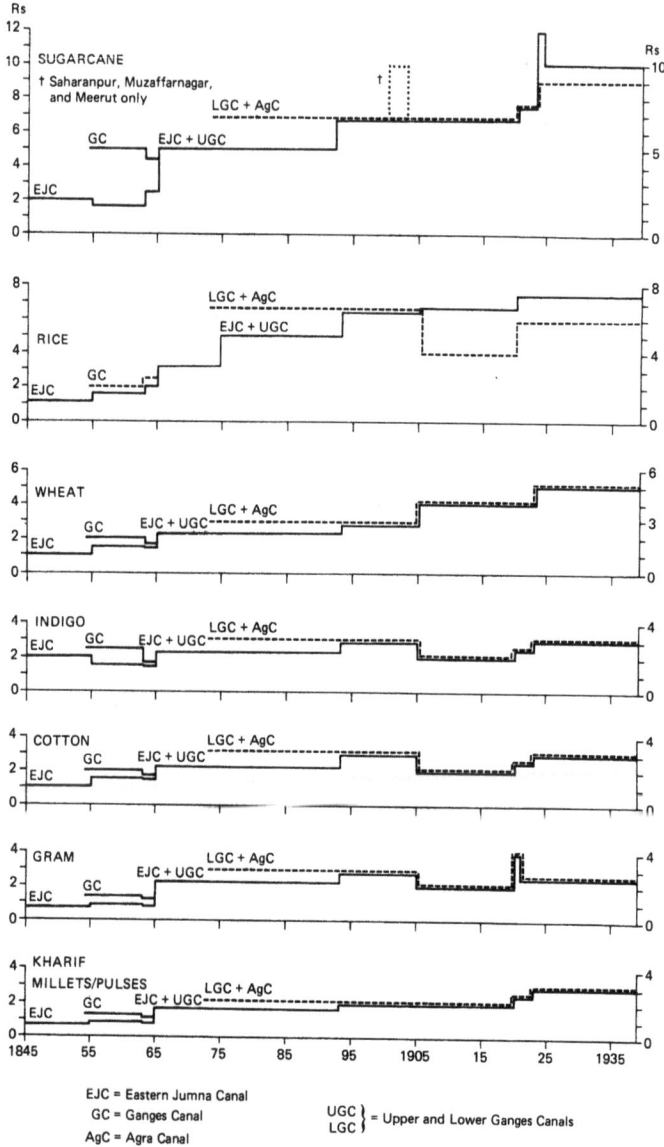

Figure 3. Crop-acre water rates for flow irrigation on major canals, 1845–1939. (Sources: *IRR*, various years; *Report of the Irrigation Rates Committee.*)

tenant as a 'stable factor in the determination of rent', the central government came back with the suggestion that the upward trend in prices still provided an opportunity to raise the water rates if it was done quickly, prior to the absorption into rents of the increased cultivating profits.[42]

The game clearly became a tiresome one for the UP Government, particularly when, in 1899, the central authorities appeared to ignore the whole basis of the local government's long-standing opposition to rate enhancements by requesting a further investigation into the profits of various crops and the comparative irrigation costs, the results of which were – along with experiments in raising rates – to be used to justify higher charges. Apart from being administratively inconvenient, the local government considered that the exercise would be unproductive beyond pointing to the 'well known facts that flush rates are comparatively low, and that irrigation from canals is cheaper than irrigation from wells'.[43] The UP Government continued to maintain that it would be 'impolitic', given the interdependence between rents and water rates, to disturb existing arrangements. The most that Indian Government pressure could achieve by 1903 was the tentative agreement of local authorities that some increases might be instituted when the crop prices improved.[44]

The chief reason, then, for the persistence of a system of low direct charges for canal water stemmed from the political priorities of a provincial government unprepared to break out of an historically derived revenue structure. But it is clear, also, that the irrigation authorities came to have little faith in the practical usefulness of a high price strategy. There were those within the department who argued in the early 1870s for water rate adjustments to influence irrigation/crop decisions, but the Chief Engineer during the key development phase of the 1870s, Henry Brownlow, oriented canal policy away from the kind of high price policy this implied. Brownlow was opposed to the idea of restricting irrigation to the best soils and cultivation, since in the Doab this would not lead to the use of the full supply. *Rabi* crops, especially, for which water was often taken to improve yields, were notoriously fickle in their need for water (as the variation in irrigation demand in Figure 4 shows), and

---

[42] UP Rev., Feb. 1904, 109.   [43] UP Rev., Feb. 1904, 86.   [44] *Ibid.*

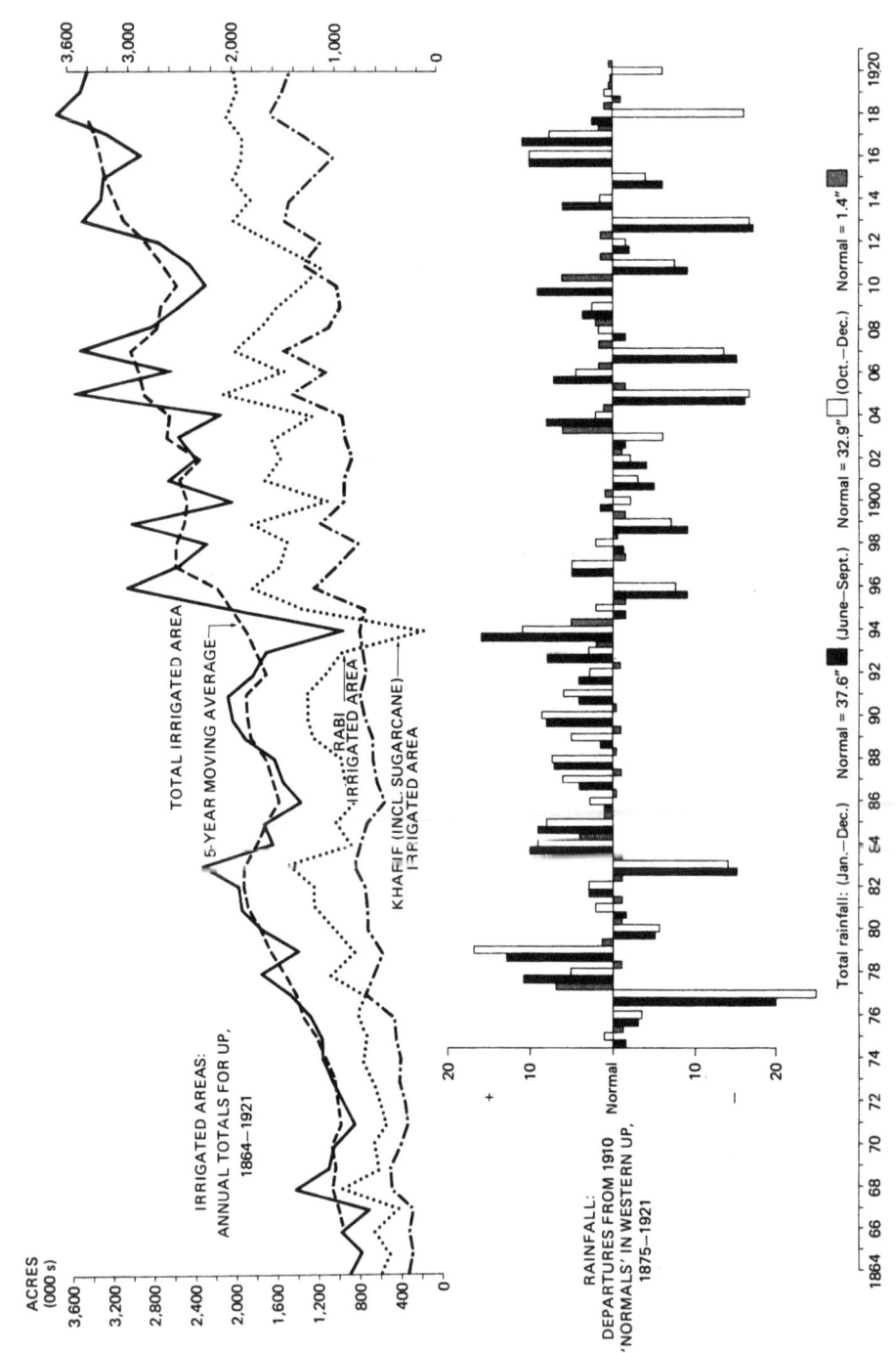

ACRES
(000's)

3,600
3,200
2,800
2,400
2,000
1,600
1,200
800
400
0

TOTAL IRRIGATED AREA

5-YEAR MOVING AVERAGE

IRRIGATED AREAS:
ANNUAL TOTALS FOR UP,
1864–1921

RABI
IRRIGATED AREA

KHARIF (INCL. SUGARCANE)
IRRIGATED AREA

3,600
3,000
2,000
1,000
0

RAINFALL:
DEPARTURES FROM 1910
'NORMALS' IN WESTERN UP,
1875–1921

+
20
10
Normal
10
20
–

1864 66 68 70 72 74 76 78 80 82 84 86 88 90 92 94 96 98 1900 02 04 06 08 10 12 14 16 18 1920

Total rainfall: (Jan.–Dec.) Normal = 37.6"   ■ (June–Sept.) Normal = 32.9"   □ (Oct.–Dec.) Normal = 1.4"

higher rates would encourage more cultivators to hold off in the hope of rain. On the cost side, Brownlow's view was that there were problems in seeking out the 'best land and highest market'; such a policy would entail not only higher distribution costs, but larger water losses through evaporation and percolation as water was led around in search of the best land.[45] Any attempt to achieve economies in water use by encouraging lift irrigation through the pricing structure would only add to the loss of water as the supplies were held up in the channels, as well as being a burden on irrigators.[46]

Brownlow firmly rejected a high price-limited coverage strategy in favour of a low price-wide distribution one. As will be shown in the following chapter, this was only in tune with the irrigation policy decisions made around this time committing the department to maximise the extent of coverage of the canal system. In effect, this meant that the irrigation service would 'look for larger revenues from kharif irrigation, as the more highly-priced crops oust the less valuable grain crops'.[47] The highly priced crops would bring higher and proportionately less variable canal revenues. Pricing was thus rejected as a means of controlling the distribution and use of canal water. Reckless cultivation – the subject of so much interest from revenue officers – would, Brownlow argued, work its own cure. Except on sanitary grounds (as in the case of rice), he was 'very averse to checking the increase of any crop by the imposition of higher water rates'.[48] Although there were several revenue officers still advocating that water use be influenced through higher charges, there were others who agreed with Brownlow. Alan Cadell, reflecting on conditions in Muzaffarnagar, came to the same conclusions as the Chief Engineer. 'The limitation

---

[45] As it was, estimates put the proportion of the water entering the system which actually reached the cultivators' fields at 50%. See Buckley, *Irrigation Works*, pp. 270–7.

[46] NWP&O Rev., March 1880, 23, Note by Col. H.A. Brownlow.

[47] *IRR 1874–5*, p. 24. [48] *IRR 1871–2*, p. 6.

Figure 4. Growth in irrigated area and relationship between rainfall variations and seasonal movements in irrigation. (Sources: 'Summary of Indian Rainfall for Fifty Years, 1875–1924', *Memoirs of the Indian Meteorological Department*, xxv:11 (1928), pp. 15–108 *passim*; irrigated areas taken from *IRR*, various years.)

of irrigation', he wrote, 'must be affected not so much by prohibitive rates, which do more injury in poor tracts than good in rich neighbourhoods, but by prudent distribution of canal water.'[49]

Cadell suggested distribution based on a fixed *proportion* of village cultivation; Brownlow preferred limiting the *time* water was available as a means of inducing economy. Both suggestions played their part in the way water was distributed and used, as will be shown in the following chapter. The important point is that price was rejected as a distributional control, and that from the 1870s there was little support from the Irrigation Department for the Government of India's ongoing attempts to achieve enhancements in the overall level of charges.

Apart from the 1892 increase of flow rates on the older canals up to the levels prevailing on the LGC and the Agra Canal, and the 1873 increase in the rate for rice, the level of direct water charges remained static from 1865 to 1905 (see Figure 3). The changes that did take place in 1905, though entailing general increases in wheat (from Rs 3 to Rs 4 for flow and from Rs 1½ to Rs 2 for lift) and sugarcane increases for Saharanpur, Muzaffarnagar, and Meerut (to Rs 10 for flow and Rs 5 for lift), were, taken overall, far from representing a victory for the central government. In fact, several rates were actually reduced so as to encourage 'approved systems of cultivation'. This was the case, for example, with the reductions in cotton and fodder rates. As the Director of Agriculture explained:

The rate for early sown cotton is now only Rs 2 and rates on fodder crops are also low to encourage stall feeding of cattle. The Agricultural Department have always urged the great advantages of cold weather fodder if animals are to be kept in proper health and condition. The ideal which has been aimed at is an early sown cotton followed by a fodder crop for cattle . . . The real disadvantages of

[49] UPA, BOR, Canal Shelf, Box 2, File 8, A. Cadell, 'Note on the Enhancement of Canal Water Rates', n.d., para. 14. Cadell pointed out that, in his view, a high rate policy would endanger the continuation of what he saw as the desirable practice, particularly noticeable among the Jats, of taking the best crops and cultivation around the entire holding, thereby manuring virtually all the land in rotation. Thus, he considered that to charge high rates 'with the special object of restricting cane and cotton to a small area of very highly manured land would be to attempt to change the system of agriculture of the Upper Doab'.

pushing a commercial crop such as cotton are thus to a large extent minimised.[50]

While the reduction was also intended to give rise to the production of 'enough cotton of the medium quality to be able at least to supply the requirements of the local industries at Cawnpore, Hathras and other places in the UP',[51] it was strongly recommended – as was the reduction in the rates for indigo – on the grounds that *kharif* irrigation deposited silt on the fields, to the benefit of yields.[52]

Nor, moreover, did even the sugarcane enhancement manage to endure. The ensuing seasons were difficult ones for the northern districts, with sugar-waterings and profits stretched thin by excessively dry conditions. Under pressure from the Muzaffarnagar-based Zamindars' Association, the local council reversed its decision in 1908 – just as it was to do again in 1924, when it had attempted to raise the water rate for sugarcane to Rs 12.[53]

The rigidity imparted to the system of direct rates by the link with the land revenue system meant also that the structure of rates changed only minimally. In fact, the broad pattern of charges throughout the period continued to be that of the rate structure determined in the 1860s. Originally crop area charges were roughly calibrated according to the profitability of crops and the amount of water they used. This rule-of-thumb guide was still operative at the end of the period.[54] Water rates for commercial crops, initially set low to encourage demand, remained low except where specific circumstances produced a consensus in favour of change. Rice, for instance, had its rate enhanced in 1873 due to its undesirable ecological consequences and to the related outbreaks of malaria; indigo rates were reduced in 1905 in an attempt to counter the dramatic fall in its cultivation around 1900. Some rates moved more into line with the original aims of scaling the charges to reflect the

---

[50] UP Rev., March 1916, 52, Note by H.R.C. Hailey (Dir. of Agriculture); see also J. MacKenna, 'Notes on the Fodder Problem in India', *AJI*, ix (1914), p. 45.

[51] *Report of the Irrigation Rates Committee*, p. 16.

[52] UP PWD(I), Feb. 1908, 9.

[53] *Report of the Irrigation Rates Committee*, pp. 18–20.

[54] *Ibid.*, p. 13; see also *Royal Commission on Agriculture, Evidence*, vii, p. 174, testimony of B. D'O. Darley (CE, UP PWD(I)).

amount of water used and the cash value of the crop: *gram* and peas, for instance, were bracketed with wheat until 1905, although they frequently took only a single watering and fetched much less in the market; *paleo*, literally a single preliminary watering given to often non-commercial crops to assist field preparation or germination, was only reduced to the lowest rate in 1923, having until that time been charged at the full crop rate. Given the constraints on raising the rates it is not surprising that the structure of the rates was altered so little.

Experiments in volumetric pricing

In theory, most of the problems arising out of the way water was used, and which many officials considered could be alleviated by price enhancements, might have been best tackled through a change in the *form* of water charge. The crop-acre charge, varying according to the general level of profits and water use associated with each crop, and divided according to whether the water flowed onto the fields by gravity or had to be lifted there, was particularly unwieldy as a tool for encouraging economy of water use. Engineers had long been aware that charging for water according to volume taken would clearly have solved many of the problems relating to the wasteful use of water and led to increased output per unit of supply. Baird Smith's early 1850s field study of Italian irrigation systems suggested that selling water by volume from the distributary bank was technically feasible, and that a modified application of the *modulo magistrale* used in Lombardy would be suitable for adoption on the new Ganges Canal.[55]

Although experiments with a local *paimana* or module system were conducted for a number of years, the innovation did not prove a success. Central to its failure was the fact that no simple sluice had been invented which gave a uniform discharge into the village channel when there was variation in the water levels in the distributary. A further technical problem was the tendency – particularly in the nearly slopeless Doab districts, where the water velocity in the *rajbahas* was often insufficient to

[55] R. Baird Smith, *Italian Irrigation: A Report on the Agricultural Conditions of Piedmont and Lombardy*, 2nd edn, vol. 1 (London and Edinburgh, 1855), p. 39.

keep the silt suspended in the water – for the module to silt up.[56] There were problems also in preventing the modules from being interfered with. On top of these technical difficulties, and no doubt partly because of the uncertainties engendered by them, cultivators not used to canal irrigation were reported to be suspicious of the module and unable to equate measures of water volume with their specific irrigation requirements. A not inconsiderable part of this consumer resistance must have originated in the absence of any attempt, for sound practical reasons, to sell to each peasant individually. This, of course, compounded the conceptual problem for the cultivator (of having to deal in water quantity) by bringing matters of village political economy into the division of the available supplies.[57]

A second-best solution, which retained vestiges of the desirable attributes of volumetric charging, was enthusiastically promoted from the mid-1860s until around 1874, when it too was effectively rejected by cultivators for reasons not very dissimilar to those which scotched its predecessor. Like the *paimana* system, the sale of water by contract had its counterpart in Italy, where a sum per measure of land formed the basis of contract.[58] A village, or part of a village, which signed a three-year contract received a reduction of between 14% and 18% on the average total sum paid as water rates over the previous five years. On payment of this annual amount, those accepting the lease were free to irrigate according to their needs, subject only to the size, number, and position of the outlets remaining unchanged.[59] As far as the canal authorities were concerned, the new system not only provided an incentive for users to economise on water, expand irrigation, and switch to more valuable crops, but promised significantly to reduce administration, since it obviated the need for detailed measurements as a basis for the assessment of water rates, minimised the need for supervision, cut the costs of rate collection, and

[56] For details, see *PRRC 1860–1*, pp. 25–6; and Swinton, *Report of the Committee on the Ganges Canal*, p. 23.

[57] *IIC*, p. 91.

[58] Baird Smith, *Italian Irrigation*, II, p. 165.

[59] Leases (for periods of up to twenty years) had been adopted temporarily on the Western Jumna Canal much earlier. These were based, however, on fixed *kolabas*, and problems inevitably arose as changes and improvements were carried out on the system. See *IIC*, p. 91.

reduced the opportunities for the abuse of power by petty officials.[60]

By 1867–8, around 65,000 acres on the EJC were irrigated on three-year contract leases. The gains to such villages appeared considerable. According to official estimates, in some years, due to economy in water use, contract villages paid rates more than 40% below what they would have paid under the measured crop rate system. So generally advantageous did the scheme appear that some officials even advocated refusing water except on four- or five-year contracts.[61]

It was over the division of the gains from reduced water charges, however, that the hopes for the contract system were dashed. 'The total want of all co-operation among the inhabitants of any one village seems now the main difficulty', wrote C.S. Moncrieff in 1866. 'They will not decide among each other, how much each person is to pay, so that despite manifest advantages to themselves, it is only after long inducement that any will enter into contracts.'[62] The problem of profit distribution was particularly pronounced among the co-partners in the large *bhaiachara* villages which predominated on the EJC. Arranging contracts with such villages – some of which contained 1,000 shareholders – necessitated making agreements with a few representatives.[63] These were the men, usually the dominant cultivators, who were able to 'appropriate, in whatever form, more than their fair share of the profits of irrigation, while the weaker shareholders find themselves called upon to pay up their share of the contract money . . . but do not obtain their share of the profits'.[64] Numerous petitions were submitted to canal officers on this score, but the matter was legally outside the jurisdiction of the Canal Department and recourse had to be had to the civil courts, where legal

---

[60] *PP* 1881, LXXI, *Further Report of the Indian Famine Commission, with Proceedings, Evidence and Appendices*, pt III, Appendix v, 'Extracts from the Records of NWP Irrigation Branch, relating to Contract Irrigation on the Eastern Jumna Canal', p. 529 (extracts from Revenue Reports for 1861–3).

[61] *Ibid.*, p. 531 (extract from SE's Review of EJC Revenue Report for 1868–9, para. 26).

[62] *Ibid.*, p. 532 (extract from SE's Review of EJC Revenue Report for 1871–2, para. 11).

[63] *Ibid.*, p. 530 (extract from EJC Revenue Report for 1865–6, para. 9).

[64] NWP PWD(I), June 1873, 10, Note by Nasir Ali Khan (former Deputy Magistrate of Meerut district), 29 March 1873. For more details on the failure of the contract system on the EJC, see NWP PWD(I), June 1873, 1–11.

expenses further bit into the profits. When the contracts of many villages expired in 1870, those with a large number of shareholders and a large area of irrigation did not apply for renewal. Moreover, while the number of contracts on the EJC fell from 407 in 1869–70 to 83 in 1870–1, the scheme was greeted unenthusiastically on the Ganges Canal, where the greater abundance of water meant that 'people get as much as they need without the necessity of economising their supply'.[65]

In the end, as a senior canal official remarked, it became 'scarcely possible to find an argument in favour of retaining it, as far as the benefit to individual cultivators' was concerned.[66] Some pondered for a time the idea that the interests of the individual cultivators in this matter should not give way to 'the more important principle, which every wise Government will strive to establish, of rendering the people, as far as possible, independent of their rulers in small matters of village government'.[67] Others were more forthright: 'If they are such a quarrelsome, impractical set that they cannot avail themselves of our offer, and prefer to pay a higher rate for the privilege of Government interference, it is no business of ours.'[68]

The failure of the contract system constituted, at least in part, a reflection of a cultural milieu centred on the EJC tract. The numerous *bhaiachara* villages generally possessed a greater degree of egalitarianism and of anarchic decision-making than *zamindari* villages. Similar, in fact, to the contrast drawn by Robert Chambers between the socially and economically more egalitarian Sri Lankan villages and the clearer hierarchical structures typically found throughout India.[69] To Dr Chambers, the anarchy present where economic power is less skewed produces a need for tighter bureaucratic controls on irrigation; whereas the more centralised power balance in hierarchical systems enables irrigation to proceed with less disruption from conflicting groups. In this context the failure of the contract

---

[65] NWP PWD(I), June 1873, 6, BOR to NWP Govt., 25 Nov. 1872.

[66] *IRR 1874–5*, p. 16A.

[67] NWP PWD(I), June 1873, 5, SE 4th Circle to NWP Govt., 12 Nov. 1872.

[68] NWP PWD(I), June 1873, 7, SE 1st Circle to NWP Govt., 10 Dec. 1872.

[69] R. Chambers, 'Men and Water: The Organisation and Operation of Irrigation', in B.H. Farmer (ed.), *Green Revolution? Technology and Change in Rice-growing Areas of Tamil Nadu and Sri Lanka* (London, 1977), pp. 360–1.

system in a region substantially composed of *bhaiachara* communities ought not to be surprising. Certainly, observers noted that the 'Mahajan proprietors' had shown their 'superior shrewdness' with respect to entering contracts, while the Meerut Deputy Magistrate was sure that 'such difficulties' as had occurred in *bhaiachara* villages 'would not occur in a pure *zemindaree* estate'.[70] This would especially be so where the dominant party possessed, in addition, the sort of quasi-proprietary right over the village *kolaba* (outlet) described in the chapter which follows. As it was, in the *bhaiachara* villages the costs of conflicts simply outweighed the concessionary element in the pricing, and so gave rise to a demand for the government to resume its former role in water distribution and rate collection as an alternative to additional leverage being placed in the hands of the powerful village factions. This rejection of the contract system marked the end effectively of the attempt to link water charges with volume alone rather than with specific patterns of use.

Effects of pricing policies

Together, the wider politico-administrative context and the specific practical concerns of the Canal Department contrived to bring about a situation where low user charges were maintained on the Doab canals. The state, as monopolist supplier, took, in direct revenue, a relatively small proportion of the typical incremental yield value. What effects did the form of pricing, and the structure and level of water charges, have upon the demand for irrigation, the way irrigation water was used, and the distribution of its benefits?

Form of pricing

There is clear evidence that the spread of canal irrigation was in places held back as a direct result of the entanglement of canal and land revenue systems. Settlement reports abound with references to *zamindars* who prohibited their tenants from taking canal water to their fields in the years leading up to

[70] NWP PWD(I), June 1873, 10, Note by Nasir Ali Khan, 29 March 1873.

resettlement, so that their revenue assessment might be kept down and they might capture the whole amount of rent enhancement after the settlement. In addition, the owner's rate appears similarly to have restrained the spread of irrigation. It may have been a convenient administrative device, but it clearly proved a disruptive influence within the countryside. An engineer on the Agra Canal described the problem:

The chief argument brought against the owner's rate is that, by the friction it has caused between the Zamindars and Cultivators, it has hindered the spread of canal irrigation. The Zamindar must either get the owner's rate from his tenants or he must pay it out of his own pocket, and recoup himself by enhancing the rents of 'dry' fields. No difficulty is experienced with tenants-at-will, as the Zamindar is free to make his own arrangements with them. But if an occupancy tenant refuses to pay either the owner's rate or a higher rent, the Zamindar must bring an enhancement suit against the man; and after the rent has been raised it is to the interest of the Zamindar to prevent his tenant from irrigating, as the smaller the area of 'dry' fields irrigated, the less the amount of owner's rate to be paid by the Zamindar.[71]

Forbes added that the situation would not have arisen had the direct charge proposals of Hume and others been adopted. Remarkably, he even recommended that, in future enhancement suits against occupany tenants based on canal irrigation, 'the court decrees simply that the owner's rate is to be paid by the tenant'.

Such a perverse ruling would only have been in line with reality. Outside some northern tracts (where tenants were frequently sufficiently powerful to make the owners pay the charge), there was 'every reason for believing that the owner's rate has been collected from cultivators instead of from Zamindars'. One engineer, Captain P. Marindin, had even seen the charge entered on the irrigation statements, or *parchas*, of individual tenants.[72]

The retarding effect of the charge on the spread of irrigation appears to have been particularly marked in villages belonging to large *zamindars*. Tarauli, in Muttra, was just such a case. The canal engineer was in no doubt that the fall in the irrigated area in this village, from 684 *bighas* in 1875–6 to 95 in the following

---

[71] Note by Col. J.G. Forbes (1873?), quoted in *IRR 1889–90*, p. 51B.
[72] *IRR 1875–6*, p. 21B.

year, was very largely due to the *maurusi* tenants' refusal to
shoulder the owner's rate.[73] Indeed, some of the most notable of
such cases in Muttra were to be found among the numerous
villages there belonging to the temples.[74] The irrigation on an
entire distributary, the Hasanpur *rajbaha* of the Agra Canal,
slumped dramatically in the *rabi* of 1876 by comparison with
the nearby Hatim *rajbaha*, due to the fact that the villages on the
former distributary were virtually all part of one estate, the
zamindar of which had demanded an extra twelve *annas* per *bigha*
to cover the levy. In Bulandshahr, too, the combination of the
owner's rate system and large estates in the eastern *parganas*
continued noticeably to retard the development of canal
irrigation through the 1880s.[75]

In what ways did the system of charging for water on a
crop-acre basis affect the manner in which it was deployed by
cultivators? Perhaps the most significant difference between
crop-acre and volumetric rates was that, under the former
system, once the cultivator had made his choice as to which
crop(s) he wished to irrigate, he effectively faced a fixed charge
regardless of the quantity of water he used. He may not have
had the advantage of being able to divert some of the water
designated for his indigo crop to perk up a nearby *juar* field
without running the risk of paying a full crop-rate for that
watering in addition, but he was able to treat irrigation in
nominated fields as a fixed cost and to use canal water to the
point where its marginal product was zero. As far as water
volume was concerned, the user outlay did not reflect its
marginal value. Inevitably this produced a tendency to
wasteful use of water by cultivators and canal villages with
ample supplies. Thus Le Poer Wynne describes how canal
water in Saharanpur, rather than being led around the fields
and permitted to enter on all sides, was generally introduced
into the field at one point and 'allowed to find its way as it best
can to all the corners. The consequence is that the soil is turned
into a quagmire at the end where it enters, is sufficiently

[73] NWP&O PWD(I), Feb. 1878, Irrigation Revenue Report 1876–7, IOR pagulation
120.
[74] *IRR 1889–90*, p. 51B.
[75] MCR, BOR, vii, 14(c), Ser. 2, T. Stoker, AR Tahsil Anupshahr (Bulandshahr),
1889, p. 12. See also *IRR 1884–5*, p. 2B; and *IRR 1880–1*, p. 3A.

watered in the centre, and quite insufficiently watered at the other end.'[76] The practice was apparently 'universal even among the Jats'.

In villages where labour was in relatively short supply compared to irrigation water, and perhaps land also, it would not be surprising to find such behaviour, given the pricing system. It is significant that Le Poer Wynne noticed that more care was taken to distribute the water evenly over the fields when wheat was being irrigated, since this crop was grown during a period when canal supplies were usually limited.[77] It should be remembered, of course, that while the crop-acre pricing system allowed private benefits of the kind described, there were social costs to such behaviour. A lower output per unit of water would result for the system as a whole, since such waste usually meant that other consumers were deprived of a full supply. There were also social costs incurred in the form of waterlogging.

One of the responses to uncertainty of supplies on the part of those consumers unfavourably placed within the canal system was to apply more water during their turn than was consistent with the achievement of high yields. This action gave some protection against losses which would result in the event of a delay or shortfall in subsequent supplies. The form of zero marginal cost pricing permitted such risk-averse responses – though inevitably they were part of the cause of uncertain supplies. Again, social costs were incurred in the form of output forgone.

Finally, with respect to the form of charges used, it should be asked whether there is any evidence from the volumetric pricing experiments of a distinct pattern of water utilisation. Senior officials were certainly quick to point out that the contract villages seemed to water a larger proportion of crops other than sugarcane, opium, and rice; and that the payments per acre irrigated in contract villages (lower than non-contract villages by one-fifth in *kharif* and one-quarter in *rabi*) could only be accounted for by the fact that 'the cultivators irrigating from contract kolabas were more careful of their water, and by such action were able to make it go further than they them-

[76] Le Poer Wynne, *Saharanpur SR* (1871), p. 139.
[77] *Ibid.*, p.140.

selves, or others, under the measurement system would have done'.[78]

The latter observation is clearly consistent with what one would expect, but an examination of the cropping patterns of contract and measurement villages shows differences in the proportions of various crops, but no systematic tendency for the differences to repeat themselves through time; indeed the year-to-year comparisons suggest no meaningful trend one way or another.[79] It is difficult to draw any conclusions on this issue since, even if we assume no bias in the measurement as between the two systems, there is practically no way of telling whether the character of contracting villages was in any way systematically different from that of the non-participants. Tenure, village size, location, reliability of the water supply – all might influence both the decision to enter into contracts and the effects of the different bases of payment. What is clear (and is dealt with in the following chapter) is that significant improvements in yields, and in net returns from canal irrigation overall, still relied upon increased reliability of supply. Without this, adjustments to the mode of pricing could only have a marginal effect.

Structure of water rates

In terms both of the water charge as a proportion of average crop value, and of the actual price paid per unit of water taken, there is no doubt that the adopted scale of charges favoured the commercial crops. In the case of *rabi* crops, for instance, during 1865–1905 wheat and *gram* bore the same canal rate, even though *gram* usually took only one or two waterings compared to wheat's four or five and despite the fact that the price difference between the crops in the market was frequently five to ten *seers* per rupee. In addition, the practice of charging the full crop rate once the *paleo* watering had been taken also worked against the poorer crops. The effectively high-threshold water charge induced cultivators to delay taking a *paleo*

---

[78] NWP PWD(I), June 1873, 7, SE 1st Circle to NWP Govt., 10 Dec. 1872.

[79] For summary tables of cropping patterns under the two systems, see NWP PWD(I), Oct. 1871, 2 (1869–70 season); *IRR 1866–7* (Roorkee, 1868), opp. p. 26 (1866–7 season, EJC villages only); and *PRRC 1865–6*, pp. 42–3.

watering, in the hope that rainfall would meet the limited moisture requirements of these relatively hardy crops. The more valuable – and more moisture-sensitive – crops, certain to need an artificial supplement to rainfall, were less subject to such finely calculated, yet yield-depressing, decisions.

In practice, however, for all the debate between officials as to what constituted an appropriate price structure, the evidence suggests that the differentials – and such adjustments as were made to them – almost certainly affected crop choice hardly at all. In 1873, for example, the rate for rice increased from Rs 3 to Rs 5, to discourage a crop which, although profitable, was often injurious to local health. Outwardly the measure would seem to have been successful, for the EJC rice area fell immediately from 37,750 acres in 1872 to 21,480 the following year. By 1874, however, the rice area had recovered to 35,430 acres.[80] Evidence suggests that a more effective method of reducing the cultivation of rice had been carried out in the late 1860s, when the previously copious water supply had been progressively cut back and the timing of canal closures for repairs and maintenance had interfered with rice irrigation.[81] Similarly, the 1905 reduction in the charge for cotton can hardly be said to have inspired the dramatic growth in irrigated cotton of this period – and the Director of Agriculture was mistaken to suggest that it did – since a clear upward trend was apparent from the late 1890s, and an acceleration from 1903–4 (see Figure 5).[82] By the same token, the accompanying reduction in the indigo rate did nothing to reverse its demise. Only changing market conditions could reverse the respective fortunes of these largely interchangeable crops – as they did, briefly, in the 1916–18 period, when indigo factories reopened, advances flowed, and uncertainty surrounded the cotton prospects. Canal officials looked closely for signs that the 1892 flow rate enhancements on the older canals had altered the flow–lift irrigation ratio, but could discern no such effect.[83] The only

---

[80] Statistics taken from *IRR 1875–6*. It can be seen from Figure 5 that the expansion of rice irrigation, in fact, followed a long-run trend similar to that of sugarcane.

[81] *IRR 1872–3*, p. xc.

[82] UP Rev., March 1916, 52, H.R.C. Hailey (Dir. of Agriculture) to UP PWD, 27 Aug. 1914.

[83] *Report on the Seasons and Crops of the United Provinces, 1916–17* (Allahabad, 1917), p. 3. (Subsequent references to these annual reports will be abbreviated to *S&C*.)

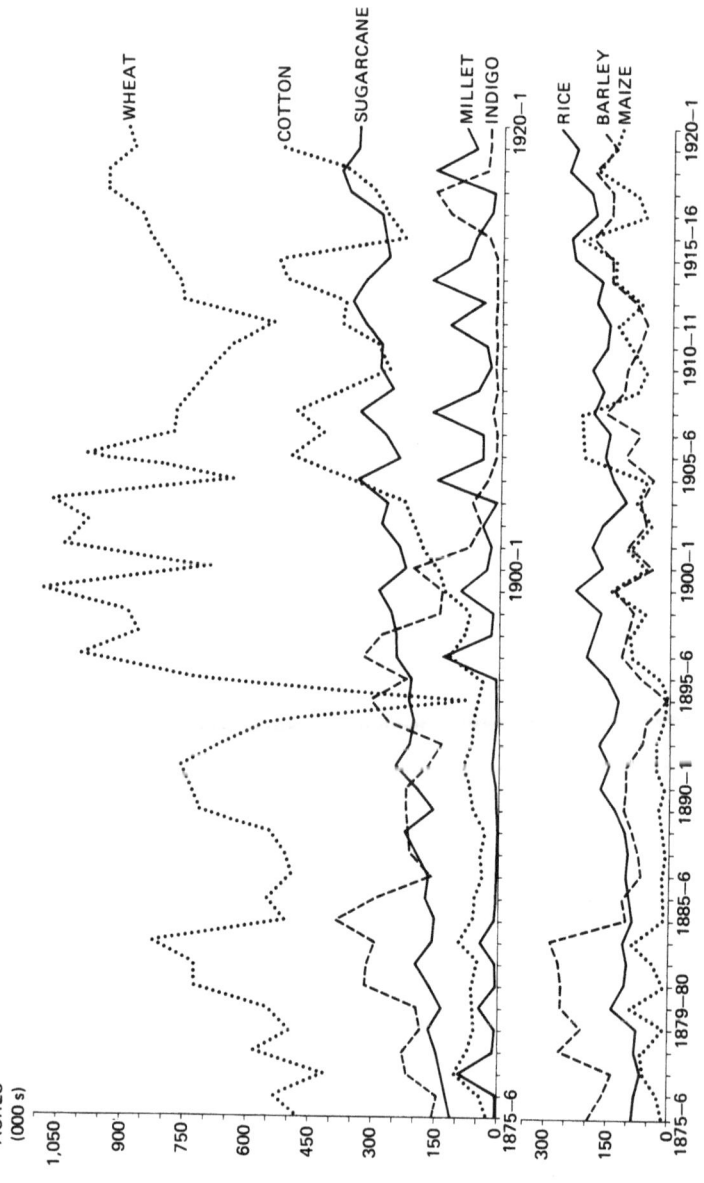

Figure 5. Canal-irrigated areas of major crops in the UP, 1875–6/1920–1. (Source: *IRR*, various years.)

explanation officials could proffer was that price rises since 1892 had nullified its impact.

There are several reasons why canal pricing is unlikely to have influenced crop choice. In the first place, the generally low overall level of direct charges meant that differentials were small, especially as a percentage of the total crop value. Canal charges were a larger percentage of total *cash* outlay, it is true, but irrigated cultivation involved a far greater overall input commitment, most of it in the form of utilising existing resources rather than the paid-out kind, and it was the organisation of these resources according to family priorities which was more pertinent to the production decision than a two-rupee water rate differential on, say, a wheat crop worth Rs 40–50. The point is really a fundamental one in that it is about peasant response to a marketed input. Government officials, in their protracted discussions on appropriate pricing procedures, were hampered by their apparent lack of understanding of the way in which the peasant producer incorporated such an input into his operations. The profit and loss accounting framework the officials continued to apply to peasant decision-making to justify their decisions and recommendations on prices had little relevance to the peasant cultivator. Just as the latter frequently paid an 'uneconomic' rent for an extra field because it allowed better utilisation of his resources – in particular his fixed stock of labour – in relation to his family's needs, so the realignment of irrigation charges was likely to be seen within the context of the peasant producer–consumer unit, and not simply in terms of a particular crop's profitability. In all of the discussions on pricing I have examined, only one revenue official, W. Irvine, appears plainly to have recognised the implications of this for irrigation pricing policies. 'Unless the water-rates are forced up to the highest point the best crops can pay, the tenant will still irrigate even his inferior crops, meeting the water-rate from the margin of profit left on his high class crops.'[84]

In other cases, price differentials proved irrelevant to the allocation of water between crops not so much for the reasons just outlined, but simply because *supply* conditions were the critical determinant of water use. This was the case, in fact,

[84] NWP&O Rev., Sept. 1881, 56, quoted in BOR to Sec. NWP&O Govt., 24 Sept 1881.

concerning the allocation not only between different crops, but also between cultivators and between localities. The discussions on the fine-tuning of the rates' structure, in fact, give an unreal impression of the extent of choice open to the irrigator. In reality, the demand for water, particularly on the intensively used Upper Doab canals, habitually outstretched the supply. From Aligarh, for example, one executive engineer reported: 'As a rule (and especially when the market prices are high) a much larger area is sown with indigo than the available supply is capable of keeping alive.'[85] The irrigation needs of sugarcane frequently interfered with later-sown crops: in 1876–7, cultivators could not get water to prepare the fields for *munji* and had to sow fodder crops instead;[86] in 1891, high *gur* prices caused a 25% increase in sugar sowings on the LGC, and such was the strain on supplies to keep the crop alive that there was a 'decrease under other kharif crops'.[87] A situation of excess demand – which arose in most years for periods of varying duration – meant that allocation became more of an institutional process: water was thus frequently not available to all those willing to buy. The irrigation decisions of some individuals or of groups within the village effectively pre-empted subsequent decisions regarding the irrigation of crops whose growing seasons followed.

Low level of direct charges

The maintenance of low rates throughout the Doab encouraged the use of canal water at the margin. Tracts or individual fields with poor soil, peasants with limited resources, and those areas and users with an unreliable canal supply could, when the chance arose, make profitable use of canal water because of the relatively small charge which was entailed. However, while low rates did not exclude the marginal users, and thus allowed the widest spread of canal irrigation, there is no doubt that these rates – together with the price structure favouring commercial crops – meant that those villages and cultivators in a position to combine fertile soil, effective organisation, and a reliable water

[85] *IRR 1873–4*, p. 25A.
[86] NWP&O PWD(I), Feb. 1878, Irrigation Revenue Report 1876–7, IOR pagulation 79.
[87] *IRR 1891–2*, p. 57B.

supply were very much at an advantage. This was particularly the case in the Upper Doab districts.[88] Theoretically, of course, a policy of price discrimination, aimed at still selling the full supply but at prices varying according to the willingness of different localities to pay, could have increased irrigation returns and permitted the increased subsidisation of 'protective' schemes in drought-prone areas which was desired by the Government of India. In practice this option was not feasible, since, even in the richer zones, marginal users – especially the smaller peasants – were to be found. As on most of today's irrigation schemes, therefore, the government had to accept that the system would be priced in a way which substantially benefited particular groups and regions.

Cheap water showed up clearly as a competitive advantage *vis-à-vis* other regions. The extensive cultivation of sugarcane in the UP, where conditions were less suitable to its growth than, for example, in the Bombay Deccan, was ascribed by the Agricultural Adviser to the Government of India to the fact that 'the facilities for cheap irrigation have been greater in the UP than in other parts'.[89] It is clear also that indigo, for the cultivation of which the Doab was clearly marginal in terms of natural growing conditions, could only have been the major cash crop it was during the second half of the nineteenth century because of the cost advantage given to it by cheap water supplies.[90]

Canal pricing also had implications for the workings of peasant society itself. For some peasant groups, the low and falling real direct state charges for canal water – in the context of an easing burden of agricultural taxation – gave rise to opportunities for capital accumulation and growth. On the other hand, however, given imperfect markets, it was feasible that the consumer surplus which attached to canal water might be siphoned off by existing or new systems of appropriation which not only redistributed the benefits but altered the way in which the canal water was used. 'Underpriced' water gave rise

---

[88] The disparities are reflected in the comparative percentages of crops irrigated by the Doab canals shown in Appendix 3 below.

[89] W. Sayer, 'Sugarcane Cultivation in Non-tropical Parts of India', *AJI*, xi (1916), p. 367.

[90] See Ch. 8 below.

to an excess demand and this, combined with a minimum of impartial official supervision over the supply, gave rise to a plethora of monopolists – working at different levels, from the *chaukidar* patrolling the *rajbaha* to the *zamindar* with his indigo factory – to prise varying shares of this surplus away from the actual user. While user charges adopted were powerless to reflect the utility of the irrigation outside areas of marginal irrigation, rent levels and other levies by intermediaries were more closely related to the size of productivity increments arising from the use of canal water.

A benign pricing strategy effectively meant that allocation of canal water took place largely through administrative and bureaucratic channels. Beyond the clumsy technical controls over use which could be incorporated into the system, and beyond the limited restraining influences of the higher administration, canal water filtered into the fields via the systems of control and influence which are part of any peasant setting. On the one hand, therefore, a range of unofficial levies and charges, demanded by those able to exert some control over the canal supply, bit into the potential or actual consumer surplus and in doing so effectively moved water supply and demand schedules more into alignment. At the same time, village politics and interest groupings operated to capture varying shares of the water and/or its benefits. Rather than the price structure being used to control the impulses to agrarian change, local circumstances were paramount in determining the use patterns arising from the canals, and in determining the beneficiaries. The following chapter examines the ways in which the design, management, and control of the Doab canals gave rise to this situation whereby irrigation was 'absorbed' according to the priorities and complexities of the peasant system.

6

Management, control, and distribution

The success of an irrigation project clearly depends upon more than merely bringing water to the fields. A whole range of qualitative characteristics – the timing and predictability of supplies, the quantity of water available, and the indirect as well as the direct costs entailed in its use – have a vital bearing upon the overall productivity of the scheme, the resulting crop patterns, and the distribution of benefits between localities, communities, and individuals. The previous chapter showed how pricing was not at any stage used as an effective way of allocating water. A valuable and increasingly underpriced resource was thus of necessity allocated through a combination of technical and institutional mechanisms.

The purpose of this chapter is to examine, from the particular point of view of water supply and distribution, the evolution of the technical design and operational features of the canals over the period, and to relate the observable phases in technical development to changes in irrigation objectives, managerial practices, and overall administrative control. Once it is established how these two elements interacted with local peasant society it is possible to understand how the very character of the water supply was affected by each phase of development, and indeed the terms on which it was available to different groups within the canal tracts.

While the major purpose of this chapter is to discuss allocation mechanisms, and the role played by different levels of the canal bureaucracy in influencing how water was distributed, it is inevitable that an analysis of management and control systems should lead to the more general issue of the relationship between this pervasive state organisation and the rural community. The Canal Department was effectively subject in the pursuit of its objectives to a minimum of outside control, and there were many aspects of canal administration, other than those relating solely to distribution, which were regarded as profoundly unsatisfactory by district officials. The final section of this chapter, therefore, extends the discussion of management and control of the canal works to consider the background and outcome of this interdepartmental conflict over 'canal oppression'.

Initial phase of development: distribution through local enterprise

For at least four decades following the opening of the EJC in 1830, the entire UP canal system was – even by the standards of the early 1900s – strikingly unsophisticated in terms of design and management. Military engineers, with little practical experience to guide them, experimented with designs and modes of operation within the tight and fluctuating bounds of 'ordinary expenditures' charged against the revenues of each year. Scarce engineering skills were naturally taken up by the main construction tasks; little time and few resources existed for the details of distribution, and only gradually did the attention of the engineers move beyond the prestigious main works to the more mundane activity of establishing an effective distributary system. The construction, operation, and management of the actual water distribution system was, by default, placed in the hands of the irrigating community.

In its most extreme form, this reliance on local enterprise saw EJC irrigation initially effected directly from the canal: water was supplied to cultivators' water-courses through cuts in the canal bank. Not only was this system wasteful, but it confined the spread of irrigation to villages directly on the main canal.[1] Wider distribution in the early years depended almost entirely upon adapting natural drainage lines (*nullas*) to the purpose through the construction of simple earthen dams. Not surprisingly, this crude expedient was beset by disadvantages, especially waterlogging problems, and the channels quickly silted up in any case. After a number of years, to make better

---

[1] *Saharanpur DG* (1909), pp. 60–1.

use of the available water supply, provision was made for a basic distributary network on the EJC – constructed at the cultivators' expense – and such a network was allowed for as a matter of course in the plans for the Ganges Canal.[2]

On both canals, the official attitude adopted towards the way irrigation was organised and carried out from the bare network of government distributaries continued to be liberal, however. To encourage irrigation, cultivators were permitted to take water through simple cuts in the banks of the raised distributaries.[3] A fixed outlet, or *kolaba*, was not insisted upon, and cultivators were free to put outlets 'where they liked, and very much as they liked'.[4] Gradually, as cultivators became used to irrigating their fields, less informal procedures were introduced, and local men, 'either individually, or as representatives of communities', then presented themselves as purchasers of *kolabas*, the size and position of which were determined by the executive engineers. On the EJC the construction costs of government distributaries were realised from the irrigators through the charge made for *kolabas*. The charge initially varied according to the cost of the particular *rajbaha* involved and to the area irrigated by the outlet. An eighty-square-inch *kolaba* could thus cost up to Rs 500. The system was subsequently modified, first in favour of a graduated scale related to the size and position of the outlet, and then, later, a fixed charge of Rs 150 was made, based on the average cost per *kolaba* of constructing the *rajbahas*.[5]

The individuals or groups whose *durkjwast* (*kolaba* application) was accepted had then to arrange for the water to be carried from the government distributary to the individual field channels within the village by constructing water-courses, or *guls*, of the required length and carrying capacity. In this early

[2] P.T. Cautley, *Notes and Memoranda on the East Jumna or Doab Canal, North-Western Provinces, 1845* (Calcutta, 1845), p. 98.
[3] NWP PWD(I), July 1873, 27, Questionnaire on Management of Private Water-courses, reply to qn. 1. Information on the operation of the irrigation system beyond the government channels is sparse. This proceeding consists of selected replies by district and canal officers to a questionnaire on water-course management in their respective areas.   [4] *Ibid.*
[5] *Ibid.* The periodic changes in the mode of charging cultivators for the construction of *rajbahas* meant that the costs fell 'in some cases with unreasonable severity on certain villages'. NWP PWD(I), March 1872, 51–8 provides examples.

phase, *guls* often bore more resemblance to small *rajbahas*, being on occasion three to four miles in length.[6] The cultivators were responsible for the maintenance of both the *guls* and (until 1873) the *kolabas*. Typically a *thokdar* was appointed for this purpose, and the canal authorities interfered only when alerted to cases where ill-maintained *kolabas* threatened the *rajbaha* banks, or where excessive loss of water was occurring from *guls*. The department would then either carry out repairs at the cost of the owners or close the outlet until such repairs had been effected. The provision and apportionment of water at the village level, therefore, was 'a matter settled by the cultivators and entirely by themselves'.[7]

The unforeseen consequence of relying upon local enterprise to facilitate the spread of irrigation, and particularly the sale of outlets, was the emergence of durable proprietary rights to the water supply, alluded to in the previous chapter. Given the terms on which the *kolaba* was sold, it 'could be transferred by the purchaser' and was indeed 'constantly transferred at a high premium'.[8] Aside from encouraging speculation, such rights gave powerful rural elements an unduly pervasive – and lasting – influence over the manner in which water was distributed. The Settlement Officer for Saharanpur, writing in the late 1880s, described the continuing effects of this system:

Originally . . . subscribers were the landowners and cultivators, but transfers [of *kolabas*] were not prohibited, and now a number of kolabas have got into the hands of speculators and others, who levy a private royalty on all those who use water. By this means several villages, which have practically no kolabas of their own, obtain water practically on sufferance; while at any time the owner or purchaser of a kolaba may apply to have it transferred to another village or district.[9]

A specific illustration of the problems which arose out of the early system of financing the spread of irrigation on the EJC

[6] Distinguished by the term 'minors' from 1874.

[7] NWP PWD(I), July 1873, 27, Questionnaire, reply to qn. 7.

[8] *Ibid.*, reply to qn. 5.

[9] Porter, *Saharanpur SR* (1891), p. 12. While Porter's remarks suggest that the problem caused by such rights still haunted canal officials around 1890, the 1904 Simla Irrigation Conference made reference to the fact that this practice had been 'broken down', although this is n t elaborated upon. *Report of the Irrigation Conference*, II, *Discussions*, p. 30.

involves a group of Bowrea villages near Bidauli, Muzaffarnagar.[10] Described by officials as 'a group of wandering thieves', the Bowreas had been settled in the early 1860s through an arrangement made between the government and estate-owner Mehndee Khan. Khan tried unsuccessfully to get the Irrigation Department to supply these and other villages on his estate with water. Irrigation officials considered the soil in the area too porous and therefore wasteful of supplies. However, when drought caused the rice to fail in 1864, and the Bowreas set out to make raids on surrounding villages, the Lieutenant-Governor intervened and it was agreed that a *rajbaha* could be constructed, at Mehndee Khan's expense, to carry fifteen cu-secs to this dry area. Khan received a Rs 13,000 *takavi* loan to cover construction costs, which he agreed to repay over eight years. In point of fact, Khan never did repay the sum. While he wrangled over the costs, the irrigation authorities came to the view that it would be unwise to bestow upon Khan full rights over such a large supply and absorbed the costs within the departmental budget. In the interim period, though, Khan had a claim over the supply, and had proceeded to redirect the available water to the growing number of non-Bowrea cultivators he was able to lure to his (now irrigated) estate. The Bowreas, undoubtedly unskilled agriculturists, no longer served Khan's best interests in his capacity of landlord, and the canal officials were unable to prevent him from depriving many of these tenants of canal water.

On the Ganges Canal and subsequent works such formal transferable proprietary rights were not permitted to emerge. The charge made for the *kolaba* covered only its cost, and ranged from Rs 2 to Rs 30, depending on whether it was a simple wooden frame, an earthenware cylinder, or of masonry construction. A specific ruling was embodied in the 1873 Canal Act to the effect that outlets were to be constructed and maintained at the expense of the government and thence no private rights to the outlet or use of water from it could be claimed.[11]

---

[10] NWP PWD(I), Nov. 1872, 14–51, Proceedings on Irrigation of Bowrea Villages from Bidauli Rajbaha, EJC.

[11] NWP PWD(I), July 1873, 30, Rules for Management of Private Water-courses.

In practice, of course, there were numerous ways in which *zamindars* and dominant villagers could ensure that the distribution system worked to their advantage regardless of such proprietary rights. For one thing, they were free to choose a line for the village *gul* which best served their fields.[12] Moreover, even where less influential villagers could find the resources to bring irrigation to their fields, the very process of arranging for the construction of a *gul* was often an impossibility for such men, as R.S. Whiteway discovered in Muttra, where he felt irrigation was being held back by leaving the construction of 'minor distributaries' to the cultivators: 'It is easy for a rich landlord to apply to the Collector to have land taken up in the next village to make his water course, but practically impossible for a petty proprietor to incur the odium of an application not only for land to be taken up in the next village, but from his neighbour's field.'[13] According to canal and settlement officers, it was very common for 'powerful elements and their agents' to obstruct even their own tenants who wished to take their water-courses along the edges of their fields.[14]

The *kolaba*-purchaser, of course, irrigated by right of purchase, and it was not uncommon for 'the applicant for a kolaba [to] use the whole of the water drawn off by it, in which case [other interested parties] do not get a share'.[15] In cases of group purchase, shares in irrigation were normally proportioned according to contributions to the original *gul* and *kolaba* outlays and non-sharers were accommodated only when there was water to spare. Where the demand for water by *kolaba*-sharers exceeded the supply, apportionment was usually based on the areas cultivated by each, but 'very often',

---

[12] Even the rudimentary system of distributaries eventually constructed by Cautley on the EJC was tailored to the needs of the 'influential landowners', and the resulting unequal distribution was a source of considerable local discontent. *Saharanpur DG* (1909), p. 61. Since the pattern of distribution was wasteful of water and frequently obstructed drainage, in 1871 the distributaries were 'completely remodelled, attention being paid to the conformation of the country rather than the convenience of the landowners'. *Meerut DG* (1904), p. 54.

[13] Whiteway, *Muttra SR* (1879), p. 13.

[14] *Ibid.* See also NWP PWD(I), July 1873, 27, Questionnaire, reply to qn. 33.

[15] NWP PWD(I), July 1873, 27, Questionnaire, reply to qn. 1.

observed Cawnpore division's Executive Engineer, 'the stronger parties generally take more than their share'.[16]

The possession of a share in a *kolaba*, particularly when combined with social and economic influence, allowed individuals and groups not only to maximise the proportion of their holding receiving canal water, but to ensure their position in the distribution of turns for watering from the *gul*. The actual ordering of 'turns' at taking water was usually determined by respective shares in *kolaba* and *gul* costs. To be high on this ordering improved considerably the reliability of irrigation. When water was in short supply *tatils*, or periodic closures, were established on *rajbahas* (and even on large *guls*, where they served several villages and disputes over them became intense). The water channel was divided into sections, each of which was designated a period of days during which its *kolabas* would be allowed open and water admitted into the village water-courses.[17] In times of restricted supply, therefore, to be high on the order for watering turns when the outlets were open for irrigation was often the difference between obtaining water and obtaining none.

It ought to be stressed, however, that it would be misleading to give the impression that control over the distribution of canal water fell at once into the hands of dominant *zamindars*. It is true that those with an interest in the manufacture of indigo often had considerable influence over the arrangements for the supply of canal irrigation, but such cases were the exception. Sometimes, a *zamindar* who was 'more active or more jealous of his cultivators than his neighbours' owned the *guls* and made a charge for their use – and it was apparently 'an understood thing that the landowner if he retains any of his own lands may water it first, even if he has no share in his tenants' kolabah' – but in general, officers found that 'few of the zemindar class interest themselves in canal irrigation'. This was especially the case with non-resident proprietors, who 'in nine cases out of ten' had nothing at all to do with the practical matters of irrigation.[18] The most frequent pattern, according to one

---

[16] *Ibid.*, reply to qn. 7.
[17] *Tatils* are further discussed below. See also Buckley, *Irrigation Works*, pp. 282–4, for technical details.
[18] NWP PWD(I), July 1873, 27, Questionnaire, reply to qn. 33.

officer, was for 'the cultivators to do everything, apply for and pay for the kolabahs and gools, and manage them independently of the zamindar'.[19]

The relevant power structures engaged in controlling and directing the supplies of water were typically those of the cultivating body within the village itself, and the picture which emerges from the enquiry into water-course management conducted in the early 1870s – as well as from other scattered sources – is one of petty elites often able to reinforce their multi-faceted advantages within the village through their control over the allocation of scarce water. [20] These village elites were not unduly affected as government control extended down the hierarchy of distribution channels and progressively confined local control to within the village boundaries. Government officers were aware that these men could exert powerful influence on decisions concerning the allocation of water, but for ideological and cost reasons they were committed to non-involvement with such village level political processes. Except where violent disagreement in a village or block caused irrigators to petition the executive engineer to act as an arbitrator – for example, to draw up a *warabandi*, or list of times at which various irrigators would have access to canal water[21] – apportionment was 'in no way interfered with by the canal establishment'. The typical official response to such problems was to place the onus for solution upon existing village institutions. Thus, although the Muzaffarnagar Collector was 'several times appealed to during ... drought', he 'never interfered otherwise than by going to the spot and getting the headmen of the village, whether zemindars or moocuddums, to take the matter up'.[22]

Operationally, this phase of canal development – though it effectively launched canal irrigation in the Doab, and in a way

---

[19] *Ibid.*

[20] Note that the extent to which these advantages could be pressed was restricted by the fact that, typically, several *kolabas* served each village.

[21] NWP PWD(I), July 1873, 27, Questionnaire, reply to qn. 33.

[22] *Ibid.* The allocative process in such cases is described in detail by F. Bion (EE, EJC) in *IIC(E)*, p. 20. The arrangement embodied in the *warabandi* would be formalised by the canal officer acquiring the signatures of the involved parties on the *ikrarnama* (agreement).

which economised upon government resources – gave rise to a noticeably crude system. The reach of the canals was limited and the practice of assigning to cultivators a large role in the construction, operation, and management of the distribution network led to the wasteful use of water in poorly aligned, ill-maintained, and environmentally disruptive water channels, which frequently served the needs of their owners only at the expense of nearby villages and fields.

The design of the canal distribution system was essentially based on the principle of 'open flow', whereby a fixed discharge flowing continuously through uncontrolled outlets would reach the fields through a hierarchy of progressively smaller channels, calibrated so as to balance the available supply with local demand. The reality by no means matched the theory and the system was so technically unsophisticated that the water supply was for many users grossly uncertain. This particularly applied to villages and users towards the extremities of the system, since these suffered from the ability of the villages and irrigators above them in the hierarchy to take water to suit their particular needs (either their normal amount when water was short, or more than normal when drought conditions prevailed) and thus deprive other users. Part of the problem stemmed from the initially permissive attitude towards the granting of *kolabas*. Up to a dozen outlets could serve the same village, since individuals and groups of irrigators were free to arrange for an outlet which best suited their needs. Their potential drawing power was, moreover, further enhanced by the simple expedient of clearing silt from the *guls* – an exercise cultivators were invariably reluctant to carry out without the prod of heightened competition for water.

The canal authorities had no control over this, and indirectly contributed to the uncertain supply situation by installing during the early decades a grossly defective distribution system. If the often quite large *guls* in the control of *zamindars* were poorly aligned and graded, leaky through lack of maintenance, and not at all calibrated in the interests of users further down the system, then much the same could be said for the government channels. The irrigation system as it stood was extremely uneven and inflexible. In its technical and managerial aspects, therefore, the system presented considerable obstacles to making more efficient use of the core water supply through improving the reliability of the supply and expanding the extent of irrigation.

Transitional phase of development: interventionalist policies

Expanding ambitions and technological constraints

The reorganisation of the administrative structure and the adoption of new methods of financing for irrigation development marked the beginning of a new phase of canal development in the post-1860s period. These changes were accompanied by a re-examination of the aims of the irrigation service, and its role within the broader administrative system of the Raj. One of the policies adopted at this juncture was of particular importance in that it influenced the whole nature of subsequent canal development in the Doab. In the words of Richard Strachey, Inspector-General of Irrigation during 1866–9, the new policy aimed at 'extending irrigation generally and so far as it is possible in a manner that shall to the utmost guard against the worst effects of severe drought'. Taking into account the rainfall conditions and the extent of well irrigation, future works were to be designed on the basis that 'a certain normal standard or proportion of irrigated area should be regarded as claimable by every district to which water can be given'.[23] The 'normal standard' settled upon during the 1870s was 42.5% of the 'cultivable area' within a village. No canal water was to be given to areas with such a percentage under well irrigation, and where the well percentage fell short of 42.5%, provision of canal water was to be limited merely to raising protection to this standard. In practice the guidelines proved impossible to adhere to,[24] but the principle was clear: equity was to be an important guide in the establishment of the distribution system; as many villages as possible were to be

---

[23] NWP PWD(I), Jan. 1869, 33, Col. R. Strachey, 'On principles to be followed in determining the capacity of the Irrigation Canals in Upper India; in regulating the distribution of the available water supply to the districts through which a canal passes, and to villages or individual cultivators; and in assessing the charge made for the water supplied for irrigation'. See also NWP PWD(I), Jan. 1869, 32.and 34–45.

[24] The practical drawbacks to this scheme of 'rateable distribution' are discussed in Ch. 3 above.

served, with a view to maximising protection rather than production.

This principle was brought to bear on another issue being debated at this time, that of the 'core' capacity to be built into the canals. The debate arose when, during the early 1870s, plans were being made to remodel the Ganges Canal so as to take account of the projected LGC system with its own intake sited in Bulandshahr. Through supplying the lower sections of the original Ganges Canal, the LGC scheme would release 1,800 cu-secs for the upper system (UGC). Some irrigation officials, most notably Officiating Chief Engineer Henry Brownlow, supported by the Lieutenant-Governor, argued for the remodelling of the UGC on the basis of a discharge of 6,000 cu-secs, a figure based on the minimum discharge at the Hardwar headworks of the three previous seasons, one of which was a drought. The doctrine behind this was that the supply should be limited to the largest possible area for which water could be guaranteed in the *rabi* season, when the supplies in the river were at their lowest prior to the spring melting of the Himalayan snows. Strachey, however, in his 1869 Inspector-General's directive on the principles to be followed relating to capacity, distribution, and water charges, had insisted that the minimum supply should be fixed from the *average* minimum – not the *actual* lowest supply, which was of rare occurrence. He argued that the extra capacity would allow the use of the ample *kharif* water supply, and that since the cultivator grew his food and fodder crops during this season, full use of the *kharif* supplies would ensure maximum famine protection.[25] Indeed Strachey – by this time no longer Inspector-General, but still influential in his position as Acting Secretary to the PWD in Calcutta – was undoubtedly behind the intervention of the PWD Secretariat in the debate over the UGC. 'The good that would result to the country at large by furnishing it for a number of years with the average supply of water available',

---

[25] NWP PWD(I), Jan. 1869, 33. It had not, at this stage, become clear that the tendency was for cultivators to decline to water such crops in drought years (see Ch. 7 below). One other influence on this decision was the recent experience of famine (1868–9). Then the canals had indeed been shown incapable of keeping alive the crops which had been put down in November with the aid of canal water; but, at the same time, late January rains had saved an area of *rabi* crops much larger than would even have been sown under a more restrictive policy on canal capacity.

wrote the Inspector-General in his 1873 memorandum on remodelling the Ganges Canal, 'seemed so to outbalance the possible loss that might accrue in an exceptional year that the government decided on taking the average and *not* the minimum supply as the standard from which to work.'[26] The eventual installed capacity of the UGC was 8,000 cu-secs.

With project design policies now explicitly oriented towards achieving a wider and more even spatial distribution of canal irrigation than had resulted from irrigation development hitherto, the authorities had to devise some way of rationing supplies, not only in order to serve the villages the system was being extended to cover, but also to enable them to distribute the available water more evenly and improve the reliability of supplies to the downstream users. The decisions on capacity merely heightened the need for such a control mechanism, since they effectively built into the system additional periodic shortages.

Manipulating price levels, it has been shown, was rejected as an impractical rationing device and innovations in the actual form of pricing had failed. Alternative methods had to be devised, therefore, to conserve and ration water supplies. This was done through a combination of piecemeal and rudimentary technical adjustments and the exertion of increased control over the distribution of canal water at the local level. From at least 1860, canal officers were engaged in reducing the numbers and intake capacity of village *kolabas*, and in taking over, remodelling, and managing the larger of the privately owned distributary channels. The technical adjustments were restricted to correcting the grosser defects, and did not at this stage significantly reduce the problems of distributing efficiently and evenly canal water within a system essentially designed according to the principle of 'open flow'. To achieve improved

---

[26] NWP PWD(I), Aug. 1874, 121, Memorandum by Inspector-General of Irrigation on Remodelling of Upper Portion of Ganges Canal, 26 March 1873. The Strachey guideline on capacity was in fact less radical than that favoured by the Governor-General, who advocated setting the capacity according to the average supply available from 15 June to 15 October, leaving 'the management of the cold-weather smaller supply and the system of distribution of the variable quantity of water . . . for the Engineers to settle in the best way possible'. UPA(R), Files of Meerut Commr., Dept. IV, File 9/1869, Ser. 17, quoted in Memorandum by Col. R. Strachey, 29 July 1869.

reliability of supply and an extension of the area of irrigation, the Canal Department thus had to intervene in a somewhat pragmatic fashion to control the flows within each distributary network. A dam here, a silt clearance there, combined with a system of *tatils* on specific parts of the *rajbaha*, were used to control the pattern of irrigation. Enhanced supervision through the canal establishment thus accompanied the increasing hold the government took over the hierarchy of distributaries during this period. Not unexpectedly, this intervention was accompanied by considerable conflict between the department and canal-users over access to water. This is examined here in two dimensions: the straightforward struggle for this vital resource which took place between the Irrigation Department and local user interests; and the attempts of subordinate officials to further their own interests by exploiting their position of influence over the supply of canal water.

Structure of the canal bureaucracy

It is useful at this stage to outline briefly the structure of canal administration during the post-1860s period. At the head of the provincial Irrigation Branch of the PWD was the Chief Engineer, who was also *ex officio* Secretary to the local government. The Chief Engineer was promoted from among the Superintending Engineers, each of whom had general responsibility for a group of irrigation divisions – usually a complete canal system – constituting a canal 'circle'. Each division contained 60–120 miles of main channel and irrigated, within an area of around 1,500 square miles, perhaps 200,000 acres annually from around 600 miles of *rajbahas*. Except when he was engaged on special duties, such as the projection of new schemes, the normal work of an executive engineer was the day-to-day administration of his division,[27] in which task he was assisted by up to three assistant engineers and perhaps a subdivisional officer (all usually trained at Roorkee), as well as a deputy magistrate, whose job it was to check the work of the subordinates and to deal with breaches of the canal regulations. The deputy magistrate, who was promoted from the subordi-

---

[27] See Ch. 2 above for some discussion of the work involved.

nate staff, was Indian, whereas the higher posts, particularly the senior positions, were for most of the period occupied by Europeans – the hierarchy closely corresponding to the military rank of the officers who dominated the service until the 1880s, when civil engineers moved increasingly into the higher executive positions.

At the lower levels of administration, which were organised on a subdivisional basis,[28] the subordinate establishment was supervised by the *ziladar*. The size and composition of the subordinate establishment differed both over time and between areas. The picture is further complicated by inconsistency in the job titles used at these levels. It seems, however, that the *ziladar* was usually assisted by a deputy, or *naib-ziladar*, and by perhaps a dozen suboverseers (*illakadars* or *amins*). Apart from their important role, described below, in relation to the levying of water charges, the latter had a general responsibility for seeing that the channels in their subdivisions were properly maintained. The actual clearance and repair of channels was carried out by the *chaukidar*, or patrol, of whom there could be 100 within a single canal division. Each *chaukidar* was assigned a section of a distributary – perhaps a length of six or seven miles – along which it was his duty to regulate the supply of water to villages or irrigating groups, to record and report gauge levels, and to note down the fields and crops which received water.

The recording aspect of the patrol's work was watched over by the *amin*, whose chief task was to measure and to report regularly on the irrigation taking place in the cluster of villages in his charge. He checked the work of six or eight *chaukidars* operating in his villages, and at the end of the irrigating season prepared the final record of irrigation in the village *khasra*. On being informed of the date of this exercise the *tehsildar* was supposed to arrange for the *patwari* to attend. In the early days of irrigation it was common for the two officials to go over the fields in question accompanied by a team with measuring chains, but as time went by reasonably reliable village maps became available and they compiled the irrigation list in the comfort of the *patwari*'s house. The *amin* subsequently prepared

[28] The subdivision was the physical area for which an assistant engineer would normally be responsible.

the *khatauni*, or final demand statement, and the individual *parchas* were distributed in the village. The *ziladar* and then the assistant or executive engineer were supposed to check the *khatauni*, and possibly inspect one or two of the villages (allowing fifteen days for irrigators to object to the assessment) before the demand statement was forwarded to the Collector and canal dues were collected from the appointed water *lambardars* through the revenue establishment. The *lambardars* received 3% on the sum if it was paid within two months. Generally the work was done by the *patwari*, the *lambardar* being 'too illiterate or too lazy to give himself the trouble'.[29]

With minor adjustments and small regional variations, this was the system through which canal irrigation was administered from the 1860s onwards.[30]

Departmental objectives and specific local interests

The conflicts which arose out of the intervention of irrigation authorities in the process of local water distribution are best illustrated by the clashes between the Irrigation Department and indigo planters. Such cases are prominent in the official proceedings of the 1860s and 70s due to the significant involvement of Europeans, who for various reasons were more prepared than indigenous landowners to pursue their grievances with the authorities.

In one case – untypical only inasmuch as the petition emanated from a non-European, Rajah Prithee Singh of Awa[31]

---

[29] NWP Rev., Jan. 1878, 17. See also NWP PWD(I), June 1870, 16, Report of Investigation into Collection of Canal Revenue; and *PP* 1881, LXXI, *Further Report of the Indian Famine Commission*, pt III, pp. 422–3.

[30] There was an attempt at streamlining the measurement process in 1902–3, partly so as to allow more time for the work of the subordinates to be checked by senior officials, but the rejigging of responsibilities among lower officials involved in this plan left gaps in the supervision of distribution; while villagers resented the loss of checks against wrong entries by the class of 'recording *amins*' (recruited from the *chaukidars*) resulting from the exclusion of the *patwari* from the process. With relations with the Canal Department having 'changed for the worse', the original system was reinstated in 1911. UP Rev., Dec. 1910, 143; and *Royal Commission on Agriculture, Evidence*, VII, p. 178, testimony of B.D'O. Darley (CE, UP PWD(I)).

[31] NWP PWD(I), Sept. 1868, 65, 'Complaint by Rajah Prithee Singh, against certain acts of canal officers in the Allygurh Division, Ganges Canal', 7 July 1867. The Rajah had apparently been urged by the Agra Commissioner in 1866 to bring 'any

– involved indigo villages along the Pilhutra *rajbaha* on the Etawah branch in Muttra district. The Rajah had four indigo factories in the upper sections of the *rajbaha*, while further down, at the Omergurh, a European named Newcomen was the agent of a similar factory. In 1867 an attempt was made to improve the distribution along the *rajbaha* and to reduce waste. This was general practice at the time, and involved reducing the number of *kolabas*, relocating those outlets liable to cause breaching from the high-banked water-courses, and making bed adjustments to the *rajbaha*, thus allowing the removal of planks used as dams to raise the water level in the distributary. As was also common practice, a *tatil* of three days per week was imposed on villages between miles 9 and 17 in order to make water available for the remaining four miles of *rajbaha*.[32]

The sudden loss of supply to the Rajah's villages resulting from the removal in April 1867 of the planks and the imposition of *tatils* prompted the Rajah's agent, Bhola Nath, to petition the Canal Department. The petition pointed out that reduced canal water supplies endangered indigo contracts with the villagers worth Rs 40,000. It also contained a number of accusations against canal officials for their harsh treatment and oppression. Assistant Engineer Lieutenant Hall's refusal to reinstall the planks was interpreted as an act purposely favouring Newcomen. It was, moreover, alleged that Hall – 'who often resides at the Omergurh Factory' – had told villagers that if they were to enter into contracts with Newcomen the water would be forthcoming. There was reference also to factory employees who were subjected 'to threats and even violent attack by canal officers, and a complaint to the effect that villagers had been deprived of alternative irrigation from wells since these had, 'agreeably to the orders of the canal authorities', been abandoned.

It is, of course, difficult to disentangle fact from the claims

inconvenience suffered to the notice of the government, since all government officials were aware of difficulties which existed in so large a department as the Ganges Canal, and heads of departments would always be ready to listen to what he had to say'. NWP PWD(I), Sept. 1868, 71A, Agra Commr. to NWP BOR, 14 Dec. 1867.

[32] For details of the technical adjustments made on this *rajbaha*, see NWP PWD(I), Sept. 1868, 70, Report by Lt. B. J. Parsons; and NWP PWD(I), Sept. 1868, 71D, Memorandum by Lt. J.G. Hall.

and counterclaims made during the enquiry which followed.[33] While the irrigation around the Omergurh factory appears to have increased in 1867, most villages along the *rajbaha* eventually received a better supply as a result of the improvements. The insensitive timing of the operations clearly caused considerable short-term disruption at a stage when indigo plants needed water at least every 10–12 days, and only a timely bout of rain limited the extent of the losses and made the complaint appear premature and exaggerated. The Canal Department did concede, however, that well closures – 26 in one village – had been ordered by the Deputy Magistrate. 'He denied this when I questioned him', recorded the Superintending Engineer, 'but I dare say he probably did (in his anxiety to increase the revenue from his sub-division) bring some moral pressure to bear on cultivators to induce them to abandon well irrigation.'[34] It emerged also that the Executive Engineer, Lieutenant Parsons, had struck the *karinda* at the Heera Nugla factory,[35] prior to threatening to withdraw altogether the irrigation to the villages supplying it with plant. The incident arose out of the discovery by Parsons that the factory establishment had been obstructive in the clearance of several feet of silt from the distributary bed, since these deposits effectively gave a better supply to the factory's indigo cultivators. Labourers from Heera Nugla who had accepted advances from the suboverseer had been threatened with expulsion from the village if they did the work.[36]

An accumulation of bad feeling between the Canal Department and a prominent European *zamindar* named Martin, who had a factory in Phaphund (Etawah), culminated in Martin's cataloguing a history of trouble with the canal authorities in an article which appeared in the *Delhi Gazette* on 3 September 1868. This move followed the department's rejection of Martin's claim for compensation for indigo lost due to failure of water supply, on the grounds that the losses arose because of bad seed rather than shortage of water. Martin, who according to the Mainpuri Collector was 'always quarrelling with the

---

[33] See NWP PWD(I), Sept. 1868, 65–7 and 70–71E for full discussion.

[34] NWP PWD(I), Sept. 1868, 52.

[35] Parsons claimed that he was provoked by the *karinda*'s 'impertinent manner and style of address'.   [36] NWP PWD(I), Sept. 1868, 71E.

canal men',[37] complained in a letter to the Agra Commissioner that 'year after year, one-third to one-half the indigo crop was suffered to be destroyed' by inadequate water due to defects in the irrigation works and 'the canal officer's refusal to keep them in order'.[38] His *ryots* had been 'so worried and persecuted by [the Executive Engineer] and his subordinates, that they have been obliged to give up indigo cultivation for a time', and for three years the factories had been 'all but closed'. Now, under the impression that the 'ill-feeling' had worn off, Martin had made heavy advances in 1867, only to incur an estimated loss of Rs 25,000 when 'the canal was again closed against my ryots both for irrigation and drainage'.[39]

As in the case of the Rajah of Awa's petition, there was substance in these claims. 'I have convinced myself he [Martin] has good grounds for complaint', recorded the Collector of Mainpuri, who had personally inspected the area.[40] His Etawah counterpart agreed with this interpretation, pointing out that although 'mixed-quality' seed was a problem in 1868 it was difficult to deny that 'crop failure in many fields was due to want of water'.[41]

It was in fact the Mainpuri Collector who made the observation that is crucial to understanding the true nature of such conflicts: Martin, he said, 'has got such a quantity of plant that he could scarcely expect perfect irrigation in all fields'. The important point here is that all such men with indigo interests were committed to both increasing and securing their supply of canal water, since this was the key to increasing the supply of plant to the local factories. Encouraged by the system of crop advances, a substantial part of the crop was speculative in the sense that it was sown with the aid of favourable showers and exceeded the area for which canal water was reasonably assured. 'I think there can be no question', the Agra Commissioner informed the NWP Government, 'that Mr Martin endeavoured by unauthorised means', including

[37] NWP PWD(I), Feb. 1869, 124, no. 3.

[38] NWP PWD(I), Feb. 1869, 124, no. 8, Martin to Agra Commr., 30 Aug. 1868.

[39] Martin was referring in this last point (closure for drainage) to the refusal of the canal officials to allow a cut to be made in the distributary banks to release floodwater held up there.

[40] NWP PWD(I), Feb. 1869, 124, no. 4.

[41] NWP PWD(I), Feb. 1869, 124, no. 5.

offering indigo advances from government *tehsil* offices, 'to inculcate a notion that he would be favoured above all others in the matter of canal water, and that he used this advantage for the purpose of furthering his speculation in indigo planting'.[42]

It was not necessarily disadvantageous to factory-owners if part of this crop failed, since in many instances this merely committed the peasant to grow indigo in the following year.[43] More important, perhaps, was the fact that the indigo-planter could use the evidence of inadequate water supplies, morally buttressed by the sentiment 'my interests and those of the people are identical',[44] to press for the provision of an additional and/or less variable water supply. Sometimes, as in the case involving Martin, the peasants themselves could be encouraged to take their complaints to the district authorities. Fifty-eight such claims for redress emanated from Martin's villages in 1868,[45] following the circulation of the idea that the government would compensate those who could prove loss of indigo due to canal water failure. It transpired that these petitions were written for the cultivators in the *tehsil* offices, where 'a good deal of the business of the indigo concern appears to have been transacted'.[46]

Thus, the sometimes acrimonious disputes over irrigation supplies to indigo villages were often simply part of an ongoing struggle for resources, and were only partly a reflection of the canal's technical inadequacies or, indeed, of 'oppression' by the canal bureaucracy. The Canal Department sought to protect its interests and pursue its broader irrigation objectives. It was inevitable that these would conflict with the ambitions of certain interests, of which the indigo factory-owners were merely the most vocal.[47] The allegations frequently made against canal officials, though not without substance by any means, were often, at base, attempts to discredit a conscientious, and therefore troublesome, local official, in the hope that he would be replaced by someone more amenable.

[42] NWP PWD(I), Feb. 1869, 124, no. 1.
[43] See Ch. 8 below for more details on the organisation of indigo production.
[44] NWP PWD(I), Nov. 1873, 84, H. Robarts to NWP Govt., 12 April 1873.
[45] NWP PWD(I), Feb. 1869, 124, no. 8, affixed statements.
[46] NWP PWD(I), Feb. 1869, 122, SE to CE, 29 Oct. 1868.
[47] See also the strong complaints of J. Michel (NWP PWD(I), Jan. 1872, 57 and 67) and H. Robarts (NWP PWD(I), Nov. 1873, 84).

Subordinate officials and corruption

The conflicts which existed between the canal administration and the local community in their respective roles as suppliers and consumers of canal services should, in fact, be distinguished from the conflicts which arose when officials pursued their own ends by exploiting their position within the canal bureaucracy. There were numerous opportunities for engaging in petty corruption, some of which were perennial and occurred regardless of the level of technical or managerial development. It was relatively easy, for example, to falsify irrigation returns, giving rise to the impression of one observer that 'a large part of the Government demand finds its way into the pockets of unscrupulous ameens, zilladhars and canal chowkeedars, entrusted with the measurement of irrigation'.[48] Another typical example is described by the Collector of Cawnpore: 'It is the custom for the canal chaukidars to seize all cattle found straying on the bank of a canal, and a great many which are not found straying there; and if they are not bribed to release them by the owners, the chaukidars hand the latter over to the Canal Magistrate to be fined.'[49]

The opportunities to engage in such practices increased markedly during the phase when the authorities attempted to control through its establishment the distribution of water, since this involved a considerably enhanced supervisory role for the subordinate officials. With local level supervision by engineering officers necessarily minimal, the subdivisional establishment gained a certain notoriety in the eyes of district officers for their corrupt practices. Indeed, an enquiry into canal corruption in Baraut *pargana* – conducted in response to A. P. Webb's allegations against the canal administration contained in his 1878 pamphlet, *Irrigation Topics*[50] – concluded

---

[48] NWP PWD(I), Feb. 1869, 124.

[49] NWP PWD(I), Sept. 1876, 25, C.A. Daniell to Allahabad Commr., 13 Feb. 1875.

[50] Reprinted verbatim in NWP&O Rev., Oct. 1879, 62. It is worth recording that Webb held what the irrigation authorities regarded as 'an animus' against them. He is reported to have admitted that the pamphlet was written as a response to news of a possible increase in water rates. Moreover, he had previously had requests for his exclusion from *tatils* refused and considered the Assistant Engineer discourteous to him. See NWP&O PWD(I), Aug. 1881, 8, Maj. C.S. Moncrieff to CE, 7 July 1879.

that the charge was 'proven beyond the shadow of a doubt'.[51]
The investigating officers found that 'unlimited opportunities' existed for making illicit gains as a result of the 'practically enormous power in [subordinates'] hands for harassing or favouring irrigators', and they concluded: 'Very considerable gratuities are given systematically to all classes of canal subordinates, sometimes in kind and sometimes in money . . . It is feared not even ziladars are exempt, and of which it is clear that the Deputy Magistrate could not have been ignorant.'[52]

Ziladars, it appears, commonly received *faslana* (harvest) payments, either at a rate per *kolaba* or at a rate per *bigha* of cultivation, and these contributions were often supplemented by additional payments to secure considerations such as timely first waterings. As well, Baraut cultivators made goodwill payments to *chaukidars* and measuring *amins*, sometimes at a rate per yoke of animals powering the sugar presses.[53] The most important additional payment, however, was for the purchase of *tatil* water – that is, for access to water when the outlet in question was officially scheduled to be closed. 'We buy tatils', admitted the *lambardar*'s agent in Mauza Jangarh (a village possessing just one half-*kolaba*). 'Sometimes we pay Rs 2, other times Rs 4.'[54] According to Webb, it was well within the power of canal employees to enforce payments or even to precipitate a demand for *tatil* water: 'Nothing is easier', he pointed out, 'than to drop a plank into the head of the rajbuha and diminish the supply.' This was a particularly common practice, he maintained, when low supplies endangered cane or wheat crops. As an alternative to such exactions, local subordinates could enter into lucrative arrangements with dominant

---

[51] NWP&O Rev., Oct. 1879, 49, BOR to NWP Govt., 6 Sept. 1879. Investigations were conducted by J.H. Fisher (Meerut Collr.) and Maj. C.S. Moncrieff (SE, EJC).

[52] *Ibid*. 'It had even turned out', Moncrieff reported incredulously to the Chief Engineer, 'that [the Executive Engineer's] own munshi had the night before been intimidating the zamindars from giving evidence.' NWP&O PWD(I), Aug. 1881, 8.

[53] NWP&O Rev., Oct. 1879, 52, J.H. Fisher to Meerut Commr., 6 Sept. 1879.

[54] NWP&O Rev., Oct. 1879, 35. Such payments were not hidden by the villagers: the *faslana* charge was invariably to be found in the *malbah* accounts (record of village cash payments); in some villages a *malbah-nahr* (water-course account) was maintained separately by the village *bania* in the same *bahi* (account book). The recorded total money payments for eighteen villages inspected came to Rs 928 for the year. NWP&O Rev., Oct. 1879, 52.

villagers – what Webb called the 'strong faction which fattens under the shadow of canal favour. These, chiefly lumbardars and their relations, in return for large supplies of water promote canal oppression in all its forms.'

The corruption was clearly embarrassing to senior canal officials, particularly since it provided yet more ammunition for their Revenue Department critics. It is important, however, to set the subordinates' activity in context. In the first place, such practices were clearly not as universal as many believed. Even Webb conceded that on the Ganges Canal, 'where irrigation by rajbuha section is, I learn, not general, and water is to be had abundantly in many localities', people were 'as a consequence tolerably free from oppression born of interference with the supply'. Secondly, much of this activity caused a net loss not so much to the irrigating community as to the canal revenues. 'It is evident enough', investigating officer C. S. Moncrieff concluded, that the cultivators 'have not been put to this expense for nothing. Measurements have been dishonestly performed, lands have been unfairly assessed, water has been given out of turn.'[55] In fact, the 'oppression' which was apparently so pronounced where *tatils* were needed to ration the short supplies of water might more realistically be seen as a 'service', through which the *chaukidar* in particular supplemented his regular pay of around Rs 5–7 per month. So crude and disruptive was the *tatil* system as a regulatory device that the operation of the 'informal market' for canal water arguably produced more positive than negative effects through imparting – for some cultivators at any rate – a much-needed degree of *predictability* to their water supplies. This seems to have been the view taken by the Chief Engineer at the time, Henry Brownlow, who refuted the idea that the *chaukidar* was 'an exceptional villain, who wanders about his beat levying blackmail on all who venture to touch canal water'. He was sanguine about the prospects of preventing the operation of an informal market for canal water given that 'the Zamindars themselves see nothing wrong in it'.[56]

Why did the *tatil* system tend to increase the unpredictability

[55] NWP&O PWD(I), Aug. 1881, 10.
[56] NWP&O Rev., May 1881, 7, letter to NWP Govt., 10 March 1881.

of supplies? Periodic closures were designed so as to ration available water by restricting the time allowed for irrigation in those villages affected. They were necessary because, even after the reductions in their outlet capacity, villages in the higher sections of *rajbahas* were still able to obtain a disproportionate share of supplies by clearing and deepening their *guls* (and thus increasing the rate of intake into village channels) or by irrigating at night. Villages at the tails were obviously vulnerable to such practices, and *tatils* were thus imposed on upstream villages with the aim of allowing a certain level of supply to reach lower parts of the distributary. It was an important assumption lying behind the system of rotating closures that the water supply in the *rajbaha* would remain roughly constant. In fact, as Webb made clear, in the crudely constructed and imperfectly maintained distribution network existing at the time, supply was anything but constant due to the silting up of channels. Level variations in the main canal also played a part in the problem. Frequently, therefore, a low supply coincided with the open period allocated to outlets along a section and irrigation was sharply curtailed. Since compensatory irrigation was prohibited by the department, it was clear – as Webb pointed out – that 'if subordinates did their duty . . . irrigation by these water-courses would be very trifling'. As it was, 'the subordinates turn this to pecuniary account and affect the desired irrigation'. It was easy to open upstream outlets out of turn and keep them running all night, since it was difficult for an assistant engineer to patrol even ten to fifteen miles of distributary out of the perhaps 200 miles in his overall control.

There was clearly a close relationship between the wealth of a particular village, in terms of the value of crops raised, and the level of payments in *faslana* and for *tatil* water. Villages which were located reasonably near to the heads of distributaries could normally expect full supplies of water, and they engaged the services of local canal officials to ensure that departmental attempts to spread supplies along the length of the distributaries did not affect the timing or quantity of water available for their intensively cultivated fields. The cultivators in these villages were prepared to pay heavily for such assurance because, quite simply, many of them could only afford to commit themselves to large areas of input-intensive crops if they could rely upon water supplies. Thus the Jat villages in Baraut appear to have paid most heavily in the form of *faslana*. Shabka Kallan village, for example, made payments amounting to Rs 12 on each of its four *kolabas*. Payments were especially heavy in villages where the water-sensitive cane crop was grown.[57]

Illegal payments decreased with distance along each distributary: 'Near the tails of the rajbuhas', observed the Meerut Collector, 'the chaukidar has less in his power and payments sink to perhaps nil.'[58] In these sections, of course, there were usually no *tatils* applied, and the influence of the subordinates was further restricted by the fact that down-stream villages had developed far less reliance upon canal water. Koil *pargana* statistics confirm that as the distance from the *rajbaha* head increased, the proportion of cultivated area under irrigation fell progressively.[59] Moreover, the area under valuable crops similarly diminished. Some villages, faced with an unreliable canal supply, constructed or maintained sup-plementary sources of irrigation (an option which was open to many cultivators in upstream villages, and which effectively impaired the bargaining position of subordinate officials). Many downstream villages responded to the scanty supplies by increasing the efficiency with which they used what water they did receive. It was in the lower reaches, for instance, that attempts at lining channels with roof tiles were to be found,[60] while *kiaris* were invariably smaller and *guls* better maintained in many cases.

Broadly considered, if *tatil* controls had been properly

---

[57] NWP&O Rev., Oct. 1879, 49.   [58] NWP&O Rev., Oct. 1879, 52.

[59] Average irrigated area as a proportion of cultivation in Koil *pargana* villages fell from 76% in villages within two miles of the main canal to 34% for villages in the four- to six-mile range. This applies also to holdings. Empirical studies show that at around 500 yards from the outlet, the percentage of the holding area irrigated is typically one-half that of holdings near to the outlets. See R.C. Repetto, *Time in India's Development Programmes* (Cambridge, Mass., 1971), p. 92.

[60] *Royal Commission on Agriculture, Evidence*, VII, p. 377, testimony of Dr R. Mukherjee (Univ. of Lucknow). For a contemporary study of how water scarcity encourages community action to safeguard available canal supplies, see R. Wade, 'The Social Response to Irrigation: An Indian Case Study', *Journal of Development Studies*, XVI:1 (1979), pp. 3–26.

enforced, there would have been a reduction in the difference in the quality of irrigation between villages located at opposite ends of the distributary system. Different expectations as to the character of the supply would clearly have affected crop choices and decisions on intensity of cultivation. As it was, illegal sales assisted the well-placed (and frequently rich) irrigators to maintain their levels of cultivation, while every time an outlet was illegally running the expected increase in the supply to lower villages as a result of *tatils* was being diminished. Inasmuch, however, as purchases of *tatils* raised the marginal cost of canal water, some economies in water use would have resulted in upstream villages. These would have been at the expense of marginal cultivators and poorly placed holdings or fields, thus producing situations such as that described by Opium Agent J. E. Hand: 'I have not been able to get all my poppy sown on account of the water being scarce; our cultivators are not the swell cultivators of the village; the Thakurs and the Brahmins get the water first, and by the time our men's turn is reached the water stops.'[61] Improved supply made available to lower villages would undoubtedly have been captured in many cases by the influential cultivators (given what we know about distribution within villages), who would thus have been given the opportunity to extend irrigation on their holdings and grow more valuable crops.

Phase of increasing technical control

The subversion of the crude distributional control embodied in the *tatil* system underscored the desirability of reducing the discretionary role of the canal establishment in the day-to-day distribution of canal supplies. Theoretically, it was not difficult for engineers to devise ways of doing this by means of technical adjustments to the system: even within the crude open flow system, refinements, properly maintained, could provide for a more even distribution of supplies. In practice, in spite of the fact that investment in project refinement was relatively cheap and quick-yielding compared with outlays on new works, it was the latter which received most of the funds, and engineers were

[61] *IIC(E)*, p. 127.

hampered in the improvements they were able to make by lack of funds for such schemes.[62] A systematic and carefully integrated attempt at remodelling only got under way, therefore, in the 1880s. Prior to that, remodelling activity was somewhat *ad hoc* and concentrated on the correction of the grosser defects.

It was advantageous, however, that by the time the improvements were seriously tackled, the government had extended considerably its hold over the channels at the lower end of the hierarchy of distribution. Most such channels still in private hands were less than one mile in length. In fact, this control over the distribution network was further enhanced by the decision made during the planning of the LGC that all principal water-courses would be constructed and maintained by the government. This was an explicit recognition of the fact that savings in the form of water economies resulting from such a policy would more than cover the costs involved.[63]

The remodelling of the distributary network reached its peak in the 1890s, with 1,400 miles (or around half) of the UGC total being adjusted during 1894–1900, and over half the total mileage of channels on the Agra Canal and the EJC. The broad aim was to reduce waste and to facilitate a maximum area of irrigation per cu-sec entering the system, while also dispensing with the need to apply *tatils* within individual *rajbahas*. The principal adjustments were as follows:[64]

(i) Channels were adapted to 'command' the country to allow

---

[62] It is an established fact – though one which, for various reasons, is not even today fully acknowledged in development budgetary allocations – that on the basis of expected incremental costs and benefits, the improvement and extension of existing capacity promises often higher (and more immediate) returns than the creation of additional capacity through new schemes. This principle is not confined to expenditure on the irrigation system alone: complementary investments in transportation especially are likely to increase the returns arising from irrigated farming.

[63] Experiments conducted by J. Ivens on the Ganges Canal indicated that water losses per mile in village water-courses due to spillage, percolation, and evaporation varied from 21% to 67% of the water entering. Buckley, *Irrigation Works*, p. 68. The costs of the construction of water-courses on the Fatehpur branch were discovered to have been recouped in a single season due to a far more rapid development of irrigation. See UP PWD(I), Sept. 1901, 5.

[64] Described in *IIC(E)*, p. 137.

the substitution of flow for lift irrigation. This was done by providing gentler slopes and masonry falls.

(ii) Distributary channels themselves were adjusted at each mile section to carry volumes sufficient to allow water to get to the tails instead of being consumed at the heads.

(iii) Both the numbers of outlets and their sizes were reduced to suit the areas to be irrigated (on the basis of current irrigation); many *kolabas* were repositioned to serve water-courses flowing through high ground to command tracts on both sides.

(iv) The lower the velocity, the more the slopes of both distributaries and minors were reduced so that the beds and sides of channels would not be scoured out but would gradually be 'puddled' over the wetted perimeters, thus reducing percolation.

The enlarged distributaries were designed so as to run full and on alternate weeks, rather than continuously. This significantly reduced percolation losses, while the combination of the adjusted distributary capacities and the limitations on *kolabas* meant that it was no longer necessary to *tatil* sections of the distributaries, since – theoretically at least – there was now sufficient water to meet the drawing capacity of each village and still supply the tails. Distribution control now lay not with the *chaukidar* but with the executive engineer, who simply controlled whether an entire channel was in flow or not. As a result of these adjustments, all villages at the tails of the EJC distributaries were reported to have increased their irrigated areas.[65]

The effects of these design improvements were not confined solely to the tail villages. The increase in the proportion of flow irrigation and the speedier execution of the irrigation process contributed to an expansion of irrigation in most villages taking water. Indeed the ability to irrigate a given area in a shorter period of time was an important aspect of the system's increased elasticity and flexibility in times of abnormal demand.[66] A breakdown of statistics for the Anupshahr branch

---

[65] *Ibid.*, p. 20.
[66] For examples of increased flexibility, see *IRR 1905-6/1907-8*, p. 26. It was the stated opinion of the Chief Engineer H. Marsh that the *speed* of irrigation had increased to the extent that as many fields could be watered in a week with remodelled distributaries as were formerly watered in a month.

of the UGC provides a good indication of the effects of the remodelling overall. The 68,000 acres irrigated on the branch during the dry year of 1877–8 were confined to the upper two divisions; the third division apparently received no water at all. Remodelling led to progressively larger areas of irrigation on the distributary as a whole, and to a dramatic increase in tail-end irrigation. In 1899–1900, an area of 100,000 acres was watered in the third division, out of a total for the whole line of 250,000 acres.[67]

Of course, these statistics of expansion also reflect certain other technical developments and design improvements which enhanced considerably the system's overall flexibility. Prior to the construction of the Jani and Kot escapes, for example, engineers on the Ganges Canal had trouble in responding to a high demand confined to the upper portion of the canal without swamping the lower reaches of the canal.[68] The flexibility of the core supply was substantially improved by the construction of the LGC, with its own headworks intake and the facility to pass supplies from the UGC system into the LGC by means of the Cawnpore terminal line.[69] This not only made water available for the expansion of irrigation capacity in the Upper Doab, but improved levels in established distributaries there and thus allowed increased irrigation. Overall flexibility was enhanced also by a link-up system with the Agra Canal. Although the latter's design capacity was fixed at 1,100 cu-secs, actual supply was sometimes as little as 700, and once, in 1892, fell to 400 cu-secs. The Hindan River cut was subsequently excavated to supplement the supply of the Jumna at Okhla, where the headworks were sited, and a further supplement could be obtained by way of the Jani escape, which connected the Hindan to the Ganges Canal.[70] The increasing integration of the main channels of supply was supplemented by the installation from the 1860s of a telegraph network, which enabled a greater degree of fine-tuning within the overall system than had hitherto been possible. Supplies in the main canals could now be more responsive to localised needs.[71]

[67] *IIC(E)*, p. 138.   [68] *IRR 1871–2*, p. 17.
[69] For details, see *Triennial Review of Irrigation*, pp. 32–64.
[70] *Ibid.*, p. 65; see also Lane, *Muttra SR* (1926), pp. 5–6.
[71] See UP PWD(I), June 1903, 1.

Increased sophistication had undoubtedly been achieved by the time the Irrigation Commission came to review the state of affairs in 1901. But a basic doctrine had not been lost sight of: water economies resulting from technical improvements ultimately took the form of extended irrigation coverage. The twin aims of maximisation of famine protection and, within the existing pricing system, of maximising direct canal revenue pushed the reach of the canal system ever outwards and ruled out the use of the benefits of increasing technical sophistication for more concentrated and intensive irrigation. The Chief Engineer, Sydney Preston, made a telling statement on this matter in 1900, when a discussion was taking place as to whether to proceed with the proposed Nandgaon distributary on the Agra Canal (a project first proposed in 1890 but abandoned due to doubts over the adequacy of the canal's water supply). One executive engineer maintained that the supply was still inadequate. Preston disagreed:

I am convinced from my own experience, and it is also common sense, that *we must make the distributary first, and then we shall find water for it* . . . if a new distributary is added, it too in turn will receive its share of the available supply. It is true this will be done by withdrawing a portion of the supply from the remaining distributaries and villages, but all share alike, and it is in my opinion advisable to give as many villages as possible the benefit of a share of the irrigation, even if the present supply is thereby reduced.[72]

This basic principle of distribution therefore continued to exert its influence on the character of canal irrigation in the Doab, and was supplemented in its effects by a process whereby the irrigated area would tend to spread in seasons when the supply was adequate, and whereby these additional fields would ultimately be recognised by the canal authorities as having a claim on the water supply.

This increased flexibility and technical sophistication lay behind the expansion in normal irrigated areas from 1.6 million acres in the 1880s to 2.6 million in the late 1890s (see Figure 4). The core supply had not been notably improved; it was simply being conveyed more efficiently and spread faster and more

---

[72] UP PWD(I), Feb. 1903, 2, S. Preston, Project for Nandgaon Distributary, Agra Canal, 18 Jan. 1900. Emphasis added.

thinly. In the process, major malfunctions had been substantially eliminated, but the policy of maximum coverage, applied within a system based on the open flow principle, militated against the achievement of supply reliability. Villages at the tails were now able to be confident of some supplies during dry weather, but still could not rely on either the quantity or the timing. Tail-enders on the Anupshahr branch, though undoubtedly better served by the system, still found the supplies 'capricious and insufficient'.[73]

Chaudhury Mukhtur Singh's testimony to the Commission on Agriculture indicates the nature of the continuing supply problems. Supplies could be delayed due to the alternate closing of the distributaries; routine closures for maintenance could often be inconvenient to cane-growers; supplies were frequently disrupted by inadequate silt clearances on the part of subcontractors; and villages at the heads could still increase their share of the water by clearing out their *guls*.[74] The most important consequence, Singh pointed out, was the discouragement given to the higher-yielding crops and varieties.

The growing technical sophistication of the canal system inevitably eradicated many of the earlier problems, particularly those of wastefully excessive use of water. As one settlement officer remarked in 1901, there was 'seldom now . . . much margin' for overirrigation.[75] Nevertheless, then, as before, the charge could be laid that the distribution policies, like those relating to pricing, took little account of water management as it related to crop needs; in this, as in earlier phases, agriculture had to fit round the irrigation branch rather than the reverse. E.C. Buck had clearly recognised this tendency in the 1870s: 'One mistake in canal administration has been that agricultural conditions have been less studied than engineering requirements, and as a consequence the state has gained less and irrigated land suffered more than ought to have been the case.'[76] Yet the same sentiment was being

---

[73] Phelps, *Bulandshahr SR* (1919), p. 3.
[74] *Royal Commission on Agriculture, Evidence*, vii, pp. 672–4, testimony of Chaudhury Mukhtur Singh (Pleader, Meerut).
[75] Gillan, *Meerut SR* (1901), p. 6.
[76] See NWP Rev., Jan. 1879, 92, E.C. Buck (Dir. of Agriculture), Note on Construction of Masonry Wells, 7 Dec. 1878.

expressed by the Howards forty years later.[77] Low irrigation charges clearly played their part in allowing – even encouraging – the perpetuation of an extensive and not particularly reliable irrigation system. Departmental policy on rate remissions and compensation (discussed next) did nothing to force official attention onto this matter.

Irrigation management and interdepartmental conflict

To aid their pursuit of departmental objectives, the canal authorities possessed a remarkable degree of independence in the management and operation of this huge and pervasive system. With the exception of the intrusive external influence in matters of canal finance, the Irrigation Department for all practical purposes functioned as a self-contained and autonomous unit in its day-to-day operations, and in many important respects this remained the case even after the administrative reorganisation of the 1860s. Indeed, the department could hardly have wished for a more favourable operational environment in which to pursue its objectives. It controlled the regulation and allocation of supplies from the canal, and could substantially determine the distribution of supplies from extensions; it possessed *de facto* powers of coercion, in that it could threaten to withdraw supplies from users who failed to comply with its terms; it adjudicated offences against its own rules; and it recorded irrigation and assessed claims for remission of water rates due to supply deficiency (as well as compensation claims for damage or loss caused by the works). Given the nature of canal schemes, and the permissive framework in which the department operated, it was inevitable that the latter would, in pursuing its objectives, come into conflict not merely with specific interests within the irrigating community, but with non-irrigating interests and with officials in other departments whose concerns and priorities were not in accord with those of irrigation engineers.

---

[77] 'Up to now irrigation in India has been looked upon largely as an engineering problem . . . but the real value of irrigation to India depends largely upon a proper appreciation of the needs of the crops irrigated.' A. Howard and G. Howard, 'The Irrigation of Alluvial Soils', *AJI*, xii (1917), p. 199.

Such conflicts can be illustrated through a brief examination of the Irrigation Department's labour recruitment practices. Substantial amounts of labour were needed to assist in the operation of the system, and to keep it in reasonable repair. Labourers were recruited for such tasks as the annual reconstruction of bunds at the UGC and EJC headworks, as well as for routine silt- and weed-clearing operations and emergency repairs, particularly to banks. For many of these tasks, the limited time available for their completion gave a certain urgency to the problem of labour recruitment.

The annual replacement of the UGC head bunds involved gathering up to 2,000 men, and canal engineers frequently had to resort to compulsion to obtain the necessary labour. 'It is useless to deny that labour employed on the headworks during the unhealthy season of the rains is not in all respects voluntary', admitted one officer, in a memorandum urging the construction of permanent headworks at Hardwar. 'Allow full liberty of action to the labourers, offering even increased . . . daily wages, and how would the head bunds be repaired?'[78] The same problem occurred on the EJC, where it was difficult to keep the number of labourers up to the minimum of around 200, particularly when malarial fever was prevalent and many labourers suffering its symptoms found the cold overnight temperatures at the heads intolerable. Labourers, procured from plains villages 'only with difficulty', often absconded at the first opportunity. Literally half the work force could disappear overnight – as happened in early November 1869, causing the Canal Department to threaten to withhold water supplies to those Saharanpur *zamindars* who, as part of their contract agreement with the department, had undertaken to supply a certain number of labourers for the headworks and had failed to do so.[79] The problem was eased on the Ganges Canal by the introduction of a system of leasing out Canal Department land in the vicinity of the headworks on condition that for every five *bighas* the tenant would provide one man for labour on the works where required, and for every 25 *bighas* a cart and four bullocks would be made available on request.

---

[78] Quoted in NWP PWD(I), April 1867, 11, Col. W. Morton, Labour Recruitment for Ganges Canal Headworks.

[79] NWP PWD(I), Dec. 1869, 68–9.

This system provided a reserve of around 300 men and 65 carts for the difficult September–October period.[80]

Engineers tended to be guarded on the subject of compulsory labour, since the use of forced labour was only sanctioned by formal legislation (the 1845 and 1873 Canal Acts) for use in coping with emergencies, such as where breaches in canal banks threatened to cause crop damage. The practice of compelling labour contributions, in fact, arose out of the period when *rajbahas* were constructed on behalf of landowners, who were granted *takavi* loans towards the cost. Since local labour was cheaper than imported professional excavators, it was in the interests of the landholders to arrange for labourers to attend the works, and the system was extended to silt clearances, which were also at that stage paid for by the *rajbaha*-owners.[81] The system began to break down, however, as prosperity increased in the canal villages, and it became more and more difficult for officers to arrange for labour to attend the various works. The low-caste Chamars from villages on the EJC, C.S. Moncrieff observed in 1866, 'will no longer go at the bidding of the zamindar to clear silt rajbahas, much as the latter would wish it'.[82] In response to this development, Moncrieff had insisted that villages taking irrigation contracts should agree to supply working parties (on the basis of two labourers per water-course) at market rates when called upon by the canal officer. 'This may seem hard', he conceded, 'but I do not know how else to carry on the work which is necessary to keep the canal open at all.'

At the root of the difficulty, according to local estate-owner A.P. Webb, was the fact that the irrigation rates of pay for such work were somewhat below what might be described as the 'market rate'. Webb's 1868 complaint about 'labour impressment' for canal works, which in the event foreshadowed his subsequent pamphlet attack on the abuse of power by the canal bureaucracy, was probably prompted by his losing exemption from *muddud* (the obligation to supply labourers) following a disagreement with a canal official, and thus subsequently

---

[80] NWP PWD(I), April 1867, 12, Note by Capt. J. G. Forbes (EE, Ganges Canal).
[81] NWP PWD(I), Sept. 1868, 10, Note by CE.
[82] NWP PWD(I), Feb. 1868, Appendix, 'Papers on the Revenue Returns of the Canals of the NWPs, 1855–6', Report by EE on EJC, pp. 35–7.

having to supply eight *beldars* to work on the Kirthal drain or risk having his water supply withdrawn. According to Webb, in order to overcome the uncompetitive rates paid for canal work,

canal officials used *once* to impress labourers, and this brought them into collision with the district authorities. They now make the Zemindars and Lumberdars of the villages impress labour for them. [The latter,] to save their crops from injury, accompany canal chuprassees to the huts of the chamars or labourers, threaten them with expulsion from the village, and eventually use violence to them and drive them to the canal works.[83]

Another constant irritant in the relations between the Irrigation Department and the local community involved the way in which claims for compensation and remission of water rates were handled. Claims were assessed by the department itself, and involved personal inspection by assistant or executive engineers. The Canal Department clearly had an interest in minimising the revenue loss from such claims, and a good many of the field inspections for rate remissions claims could be dismissed fairly easily on the grounds that supply deficiency was not the sole factor contributing to a particular crop failure. In general, as A.P. Webb pointed out, remissions were not granted 'unless crops are destroyed', and even then only to the level of the lowest crop rate. There was certainly no remission according to the degree of damage, on the practical grounds, as one officer put it, that the 'notion was absurd that engineers should be expected to discuss the outturn of every field ... if the yield does not come up to the owner's expectations'.[84] The Irrigation Department's 'tendency to minimise the loss arising from deficient, irregular or untimely supply of water' was, in fact, well known to revenue officials.[85] The Cawnpore Collector had in 1863 taken up personally three seemingly sound complaints of crop failure due to inadequate water supply (and a further case of a charge having been levied on a wholly rain-fed crop) which had been dismissed by the

---

[83] NWP PWD(I), Sept. 1868, 31, 'Complaint by Mr A.P. Webb, Zemindar of Sowntee, Meerut District, against impressment of labour for canal works'.

[84] NWP PWD(I), Oct. 1869, 157, Lt. J.G. Hall, Inspection Report, Kulda Rajbaha, Dasna. The statement was made in response to a complaint made by Michel, a *zamindar* from Dasna (discussed below).

[85] UP Rev., April 1908, 45, H.R.C. Hailey (BOR) to UP Govt., 27 Nov. 1907.

Assistant Engineer. More than two years later he was to report to the Board of Revenue that the claim was still unsettled, despite his representations to the Executive Engineer.[86]

Claims for compensation usually involved crop damage resulting from breaches of the distributary banks or from the use of escapes. The use of escapes to reduce the supply in the canal when a sudden fall of rain caused demand to cease, or when a breach occurred, involved a particularly unfortunate conflict in that the cultivators affected were frequently those who did not benefit from the use of canal irrigation themselves. As the Executive Engineer for Etawah division pointed out, the beds of most rivers and streams which could be used as escapes had been for revenue settlement purposes 'declared to belong to engaging proprietors'. Many were cultivated right across during the *rabi* season and frequently they were dammed and stocked with fish, 'rendering the working of escapes . . . a matter of serious responsibility'.[87] The choice was often between allowing a breach to damage high ground or opening the escapes and damaging crops along the rivers – as happened, to take one instance, in 1869, when the Moondakhera escape at Aligarh was allowed to disgorge into the Kali Nadi for a period of ten days, swamping *rabi* crops worth Rs 50–100,000 growing alongside the dry-season channel.[88]

Naturally the canal authorities sought to minimise compensation payments, and in this they were aided by the understandable reluctance of claimants to become involved in costly and drawn-out legal proceedings. A notable exception, however, was Mr Michel, of Dasna in Meerut, who owned what was reputedly the largest indigo factory in India and whose suit for damages is particularly illuminating. Michel's suit was for the recovery of Rs 22,736 as compensation for flood damage to standing indigo crops in July 1868 caused by the negligence of canal officials. He alleged that defective drainage

---

[86] NWP PWD(I), Aug. 1869, 101, letter dated 6 May 1865.

[87] NWP PWD(I), Nov. 1867, 15.

[88] NWP PWD(I), Aug. 1869, 142–7. It should be noted that water entering the Ganges Canal at Hardwar took roughly ten days to make its way to Cawnpore through channels steadily diminishing in capacity. Escape installations were essential for dealing with the inevitable variations in demand which could take place during that period.

arrangements and the overflow of *rajbaha* banks caused a loss in outturn on 1,500 *bighas* of 38,000 *maunds* of indigo, worth Rs 22,736 (net of production costs) as manufactured cake.[89] The suit had been brought as a result of Michel's dissatisfaction with the normal procedures for compensation claims. The department's view was that natural drainage defects were the responsibility of individual landowners and that its liability for the damage extended to only 76 *bighas*, and on that area was restricted to the value of the crop lost.

Although the Lieutenant-Governor made known his view that the department should have come to an arrangement with Michel prior to this stage, canal officers were less inclined to seek a compromise. 'Had Mr Michel not been a European landholder', a senior canal official noted in a letter to the NWP Government, 'it is doubtful if his claims would have been admitted by executive and superintending engineers as far as they have been.' He suggested that Mr Michel 'be left to try his remedy in the civil court'[90] and was backed fully by the Chief Engineer. 'If we give way to an unsound claim', the latter pointed out, 'they will arise all over the country.'[91] Nonetheless, though they accepted that the government's case was a 'very good one', the irrigation and revenue officials who met in September 1868 to discuss how best to proceed with the case were agreed that decisions of courts of justice were inclined to be 'erratic', and that under the circumstances it was best to seek a compromise.[92] It was decided that the government should offer to execute the drainage works at its own expense, and that a payment of Rs 1,000 should be allowed to Michel for 'anxiety and mental wrong'. 'That admits nothing', one of those present remarked, with an eye to future liability claims. Michel subsequently withdrew the suit.

As was explicitly acknowledged, however, the innumerable compensation claims of the less influential received more cursory treatment. Take a typical claim, such as that for

[89] NWP PWD(I), Dec. 1869, 16–38, 'Mr Michel's suit against Government'.
[90] NWP PWD(I), Dec. 1869, 26, Maj. H. A. Brownlow (SE) to NWP Govt., 6 Aug. 1869.
[91] NWP PWD(I), Dec. 1869, 28, Note by Col. W. Greathed.
[92] NWP PWD(I), Dec. 1869, 36, Maj. H.A. Brownlow (SE) to NWP Govt., 13 Sept. 1869.

damage to fields temporarily taken up for purposes of repair to *rajbaha* banks. C. A. Daniell, Collector of Cawnpore, in a letter to the Allahabad Commissioner described the damage he had observed locally, pointing out how contractors found it convenient to take the manured soil to a depth of up to three feet over an area thirty to forty yards square, effectively returning to the cultivator 'a saturated pit'.[93] Since it was difficult for a canal officer to inspect and settle all such compensation cases, the matter, Daniell wrote, 'falls into the hands of some native subordinate, who knows that if he gets the job over quickly, and silences complaints, his conduct is not likely to be enquired into. A little bullying, a strong assertion that it is no good holding out for higher terms and a compulsory razinama taken from the man whose cultivation is destroyed, settles the matter.' The 'trifling sum' received in compensation for such damage was undoubtedly the typical experience.

Concern over the pervasive and unaccountable exercise of powers by the canal bureaucracy lay behind attempts in the 1860s and 70s to regulate the activities of the canal service, and indeed to transfer some of the functions of the irrigation branch to other authorities. The main pressure for this change emanated from the Revenue Department, whose officers in the Doab districts daily came into contact with the operational and administrative defects of a service which seemed to impinge on almost all aspects of their local duties. The post-Mutiny settlements in the canal districts kept in the official gaze a catalogue of problems arising out of the technical and administrative shortcomings of canal irrigation. These have been discussed in earlier chapters, and notably included the conflict between water distribution policies which maximised immediate canal revenue as against those which would have increased long-term land revenue. There was as much concern over the effects of the canal on standards of husbandry, and over the lack of agricultural knowledge among canal engineers. In addition, petitions and complaints received by district officers on their rounds kept them in mind of the numerous petty annoyances associated with a bureaucracy operating a technically crude instrument of irrigation and supervised in its

[93] NWP PWD(I), Sept. 1876, 25.

official and unofficial activities by a tiny corps of overstretched European officers. Moreover, even these officers could be found perpetrating 'arbitrary acts', such as the stoppage of supply to whole villages for transgressions committed by one or two of its inhabitants (as, for example, in cases of forced labour recruitment, which revenue officers considered to be 'against the idea of free markets and the liberty of the subject').[94]

This was the background to an experiment, conducted in the face of 'vigorous resistance' by the canal authorities, in handing over the irrigation measurement duties to the revenue authorities in the early 1860s. Under the existing system, the handling of the revenue side of canal irrigation was divided between the Revenue and Irrigation Departments, with the irrigation officials measuring and recording the irrigation and compiling a list of payments due from individual irrigators, and the actual collection being undertaken through the revenue authorities. It was, of course, unsatisfying to district officers to have to enforce final demand statements which, they knew from experience, sometimes included dues on fields which had been inadequately irrigated or not irrigated from the canal at all.[95] Revenue officials thus welcomed the scheme, which promised also to reduce establishment costs and introduce a check on corruption and inaccuracies. In practice the new scheme was a failure and was dropped after two years' trial in Meerut and Etawah. Far from reducing costs, in a heavily irrigated district like Meerut, it severely disrupted the work of the *tehsil* revenue establishment; it was, moreover, particularly disruptive to the Canal Department, since the requirement that the *ziladar*, *naib-ziladar*, or suboverseer had to accompany each measuring party effectively meant stopping 'the entire current work of the canal for about three months every year, as each sub-overseer's whole time would be employed with the measuring parties'.[96]

---

[94] NWP PWD(I), Feb. 1869, 101, proposed North India Canal and Drainage Bill, marginal note to para. 82.

[95] *PRRC 1864–5*, pp. 23–4. The intended use of the village maps (where they existed) instead of the measuring chain to speed up the operation was hampered by the frequent non-attendance of the *patwaris*. On the problems caused for *tehsil* revenue operations by the extra burden, see NWP Rev., May 1866, 23.

[96] They were also annoyed by such instances of bureaucratic insensitivity as that complained of by the Cawnpore Collector in 1865, where the water rate demands

It was, in fact, the judicial arrangements which particularly upset the district officers, since these were such as frequently to deny the cultivator effective right of redress for injury or abuse of his rights. Critics pointed out that the system whereby offences against canal regulations were handled internally – by executive or assistant engineers, or by deputy magistrates – was unsatisfactory on several counts. They argued that the officers assigned to deal with such cases, particularly the young assistant engineers, often possessed insufficient understanding of and sensitivity towards local customs, tenures, and rights. More importantly, it was asserted that the officers concerned often lacked the necessary legal knowledge. This was clearly evident in the case involving the villagers of Byree, described by the Cawnpore Collector in a letter to the Board of Revenue. Some of the villagers had been reported by the suboverseer for stealing water during *tatil*. All were brought before the Deputy Magistrate by the local *chaukidar*, without formal summonses, and eventually fined or imprisoned for fifteen days in default. No evidence for the prosecution had been examined, and, in fact, from their statements the men appear merely to have used water which had escaped from a breach in the distributary. 'That is', the Collector summed up, 'without evidence for the prosecution, and brought into court perfectly illegally, they were convicted of what is not an offence under the act.'[97]

Most of the criticism, however, was directed at what was described as the 'abnormal and inexpedient arrangement (and one peculiar to the canal department) to have the Canal Act offence reported, prosecution conducted and judgement given by canal employees'.[98] The Board of Revenue complained strongly to the NWP Government in June 1869 that the existing system whereby 'the judge, and plaintiff or defendant are virtually one and the same thing' violated the principle embodied in the maxim *nemo debet esse judex in propria causa*.[99] Moreover, to add to district officers' dissatisfaction with the

distributed in one area had been drawn up not in Hindi but in Hindustani characters, which 'not one in a hundred can read'. NWP PWD(I), Nov. 1869, 101.

[97] NWP PWD(I), Aug. 1869, 101, letter dated 6 May 1865.

[98] NWP PWD(I), Oct. 1876, 3, Report by Civil Administration Committee on Irrigation Administration. Moreover, the fines imposed by such officials were credited to the Irrigation Department as 'miscellaneous receipts'.

[99] NWP PWD(I), Sept. 1869, 95.

state of affairs, an appeal against a canal deputy magistrate's decision in a criminal case was adjudicated by his departmental superior and not by the magistrate of the district to whom this official, *quoad* his criminal jurisdiction, was subordinate.

Considerable pressure built up during the 1860s in favour of reforming the system. To the 1867 recommendation of the NWP Judicature that an independent tribunal be established to handle appeals from the decisions of canal magistrates was added a more comprehensive proposal from the Board of Revenue in 1869. The latter suggested that canal officers should not be 'entrusted with the decision of disputes among consumers of water; the cognizance of complaints regarding a deficiency or total deprivation of the supply of water; or regarding incorrect measurements and charges made for water taken from sources other than the canal'. It was argued that these matters should be handled by the *tehsildar*, with appeal to the Collector.[100]

Irrigation officials acknowledged the need for some change. 'The development of irrigation has outrun its administration, which needs to be conformed to the requirements of the time', admitted the Chief Engineer, when responding to the board's proposals.[101] But the department steadfastly resisted attempts to reduce the extent of its administrative responsibility. It based its case against the board's proposals to deprive it of its judicial role with respect to canal offences, and its role in relation to complaints and claims for compensation, on the practical advantages of the existing system. It insisted that 'on-the-spot' justice administered by canal officials, involving fines of less than Rs 50 (and averaging in fact only Rs 3–4), allowed cases to be dealt with promptly, without tying up canal officials as witnesses, and was actually preferred by the people 'to the more tedious process of the law'.[102] The department also

[100] NWP PWD(I), Sept. 1869, 95, BOR to NWP Govt., 14 June 1869.
[101] NWP PWD(I), Sept. 1869, 107, Note by Col. W. Greathed.
[102] See arguments in favour of the existing system by successive Chief Engineers, Greathed and Brownlow, in NWP PWD(I), Sept. 1869, 107 and NWP PWD(I), Oct. 1876, 6, respectively. The Allahabad Commissioner supported the Canal Department argument on this point: 'Nine men out of ten prefer to pay a small fine and not suffer detention to being kept from their work until witnesses can be summoned and their cases decided at regular trial.' NWP PWD(I), Oct. 1876, 4.

pointed out that there was often little to choose between a *tehsildar* and a deputy magistrate in handling such cases.

In the event, the Canal Department retained its judicial role and its right to adjudicate claims for rate remissions and compensation. The only significant adjustment in functions was that canal officers were made subordinate in their magisterial role to the magistrate in the district in which the case was tried,[103] rather than to the Sessions Judge, as had been the case previously. The change had been vigorously opposed by canal officers, one of whom argued against the proposal in a manner which illustrates the distrust which existed between the two departments:

I think that the appeal would very often be taken up by a District Officer with an unconscious bias against the Canal Officer. The latter belongs to a different and independent department, of which the Magistrate (being also Collector) has never heard much good. I think I am only stating a fact when I assert that most Collectors have an undefined feeling that every Canal Officer abuses his power.[104]

That the 1860s proposals for the transfer of functions to the Revenue Department were unsuccessful is partly explained by lack of interest in radical change on the part of the Lieutenant-Governor, Sir William Muir. The latter's role in resisting significant changes in relation to pricing have already been discussed. The 'Report by the Civil Administration Committee on Irrigation Administration' shows clearly that the extent of change the Lieutenant-Governor was prepared to countenance was restricted to the requirement that engineers should pass examinations in criminal procedure and the Canal Act (and its circulars) before being invested with magistrate's (second-class) powers.[105] Administration of irrigation was tidied up, regularised, rather than being significantly altered.[106]

There was another argument in favour of what was described as 'professional' (i.e. engineers') management of irrigation as

---

[103] NWP PWD(I), Oct. 1876, 3, Report by Civil Administration Committee on Irrigation Administration.

[104] NWP PWD(I), June 1868, 63A, Note by SE 1st Circle, 1 Nov. 1867.

[105] NWP PWD(I), Oct. 1876, 3, Report by Civil Administration Committee on Irrigation Administration.

[106] See in particular the rules clarifying administrative procedures issued on 22 June 1869. NWP PWD(I), Aug. 1869, 28.

against the 'non-professional' control of revenue officials. This argument was first articulated by the Inspector-General of Irrigation in 1869, and served as a standard objection to change whenever administrative arrangements concerning the canals came under official scrutiny.[107] This view did not dispute that 'non-professional' management could offer certain advantages arising out of knowledge possessed by district officers in matters relating to village customs, agricultural practices, land rights, and so forth. Such knowledge was conceded to be of value in the management of irrigation at the ground level. In favour of 'professional' management, however, was the need for the technical skills of the engineer, not only in the operation and maintenance of the system but in all manner of decisions on drainage, water distribution, extensions, and the arbitration of disputes. In particular, the efficiency and long-term development of the system overall was more likely to be brought about through 'professional' management. The improvement of the system, by this argument, relied upon the engineer's close association with its operation and his involvement with the whole process of irrigation.[108] Deprived of the control of his irrigation, one official made clear, the interest of the engineer would vanish, and those forced to devote their time to 'silt and jungle clearances and to repairs of bridges and culverts [would] merit a place among Engineers no higher than the much despised whitewasher of barracks'.[109]

As things transpired, the administrative adjustments made in the 1860s and 70s were oriented more towards increasing consultation between the Revenue and Irrigation Departments – notably in planning new canals and extensions – than towards significantly altering the balance of administrative responsibility and control. Subsequent experience would appear to support the Canal Department's view that the long-term development of the canal system was best assured

---

[107] *PP* 1870, LIII, *Report on Irrigation Works*, pp. 55–6, R. Strachey (Inspector-General of Irrigation) to Sec. GI PWD, letter 57, 8 May 1869.

[108] It would be unfortunate, Strachey argued, to exclude from management those persons best able to improve the canals' usefulness simply because 'they are not likely to have sufficient respect for the ignorance of the past'. *Ibid.*, p. 56.

[109] See *PP* 1881, LXXI, *Further Report of the Indian Famine Commission*, pt III, Appendix C, Maj. C.S. Moncrieff, 'Note on the Administration of Canal Irrigation'.

through 'professional' management. The Irrigation Commission pointed out in 1903 that the 'dual control' system in operation in Madras and on the older systems in Bombay was in many respects unsatisfactory, and concluded that it was doubtful 'whether any of the improvements in distribution which have been made on the Northern India canals, would have been proposed if the canal officers had been responsible only for the maintenance of the works. Under a dual control it would have been no-one's business to initiate them.'[110]

Canal irrigation in local society

This chapter has shown how a vital agricultural input was, at the outset, delivered into the hands of the individuals and groups of men prepared and able to put resources into spreading the canal water through the countryside. It was used to accumulate wealth, and as an important element in the competition for local influence and power. Gradually, as the state fixed upon clearer aims with regard to irrigation development, it was necessary to ease out local *zamindars* and cultivating groups from their roles as financiers, builders, and controllers of local distributary works. The chapter has dealt with the complex interaction between a bureaucracy, the size, power, and influence of which expanded significantly as the government extended its control over irrigation, and the local community. Finally, it has shown how a subsequent phase of irrigation control through greater technical sophistication tended increasingly to confine the struggle for influence relating to canal irrigation to within the boundaries of the village. The power of the subordinate establishment, though it by no means disappeared, was significantly diminished during this phase.

Within the village, of course, the canal water represented a vital input which was to weave itself inevitably into the complex patterns of village relationships. Because in most villages the water was often relatively scarce as a result of policy decisions, and because it was relatively valuable (underpriced) as well, it was inevitable that – with the government reluctant

[110] *IIC*, p. 103.

to intrude in village level distribution – allocation would have to take place through institutional means, and that such structures would favour the influential villagers.

The dramatic effect the canal had in most areas obviously· produced new opportunities for families to rise in the economic and social order – opportunities additional to those arising out of the cycles of demography and fortune which affected all rural households. But because it was ultimately allocated by institutional means, established social and political channels tended to exert a major influence over the way it was used and who reaped the benefits. If the dominant cultivators chose to use these supplies primarily for their own sugar and wheat fields – and undoubtedly it was the more substantial cultivators who had the capital and land resources to obtain the highest returns from canal irrigation – then they derived important advantages with respect to such matters as *gul*-siting and the ordering of watering 'turns'. If the dominant villagers – particularly the large landholders unable or unprepared to mobilise the resources necessary to exploit the canal fully through direct production – preferred to add canal water to the list of factors (such as land rents and crop loans) from which they derived their 'intermediary' income and influence, then they usually possessed the control (such as contacts with the canal bureaucracy) to reinforce this. Either of these groups would be likely to respond to new opportunities according to their perceived interests, and the precise effect of the canal would vary accordingly.

Institutional control over the allocation of canal water thus meant that the input was 'absorbed' into the local economic, social, and political environment. This would appear to have important implications in two areas. The first of these relates to the role of canals during famine, and the nature of the protection they provided; the other involves the whole question of differential response as reflected in patterns and rates of growth. The final two chapters deal with these issues in turn.

7

Protection against famine

One important purpose for which the canals were constructed was famine prevention – a purpose which not only provided the humanitarian or 'higher ground for advocating the works', but also aimed more pragmatically at avoiding the 'sacrifice of money' which famines entailed, as well as increasing taxable capacity.[1] Priority was given to those works calculated to prevent or mitigate the severity of drought, subject to the condition that the works should be remunerative: 'No project', an 1866 PWD circular ran, 'should be selected which is not likely to satisfy both these conditions.'[2] Moreover, as was shown in the previous chapter, all new projects and modifications to existing works planned from around that time incorporated specific design features – such as capacities based on the copious *kharif* water supplies and extensive rather than intensive coverage – in order that drought protection should be enhanced.

How effectively did canals counteract famine? The precise role of canals in drought seasons was never subjected to close scrutiny;[3] such analysis was not deemed necessary since there

[1] Cautley, *Ganges Canal Works*, I, p. 18; Court of Directors Despatch, 1 Sept. 1841, quoted in *ibid.*, p. 25. There had been a severe drought in the Upper Doab in 1824–5, but the famine experience in the official mind when the Ganges Canal was sanctioned was that of 1837–8, when 800,000 lives were estimated to have been lost and the cost to the government in revenue remissions and relief expenditures was later put by C.E.R. Girdlestone at around £2 million. According to the *Meerut DG* (1878), p. 37, 'more than twenty years elapsed before the revenue regained its former standard'.

[2] *PP* 1870, LIII, *Report on Irrigation Works*, p. 3, GI to PWDs, Circular 65, 27 April 1866.

[3] The impressionistic accounts of C.A. Daniell (Bulandshahr Collr.) and F. Williams (Meerut Commr.) are as informative as anything on the way canals functioned during drought. NWP Rev., Dec. 1868, 24, 44, and 46, Memoranda to NWP Govt., Oct. and Nov. 1868.

was no shortage of testimony to the canals' success in coping with drought. Even in the testing 1896–7 famine, according to the Agra Commissioner, it was only 'where canal irrigation was absent . . . [that] the pressure of famine made itself severely felt, and it was there that relief had to be extensively given'; the canals had been the 'salvation of the Division'.[4] Other observers refer to the 'complete immunity from famine' of canal tracts[5] and of the canal's value in such years being 'all but incalculable'.[6]

Part of Elizabeth Whitcombe's critique of the canals, however, consists of doubts about their value in such seasons. She points out that the very crops the 'overwhelming majority of the population' relied upon for consumption – the *kharif* millets and coarse *rabi* mixed crops – were those which received little protection from the canal during drought; the water was, as in other seasons, devoted to cash crops instead. Indeed, she argues that during drought the usual cropping pattern imbalance in canal tracts in favour of cash crops became even more pronounced:

Generally speaking, canal irrigation did, and could do, little to decrease the ravages of scarcity by expanding the sources of staple food supply; indeed its effect seemed to be the reverse, to contract them – a process which tended to worsen with the stimulus of the export trade in grains, particularly wheat, beginning in the late 1870s.[7]

'When the rainfall failed entirely or almost entirely', Dr Whitcombe sums up, 'canals . . . could not take its place.'[8]

In order to understand the role of the canal during famine, and to resolve the discrepancy in opinion as to the effectiveness of canal irrigation in such periods, it is necessary to be more

[4] *RAR 1896–7*, p. 19.

[5] McGeorge, *Works and Ways in India*, p. 17.

[6] *RAR 1896–7*, p. 19. 'No famine [relief] operations have been undertaken here since the canals were built', the Bulandshahr Collector commented in January 1908 (adding that none would be required in that year provided that drought was the only calamity to face). Government of the United Provinces of Agra and Oudh, Scarcity Department Proceedings (hereafter UP Scarcity), Jan. 1908, (proceeding no.) 99. (In the IOR.)

[7] Whitcombe, *Agrarian Conditions*, p. 75.

[8] *Ibid.*, p. 74. Similar reservations are to be found in Thornton, *Indian Public Works*, pp. 116–25.

precise about the phenomenon of famine itself, its processes and effects. Each famine – or bout of 'scarcity', the less extreme condition – was unique, and varied in its geographic extent, antecedent conditions, and climatic characteristics; while changing external influences (official action, trade, and transport developments) themselves added distinctiveness to each visitation.[9] Whilst unable to do justice to the complexity of the phenomenon and the range of its forms, this chapter will outline the important processes impinging upon the peasant production system in times of drought and will assess how the outcome was modified by the presence of the canal.

The scarcity/famine phenomenon was made up of an intricate pattern of chain reactions, reflecting the fact that abnormal circumstances meant different things to different groups, with each group reacting in accordance with its own perceptions of the new set of gains, losses, and risks confronting it. Its seeds lay in the disruption of normal agricultural operations; its effects, however, proved out of all proportion to the actual food situation as caution and speculation together contrived to interrupt the normal flows of foodgrains. Famine amounted, at root, to a disruption in production and thus in the pattern of relationships and terms of exchange surrounding that activity which in normal years was sufficient to provide a role for component social groups at varying, but for the most part adequate, levels of consumption.

For agricultural labourers, as for many artisans, the drastic decline in demand for their services in the face of escalating food prices made it difficult for them to earn, in Professor A.K. Sen's words, sufficient 'exchange entitlements' to subsist;[10] if neither charity nor migration could effectively relieve this crisis starvation ensued. The labourer–cultivator confronted a composite problem of diminished 'exchange entitlements' for his labour and a production problem exacerbated by scarce credit and seedgrain. Most cultivators had this fundamental problem whereby their connections with the market meant, in a

---

[9] This is made clear in the descriptions of successive famines in Bhatia, *Famines in India*.

[10] A.K. Sen, 'Starvation and Exchange Entitlements: A General Approach and its Application to the Great Bengal Famine', *Cambridge Journal of Economics*, I:1 (1977), pp. 33–59.

year of partial crop failure, the possibility that their combined resources were insufficient to cover current consumption and overheads as well as enhanced operating costs and costly credit. Substantial farmers had the less vexing task of choosing precisely how they would gain from such a year: whether through realising the cash value of their surpluses or seeking political, social, and economic advantages within the village or locality. Quite simply, the severity of the effects of any famine experience was related to the proportion of the population exposed, in the sense that their realisable stored wealth plus current income (allowing for migration and other contingency options) were inadequate to cover basic needs for the duration of the crisis. For labourers, such a shortfall was felt in terms of debilitation and loss of life; for cultivators this usually was felt in terms of compromised independence of cultivation and loss of control over their surpluses in future seasons. The larger the proportion of the population able to insulate itself from the disturbance of normal-season relationships, and from the adverse movements in the terms of trade between what they as families had to buy and sell, the less effect the famine had.

The fundamental question this chapter will address, therefore, involves the extent to which the canal operated to insulate the community it served from the damaging sequences set in train by periodically severe climatic conditions. The first part of the chapter will examine the proposition that in the apparently prime area of protecting food staple crops the canal's record is dismal. It will be shown, in fact, that the evidence does not altogether confirm this view. Notwithstanding this, however, the evidence is clear that neither foodgrain production nor food availability is the key determinant of famine, as studies of the 1943 Bengal famine plainly show.[11] The ability of the various social and economic groups to insulate themselves from the complex of forces making up the famine phenomenon revolves around both production *flows* – not merely foodgrain flows, but income-earning ones too – and *stocks*. It will then be shown how damaging was the loss of

[11] The food supply in 1943 was around 5% lower than the average of the preceding five years and an estimated 13% higher than in 1941, when there was no famine. *Ibid.*, p. 39.

cash-earning production flows, and the ways in which canal cultivators were able to counter the mechanisms of famine which interrupted these flows. The next section will examine the role of stocks, the types of checks upon their rundown, and the consequent exposure of cultivators to inflated famine-year market prices. While the section on production flows will be concerned with production, and thus with local growing conditions, the section on stocks will incorporate some of the less localised processes increasingly coming into play – particularly the influence of wider markets upon prices. The insulating role of the canal with respect to these local stocks will also be considered.

Flows and stocks are, of course, linked through the undifferentiated nature (i.e. among consumption, investment, and reserves) of a large part of peasant output. The interaction of the two, in fact, figures prominently in the process by which canal irrigation was able in some degree to insulate a large part of the rural population from the severe disruption of normal-season relationships caused by drought or regional famine. As the chapter will proceed to demonstrate, nowhere is this integration better demonstrated than in the case of the farm labourers, the group traditionally most affected by scarcity: the production flow (its level and character) affected employment; and the stocks of food affected the real wages, since payments in kind could insulate the labourer from deficient 'exchange entitlements' resulting from his having to deal with the market.

Two important additional questions also need to be considered. First, to what extent did the concentration of irrigation investment in such localities as the Doab prove to be of broad protective benefit to the wider region? Secondly, to what extent did natural forces cause the comparative security of canal villages to be erased over time through migration and the operation of demographic forces?

Continuity of production flows

Irrigation and foodgrains

It was shown in Chapter 4 how empirical evidence does not

support the notion that the canal caused staple food production to lag behind population growth. It is very clear, however, that the canals were limited in their ability to maintain the normal pattern of food production in the hot dry spells marking a failure of the *kharif* rains. Part A of Table 7.1 indicates the typical pattern. In 1896 the rain ceased prematurely in the third week of August and no rain of significance fell during September and October. While the summer crop acreage under canal irrigation increased by 511,469 over the average of the previous ten years, 221,507 acres of this increase were taken up by sugar, indigo, and cotton. In the case of sugar and indigo, the increase partly reflects larger sown areas in response to market conditions, while the growth in the irrigation of cotton mainly reflects the exceptional dryness of the season. Though the irrigation of maize and millet showed an impressive nine-fold increase, from 27,423 to 235,591 acres, this still represented only 25% of the total *kharif* irrigation. Even when rice is counted (much of which was grown for the market), food crops covered little more than one-third of the *kharif* canal-irrigated area.

The same pattern emerges for the UGC during the 1907–8 famine (Table 7.1, part B), when the monsoon had a local duration of five weeks instead of the normal twelve to fourteen. The seriously deficient *kharif* rainfall was followed by unusually delayed winter rains. Once again, maize and millet showed an increase (from 32,191 to 143,459), but this amounted to only 24% of the *kharif* total, and the rice area, down slightly, amounted to just over 6%.

What can account for this pattern whereby the canal protected *kharif* food staples to only a limited degree? One possible explanation is that the commercial crop bias affected the ability of the canals to cope with the special demands of a drought season. It has been demonstrated how the Doab's agricultural conditions made it far less reliant on the canal for its ordinary food crops than was the case, for example, in large parts of the Punjab, and that, under pressure to be remunerative, the canal authorities made a significant commitment to the irrigation of commercial crops, which could bear higher charges and needed irrigation for the most part irrespective of normal rainfall. Such crops dominated the *kharif* irrigation:

Table 7.1. Kharif crops under canal irrigation in (A) the UP and (B) the Upper Ganges Canal: comparison of typical years with drought years (in acres)

Season	Sugarcane	Rice	Maize	Millet	Indigo	Cotton	Other crops	Kharif total[a]		
Typical years 1886–95 (average)	204,239	136,213	19,372	8,051	215,923	57,563	47,803	484,925	A (the UP)[b]	
Drought year 1896	247,119	200,671	90,900	144,691	325,406	126,707	65,139	953,514		
Typical years 1902–4 (average)	153,324	34,096	23,121	9,070	26,236	78,833	31,331	202,687	B	(Upper Ganges Canal)
Drought year 1907	153,168	32,993	114,316	29,143	10,828	187,301	47,501	422,082		

[a] Sugarcane, although primarily a kharif crop, was always classified separately, and is therefore not included in the kharif total.
[b] Part A consists of figures for total government canal irrigation in the UP. Since over 90% of such irrigation was in the Doab, the inclusion of the canal area outside the Doab does not affect the overall results, apart from slightly exaggerating the increase in rice. See also Figure 5.
Sources: IRR 1896–7, p. 15 [part A]; IRR 1905–6/1907–8, pp. 68–9 [part B].

sugarcane, indigo, and rice accounted on average for 556,375 acres out of a total of 689,164 during 1886–95.[12]

Understandably, the system adapted itself to the consistent needs of the lucrative commercial crops rather than the occasional convenience of coarse staples, and there is evidence that the thirsty cash crops frequently monopolised the very capacity installed for purposes of watering food crops during a deficient rainy season. Since sugarcane, indigo, and rice were all sown before the rains, the cultivators – and the Canal Department – were committed to them by the time the monsoon was found to be deficient. As the *kharif* season advanced, and as river levels fell and canal supplies tightened, the 'heavy engagements for watering first class crops did not leave much opening for other irrigation'.[13] Although many canal officers felt – as one remarked in 1877 – that it would have been 'more satisfactory' had there been 'a larger proportion of food-grains' in the irrigation undertaken during drought years, the department saw it as incumbent upon itself to give priority to crops which had already taken water.[14] While the growing season of indigo was such that the irrigation of the crop did not unduly interfere with the watering of food crops, this could not be said of the sugarcane crop, with its longer and more obtrusive growing period: 'The increase of sugarcane . . . doubtless imperils the safety of the grain crops in a year of famine', admitted the Irrigation Department.[15]

Owing to their various technical limitations, the canals often lacked the elasticity of supply necessary to cope adequately with the demands of a drought year, and were frequently unable to do more than provide water for the crops already established. Sometimes even this was difficult, such as when a prolonged break in the rains coincided with delays in the placement of head bunds caused by high river levels. When this happened in 1880, supplies to the hard-pressed lower divisions were considerably disrupted for the several weeks needed to carry out the construction of bunds.[16] Uneven distribution of

---

[12] *IRR 1896–7*, p. 15. Not surprisingly, it was said that in the area below Meerut the importance of indigo to the 'interests of the canal can hardly be exaggerated' (*IRR 1871–2*, p. 6); and that in Meerut and above, sugarcane was held to 'create the financial success of the canal' (*IRR 1874–5*, p. 15).

[13] *IRR 1877–8*, p. 3A.   [14] *Ibid.*, p. 17A.   [15] *IRR 1874–5*, p. 14.

[16] *IRR 1880–1*, p. 1A. See Ch. 2 above for details of the head bunds system.

supplies when demand for water was high frequently caused problems for specific areas. Complaints were made in 1868–9 that Saharanpur cultivators had increased their share of EJC water at the expense of those further down in Muzaffarnagar and Meerut;[17] while in 1871–2 the Ganges Canal users in the same two districts obtained only inadequate supplies during a dry spell in September and October because the districts some 200 miles further down the canal had had bountiful rain and it was not possible to allow a full supply into the canal without running the danger of swamping the lower divisions where water was not required.[18] Even with a good supply in the main lines, the distribution system was often incapable of conveying sufficient water to meet the needs of crops other than those primarily commercial. In Cawnpore division especially, the badly aligned *rajbahas* were 'inelastic in their powers to meet an abnormal demand'.[19]

As was shown in the previous chapter, technical improvements over time imparted greater flexibility to the system, and this undoubtedly lessened the marked disparities in the availability of supplies. The extremities of the system were no longer so uniformly starved in times of heavy demand. Even so, as was also shown, the fact that the improvements were accompanied by an extension of coverage meant that the thinly stretched water supplies continued to be a cause for concern during drought. In Agra, canal irrigation still tended to diminish in such years due to the water having been taken further up the system;[20] and in Muzaffarnagar during the 1907 drought it was a 'universal complaint' that the supply was inadequate, 'especially towards the end of the distributaries'.[21]

While water supply conditions and valuable crops did conspire to squeeze out the *kharif* staples, it is clear that at the same time the cultivators made the choice to favour their cash crops: 'Until the more valuable crops in any village had received as much water as they required, not a drop was taken even for Indian corn', it was noted of Bulandshahr canal

[17] *IRR 1871–2*, p. 18.
[18] *Ibid.*, p. 17.   [19] *IRR 1872–3*, p. 23A.
[20] Mudie, *Agra SR* (1930), p. 8.
[21] UP Scarcity, Jan. 1908, 95, P. Prasad (Muzaffarnagar Collr.), 19 Oct. 1907.

division in 1868.[22] The Jats in the area, too, 'refused to irrigate poor crops till they had saved their enormous stocks of sugarcane'.[23] In 1907, to take another example, 86% of the Agra Canal's *kharif* irrigation was expended upon one crop, cotton, while fierce winds dried up the unirrigated crops and insufficient time was left for a large area of new irrigation.[24] But it is equally clear that the cultivators were reluctant to water their coarse food crops, even when water was available. As one canal officer put it in 1868:

It is almost incredible to what an extent, even in a year of drought like the present, the cultivators hold back, often till it is too late to derive much benefit, from taking the water which actually flows alongside their fields sown with the coarser autumnal crops. Thus they wait on in expectation of rain which the experience of average years leads them to look for, and so the water runs to waste.[25]

Invariably, then, the cultivators waited until the last possible moment for the rains: 'A cloudy sky was sufficient to stop all irrigation of inferior crops, when the next day, perhaps, they would be fighting fiercely over every drop of water.'[26] The rush for water commenced only when the crops would otherwise be lost, 'too late in many instances to secure a sufficient supply, and in other instances too late to save the crops'.[27]

The authorities were well aware of this seemingly perverse situation and looked eagerly for signs that the irrigation of coarse crops in dry years was on the increase. The Government of India even instructed the UP Government to ensure that they did not grant revenue suspensions where cultivators had been 'short-sighted enough' to abstain from taking canal water for their autumn crops till it was too late.[28] But how can this response be accounted for? It is not hard to understand the preference given to the commercial crops: they were sold for cash, which was used to cover rent, accumulated debts, items of

[22] NWP PWD(I), June 1871, 2, Irrigation Revenue Report 1868–9, IOR pagulation 64. 'Indian corn' = maize.

[23] *Ibid.*, IOR pagulation 63.   [24] UP Scarcity, Jan. 1908, 83.

[25] UPA(R), Files of Meerut Commr., Dept. IV, File 19/1869, Ser. 17, Enclosure to PWD nos. 236–43, Paper by Lt.-Col. W. Greathed, 8 Oct. 1868.

[26] NWP PWD(I), June 1871, 2, Irrigation Revenue Report 1868–9, IOR pagulation 63–4.

[27] NWP Rev., Dec. 1868, 44, memorandum by C.A. Daniell (Bulandshahr Collr.), 7 Oct. 1868.   [28] GI PWD(I), Oct. 1877, 39.

capital, and household goods; they occupied the best land and represented a relatively substantial investment of resources. In contrast, the millet and maize crops occupied poor soils; they involved very little investment since the seed was invariably the cultivators' own, tillage was nominal, and irrigation was not normally required. Such crops were grown mainly for consumption by the family, their hired labour, and the farm animals, and were usually worth little in the market. Moreover, because of their low input requirements, these crops were highly flexible. Indeed they had to be, since the weather was the main influence on this class of crop; if the weather was unsuitable for *juar*, for example, the cultivator could wait for the right conditions for sowing *bajra* or some other tall millet or low pulse. The farmer was, therefore, perfectly rational in his behaviour when he delayed giving water to sow inferior crops in favour of his cash crops: his experience made him aware of the high probability of sufficient rain falling to mature his standing crops or to enable a late crop to be sown.

There were other reasons why the farmer held back from irrigating his poorer crops. The millet fields were invariably out of reach of the water-courses, and hence an additional cost in the form of their construction was entailed – always assuming it was physically possible to bring the water to these fields. Furthermore, such land was often undulating, unlike the flat, compartmentalised fields in regular receipt of canal water; it was therefore difficult to irrigate such land. It must also be remembered that the yield of such crops, even under irrigation, was not likely to be good in such a year, as C.A. Daniell noticed in 1868: 'Taking it as a whole the *kharif* outturn of grains grown under canal irrigation [in Bulandshahr] will be indifferent.'[29] 'Against severe and prolonged drought', the provincial government informed the Famine Commission, 'even the canals are but a poor substitute for rain in the autumnal season, especially if, as happened in 1877, a scorching hot wind continues blowing when crops are on the ground that depend for nourishment as much on the moisture they imbibe from the air as on what is supplied at the roots.'[30] In such conditions the

[29] NWP Rev., Dec. 1868, 24.
[30] *PP* 1881, LXXI, *Further Report of the Indian Famine Commission*, pt III, Appendix V, p. 521.

cultivator was entirely justified in giving canal water 'to save and not merely to improve' his inferior crops.[31]

In respect of coarse crops, therefore, the canal villages seem to have been little better off than the villages without irrigation. In dry tracts, where early-season conditions led the cultivators to sow the usual crops, later drought withered and destroyed them; where the early rains were too light or short-lived to enable the full areas to be sown, subsequent lengthy spells without rain meant that the *kharif* area remained short. 'Of the usual khureef crops grown in dry lands, I do not believe more than one-twelfth have come to anything', remarked the Bulandshahr Collector in 1868.[32] The well areas often fared no better. Even where wells were thick upon the ground they proved of 'little value in affording protection to the kharif crop'.[33] The NWP Government made this point clear in a communication to the Government of India late in the famine year of 1877:

A well that [during the cold season] is able to protect five bighas could not in the scorching wind . . . and under a blazing sun do the same office for one bigha. In fact, both the Commissioners of Rohilkhand and Meerut reported that even constant irrigation from wells had failed to keep the crops alive, and many cultivators abandoned the attempt in despair when their bullocks were worn out by constant work. [Well] irrigation in the kharif . . . is practically inconsiderable in extent, and must always more or less be so, and therefore when it is said that the kharif has failed except on irrigated lands, this is tantamount to saying that the whole, with a very trifling exception, has failed in all but canal districts, and there it is true of the grain and fodder crops.[34]

---

[31] NWP PWD(I), June 1871, 2, Irrigation Revenue Report 1868–9, IOR pagulation 63–4. The reason for the reduced yield is that high temperatures and dry conditions accompanying *kharif* rain failure have the effect of causing the plant stomata to shut off the evapo-transpiration current through the plant. This restricts the access of air into the plant's intercellular spaces, and thus interrupts the process whereby carbon dioxide is absorbed and turned first into starch and then into growth-sustaining sugar. It was believed in the 1880s that the ordinary autumn crops, 'particularly the millets', required so much moisture that, even if enough was supplied to the roots, the grain would be small. In fact, it is more likely that the evapo-transpiration rate was so high that water available to the root zone of the plant was, despite ample canal irrigation, restricted (either because the soil texture was poor or because irrigation was not available frequently enough). See H.O. Buckman and N.C. Brady, *The Nature and Properties of Soils* (New York, 1968), pp. 194–7.

[32] NWP Rev., Dec. 1868, 44.   [33] *IIC*, p. 201.   [34] GI PWD, Dec. 1877, 13.

In practical terms it does appear that the canal villages could claim little if any advantage over dry and well villages in maintaining the area of *kharif* food staples when the rains failed. To conclude from this that the canals were of little use, however, is to overlook the importance of *rabi* crops. The plain fact is that the failure of the *kharif* food staples did not in itself imply famine – at least, not in the western parts of the province.[35] Admittedly it could create problems for certain sections of the populace, but its effects can be easily exaggerated: 'The crops are withering on dry land', observed the Aligarh Collector in 1907, 'but everyone makes light of that if they can get rain for the rabi.'[36] While defence mechanisms could in the main cope with a short-lived interruption of this kind, very great importance was placed in such a year upon the *rabi* crop; a large winter crop was needed, making use of the lands forcibly laid fallow and the resources freed by diminished *kharif* duties. Thus the cultivator with a well, while salvaging a plot or two of produce from the *kharif*, would, 'rather than attempt to save his millets and other dry kharif crops . . . prefer to devote all his energies to watering and preparing his fields for the more valuable rabi crops, hoping for help from the rains'.[37]

The danger, of course, was that the winter rain would be unsuitable in its timing or quantity to assist the farmer in putting down his *rabi* area. The dry cultivator had obvious need to fear a dry rainless autumn, for under such circumstances, even where the baked soil could be stirred, cultivators – and creditors – hesitated to risk their scarce seed. Thus in 1860–1, on top of *kharif* failure in Meerut division, 'a very large proportion of the ground was unsown for the Rubbee' and 'scarcely an acre of unirrigated land was ploughed'.[38] Even the

---

[35] The situation was different in some rice districts. While canals in Rohilkhand could save the rice crop during prolonged breaks in, or early cessation of, the rains, the entire sweep of country from Bahraich through Gonda, Basti, and Gorakhpur – where the *kharif* area was much greater than the *rabi* (and was almost entirely rice) – was vulnerable to a failure of the rice crop since it could not, as W.H. Moreland observed, be adequately compensated by a larger *rabi* 'as the land cannot be tilled'. UP Rev., Sept. 1905, 2, W.H. Moreland, Report on UP Agricultural Defects.

[36] UP Scarcity, Jan. 1908, 80.

[37] *IIC*, p. 201.

[38] UPA(R), Files of Meerut Commr., Dept. xii, File 5/1861, Ser. 125, File on Revenue Administration Report 1860–1, Meerut Collr. to Commr., 1 Oct. 1861.

well villages, however, were often in tremendous difficulties in such circumstances. Many wells, particularly *kachha* wells, simply dried up due to the effect of the drought upon the water-table. Thus in Bulandshahr in 1868–9, very many wells 'failed altogether';[39] while those in Etawah frequently 'did not give their usual supply, and many yielded water for 2 or 3 acres instead of 5 or 6'.[40] In Etah, wells were 'always liable to dry up more or less completely in a year of drought'.[41] *Pakka* wells were more reliable but still had capacity limitations, as in 1878, when 'only a limited area could be watered for the rabi ploughings and sowing'.[42] This was partly due to the slowness of well irrigation and the fact that productive factors were tied up in the process – and in a season like that of 1896, when all the *rabi* area had to be watered prior to sowing, intense pressure was placed upon well villages. The problem was exacerbated, however, by the state of the draught animals. 'Irrigation from wells will be limited', L.C. Porter remarked of Meerut late in 1907. 'The bullocks have been overworked from watering the kharif crops, and have become weak. They are not fed well as fodder is dear.'[43] The situation had been worse in 1860–1, though, when crops 'failed on well lands due to the bullocks dying from overwork and starvation'; then, with the exception of some valley land, the 1860 Meerut division *kharif* crop and cereal crop of 1861 were 'total failures' outside the canal tracts.[44]

The sheer speed with which canal irrigation could be accomplished made the canal invaluable for laying down a good area of *rabi* cereals. Not that conflicts between commercial and food crops were altogether absent even then: on the EJC particularly, *rabi* sowings were frequently delayed and sometimes prevented as the already falling water supply was used to

[39] *Bulandshahr DG* (1876), p. 76.
[40] C.H.T. Crosthwaite, 'Report on the Average Rent-Rates for part of Pergunnah Phuppoond', 20 Sept. 1870, *Revenue Reporter*, v (n.d.), p. 4.
[41] *Etah DG* (1911), p. 32.
[42] *RAR 1878–9*, p. 100.
[43] UP Scarcity, Jan. 1908, 194.
[44] UPA(R), Files of Meerut Commr., Dept. xii, File 5/1861, Ser. 125, File on Revenue Administration Report 1860–1, Meerut Collr. to Commr., 1 Oct. 1861. It was reported that 'fully one-half of the farming cattle' had died. C.E.R. Girdlestone, *Report on Past Famines in the North-Western Provinces* (Allahabad, 1868), p. 87.

bring sugarcane and rice to maturity.[45] But there were factors which compensated for such conflicts and the failing supplies. Climatic conditions were such during the winter season as to ease the pressure upon irrigation. Lower evapo-transpiration rates allowed plants to survive with less water and for longer periods between waterings, and this, along with reduced evaporation losses, allowed the supplies to be stretched further. This is reflected in the fact that the *rabi* irrigated area was consistently in excess of the *kharif* area and exhibited greater elasticity (see Figure 4). At the end of the year, when canal supplies were reaching their lowest point, the cool temperatures, the fairly reliable winter showers, and the backing of some well irrigation sustained much of the crop; and while the *kharif* millet yields responded poorly to irrigation, in the mild *rabi* conditions artificial irrigation permitted full yields to be realised. Thus in 1868–9, when from late July to February hardly any rain fell in the Upper Doab, and when valuable *rabi* seed hurriedly sown in dry areas in the wake of September showers was lost, cultivators in eastern Muzaffarnagar put down with the aid of Ganges Canal water over 90,000 acres in the *rabi* alone (25% more than the normal area irrigated in a full year). Some of it subsequently withered, but the bulk was sustained by the eking out of what canal water was available until the February rains not only saved the cereals but allowed in the end a good yield.[46]

It was in the *rabi*, therefore, that the canal had a major impact on foodgrain production in dry years. Table 7.2 confirms this: only 7% of the area irrigated by canals was devoted to non-food crops; over 1.7 million irrigated acres were devoted to food production. Moreover, this concentration of food crops in the *rabi* was not in any important sense a 'fall-back' situation necessitated by drought-disrupted *kharif*

---

[45] In the spring, too, conflicts could arise between the need to mature the wheat crop and the water needs of the new sugar crop. In 1861 this conflict appears to have led EJC cultivators to sell their cane seed to growers on the Ganges Canal, who had plentiful canal supplies; while on the Ganges Canal, in 1869, cultivators were reported to be favouring their sugar sowings to the detriment of maturing *rabi* crops. *PRRC 1861–2*, p. 41; NWP PWD(I), Oct. 1871, 2, Irrigation Revenue Report 1869–70, IOR pagulation 234.

[46] *Muzaffarnagar DG* (1903), p. 56. See *IRR 1896–7*, pp. 12–13 for a brief description of the role of the canal when the rains ceased in late August 1896.

Table 7.2. *Rabi and total crops under canal irrigation in the UP: comparison of typical years with the 1896–7 drought year (in acres)*

Season	Wheat	Barley	Gram	Peas	Other food-grains	Poppy	Other crops	Total rabi	Rabi + kharif
Typical years 1886–7/1895–6 (average)	577,945	77,852	44,230	68,025	222,333	11,798	34,797	991,029	1,680,893
Drought year 1896–7	1,015,120	118,632	22,074	43,999	459,367	26,534	91,574	1,823,251	3,023,884

Source: IRR 1896–7, p. 15.

staple production, since – as was shown in Chapter 4 – canal villages in normal seasons grew most of their food in the *rabi*. Finally, on this question of the apparent helplessness of canal cultivators to preserve their basic food supplies, it has to be stressed that meeting the food requirements of the Doab did not primarily hinge upon the precarious *kharif* staples. It has already been noted how, over time, the better-quality crops became the staples. Even in Muttra – one of the poorest of Doab districts – it was observed, following the extension of canal irrigation, that 'the better crops are more largely consumed by the bulk of the people'.[47] 'While the population of South Oudh is maintained on barley, coarse millets and pulses and the kind of grain consumed by the masses varies closely with the state of the markets', W.H. Moreland concluded in 1901, 'the Meerut Division . . . is, as a whole, a wheat-eating tract.'[48]

Cash crops

For most cultivators the problem of drought was not so much one of food provision for the family as one of meeting short-term cash commitments and current production needs without jeopardising the future viability and independence of the production unit. An enquiry into the 1877–8 relief operations made it clear that 'long before famine reaches even the agricultural tenant it becomes an impossibility for him to pay a large portion of his rent'.[49] Bijnor villages, for example, had in 1896 sufficient grain in store to pay the labour they required, but were not able to pay their rents.[50] In addition, the cash requirements of a normal season (for example, rent and debt repayments) were in a famine year accompanied by enhanced operating costs – especially with respect to *rabi* seedgrain and credit – which had to be covered at a time of production losses and income reductions.

[47] *Muttra DG* (1884), p. 138. See also Holderness, *Memorandum on Food Production*, p. 14.
[48] MCR, Scarcity Dept., xii, File 2/1900–01, Ser. 127, W.H. Moreland to NWP Govt., 24 June 1901.
[49] W.C. Benett, *Report on Scarcity and Relief Operations in the North-Western Provinces and Oudh during 1877–8 and 1879* (Allahabad, 1880), p. 382.
[50] Government of the North-Western Provinces and Oudh, Scarcity Department Proceedings (hereafter NWP&O Scarcity), Jan. 1897, (proceeding no.) 13. (In the IOR.)

Drought often forced peasants into buying fodder, something which was not necessary in normal seasons. Also, they would usually need to replace well ropes or buy additional hide-buckets; the prices of these items were said to be 'causing an outcry in the well tracts' in 1907–8.[51] The chief item of expense, however, was seedgrain for the vital *rabi* crop. Unlike the coarse *kharif* millet and maize seed, *rabi* seedgrains – and especially those of the better crops – were not widely kept by the cultivators themselves, and generally had to be purchased when sowing time came round. In famine years they were usually only available in exchange for cash. In Muzaffarnagar, many of the cultivators who had managed to keep their cattle alive and had retained their labourers during the scorching 1860 *kharif* found they could 'only with much difficulty procure seed, for the Bunyas, with devilish forethought, refused to advance it'.[52] The 'devilish forethought' was, of course, nothing more than a reflection of the risk involved in making seed loans when grain could be readily converted into cash. Thus in Unao, in October 1907, the credit supply had 'practically wholly stopped', and only the 'most well to do cultivators' – presumably with good collateral – could get seedgrain for other than cash.[53]

While the cost of seed alone for a wheat crop might cost the cultivator Rs 5–6 in cash (and other *rabi* seeds sometimes not markedly less, owing to the trend for price differentials between food crops to narrow sharply during drought),[54] those cultivators able to draw upon credit faced hard terms. While the cultivators commonly obtained credit at *siwai* rates (25% over six months) when borrowing and repaying in kind in normal seasons, the grain deals of a drought year were adjusted to take account of the extraordinary circumstances. Grain loans made in such seasons were calculated according to money rates rather than simply in terms of grain volumes: the grain borrowed in the autumn was measured in terms of prevailing prices, and the amount to be repaid was the equivalent value of

[51] UP Scarcity, Jan. 1908, 107, F.S. Swan (Unao Collr.); UP Scarcity, Jan. 1908, 195, A.B. fforde (Bulandshahr Collr.).

[52] *PRRC 1860–1*, p. 16.   [53] UP Scarcity, Jan. 1908, 107.

[54] W.H. Moreland, *Conditions Determining the Area Sown with Crops in the United Provinces* (Allahabad, 1907), p. 22.

grain reckoned in terms of the much lower post-harvest prices.[55]

Cultivators, of course, had to allow for the periodic loss of much of their *kharif* and particularly their cash-earning crops. Their cash needs in such seasons would be obtained through drawing upon future output (credit) and through running down wealth stocks accumulated during better seasons. Thus the womenfolk of Unao were, in October 1907, beginning to sell their jewellery 'as money is needed for the rabi sowings'.[56] The terms of such arrangements were hardly in the cultivator's favour, but they enabled him to get by. There are limits, however, to the ability of low-productivity peasant agriculture to withstand a series of poor harvests. The resources of the mass of the cultivators are quickly dissipated when difficult years hit production. 'Drought', recorded the 1880 Famine Commission, 'will be more or less fatal, according as it follows a season of bad or good rainfall; and it has commonly happened that severe famines have been caused by a severe drought following one or two years of indifferent rain.'[57] Muttra's *zamindars* and tenants 'cannot be otherwise than poor', reasoned Agra's Commissioner in 1908, 'having had no chance of recovering from the [1905–6] famine . . . and the impaired rabi of last year'.[58]

The critical shortage of cash and credit in a drought year had important long-term implications in that it often necessitated the rundown of capital and other assets. Several years after the 1868–9 famine, the Settlement Officer found that the well villages north of the Kirsani, in Meerut, had 'hardly yet recovered from the effects of that calamity. The sugar plantation received a great check. Loss of cattle and men, and want of capital, obliged cultivators in many villages to curtail their sugar production. So hard-pressed were they in some villages that they were obliged to sell their sugar presses to the canal villages of Chuprowlee in order to raise money.'[59]

[55] A.B. Patterson, *Fatehpur SR* (1878), p. 11. In terms of output, the typical burden of repayment for a *maund* of grain borrowed in October, after a poor *kharif* such as that of 1874 (when the grain sold at twenty *seers* per rupee), almost doubled by repayment time in April–May, due to the fall in price (to thirty *seers* per rupee) and the 25% interest. *RAR 1874–5*, p. 3   [56] UP Scarcity, Jan. 1908, 107.

[57] *PP* 1881, LXXI, *Further Report of the Indian Famine Commission*, pt 1, p. 5.

[58] UP Scarcity, Jan. 1908, 112.

[59] J.S. Porter, Burnawa Pargana Report, 1870, p. 46, in Buck, *Meerut SR* (1874).

Similarly, in Agra the loss of capital, particularly in the form of bullocks, and the subsequent inability of cultivators to generate and retain a surplus over consumption requirements meant that 'even such facilities for irrigation and good cultivation as the district affords cannot be made full use of after famine owing to the crippling effect on cultivators' resources'.[60] Although this curtailment of Agra's *rabi* area in the seasons following drought apparently became less marked with each successive bout as the higher agricultural prices attracted capital, the inflow of credit was not without its costs to the more weakly positioned cultivators: the 'improvement . . . in the financing of agriculture', wrote the Agra Settlement Officer in 1930, 'has been accompanied by the transferrence of a large part of their property to the professional moneylenders'.[61]

Less tangible assets than physical capital and land were also lost at such times. Occupancy rights clearly declined in Agra as a result of the 1877–8 famine.[62] On estates in Muttra belonging to temples and powerful landlords the custom prevailed for the landlords to compel their tenants to file notices of relinquishment in order that they could not lay claim to having occupancy rights in the land they rented. This task was made easier in times of hardship, when rents especially could not be paid.

Significantly, just as it was the canal villages of Chaprauli which had picked up cheaply the sugar presses of the necessitous villagers to the north, so it was the canal village tenantry on the large Muttra estates which was conspicuously able to resist the pressure to file relinquishment notices: the large increase in notices during the hard times of 1897 was chiefly in the parts of the district not protected by canal irrigation. This reflects the vital advantage possessed by canal villages during serious drought: the ability of cultivators using the canal to secure *kharif* cash crops and hence make cash

---

[60] Mudie, *Agra SR* (1930), p. 14. The slow recovery of the wheat area in post-drought years demonstrated in Moreland's study of Cawnpore and Fatehpur reflects the reduced prosperity of producers in the wake of a famine. Moreland, *Conditions Determining the Area Sown with Crops*, pp. 20–5.

[61] R.F. Mudie (?), *Agra Pargana Handbook* (Allahabad, 1930), p. 12. (A copy is available at the Uttar Pradesh Board of Revenue Library, Allahabad.)

[62] *Ibid.*

available to meet outgoings and finance current operations. The importance of a successful commercial crop in a year of famine was demonstrated repeatedly during the period, as in the case of Muttra and Agra in 1913–14: 'One of the principal causes that so little signs of stress were shown . . . in spite of the failure of food crops on unirrigated land, was the comparative excellence of the cotton, which is less affected by drought than almost any other crop, and the high prices realised for the produce.'[63]

While the area and yields of cash crops in the canal tracts enjoyed a fair degree of security from poorly distributed or inadequate rainfall, the incomes from commercial production in the non-canal tracts were strikingly more variable. When buoyant cotton prices promoted an increase in the cotton area of canal districts from 265,000 acres in 1910 to 410,000 the following year, the fact that the rainfall was not suitable to the crop's needs actually meant that there was a province-wide fall overall of 32% over the 1910 area. In the districts adjacent to the Doab, large falls were registered: Moradabad's cotton area fell from 74,250 acres to 30,000, and Budaun's from 80,000 to 31,000.[64] Similarly, in the aftermath of the severe 1907–8 drought, it is noticeable how the Meerut division was able to increase its sugarcane area substantially, while in Rohilkhand, Benares, and Oudh, sugarcane cultivation dropped off in the face of high production costs and need to replenish food stores.[65] Yields of *kharif* commercial crops could be fairly well maintained, too, with the aid of canal water. Even in the especially harsh conditions prevailing in 1907, when hot winds damaged even irrigated crops, sugar and cotton outturns were still put at around eleven *annas*.[66] While rice in Saharanpur's dry tracts during the *kharif* of 1899 was described by the Collector as 'absolutely gone', cotton 'very sick', and sugar, without rain, likely to fall below four *annas*, in the adjoining

---

[63] *S&C 1913–14*, p. 4. It was a common saying among Aligarh's cultivators that the cotton crop paid the rent of the entire holding. Smith, *Aligarh SR* (1882), p. 47. In Muzaffarnagar around 1920 there were at least three villages with a cane crop worth in excess of Rs 2 *lakhs*, and in at least twelve others the crop was worth more than Rs 1 *lakh*. *RAR 1921–2*, p. 30.

[64] *S&C 1911–12*, p. 3.

[65] *RAR 1909–10*, p. 3.   [66] UP Scarcity, Jan. 1908, 96.

canal tracts the thirsty rice fields promised a twelve-to-sixteen-*anna* outturn, and sugar was 'perfectly right'.[67]

With the exception of *munji, kharif* cash crops were generally not in the form of food substitutes and their prices did not automatically move upwards in such a year. Conditions did not, however, rule out favourable price increases, as 1878 proved, when Muzaffarnagar cultivators were getting Rs 6 per *maund* for *gur*, the ordinary rate being Rs 3–4.[68] In 1907–8, too, despite the squeeze on incomes, the reduced supply of *gur* due to the drought raised prices to the extent that 'independent growers of cane . . . recouped themselves in part for the loss due to the season', and in Rohilkhand the unusual phenomenon was observed whereby cultivators broke their contracts to deliver juice to the manufacturers and made *gur* for their own profit.[69]

It was mainly from the sale of part of the spring harvest that canal villagers were able to cash in on famine prices, since these typically remained high due to exports or stock replenishment until the harvesting of the cheaper grains in the following *kharif*. While landlords in most districts declined during the 1896–7 famine to bring arrears suits or to issue notices upon their statutory tenants, this reticence was not to be found in Meerut and Agra divisions, where, due to the effect of the canal, many tenants had 'reaped exceptional profits from their produce'.[70] Only the canal villages were excluded from the dismal picture of Muttra painted by the Collector late in the drought year of 1899 – a picture of depopulation, high prices, the lingering effects of the previous (1896–7) famine, and the unsettling mournful prophecies of the *pandits*. The reason, according to A. B. fforde, was that 'the high prices in [1896–7] enabled the cultivators to redeem their mortgages and make ornaments, and they are free from apprehension as to the coming rabi, so that, in spite of the poor kharif, they are hopeful'.[71]

---

[67] Uttar Pradesh Secretariat Library and Records, Lucknow (hereafter UP Sec.), Scarcity Dept., File 290/1899, Ser. 12, File on Agricultural Prospects for Kharif Crops, G.A. Streatfield (Saharanpur Collr.) to NWP&O Govt., 16 Sept. 1899.

[68] NWP&O Rev., May 1878, 77.

[69] *RAR 1907–8*, p. 7. See also Ch. 8 below.

[70] *RAR 1896–7*, pp. 27, 34.

[71] UP Sec., Scarcity Dept., File 290/1899, Ser. 37, File on Agricultural Prospects for Kharif Crops, A.B. fforde (Muttra Collr.) to NWP&O Govt., 16 Sept. 1899.

Canal villages, therefore, derived much of their strength in times of widespread drought from their ability to minimise the season-to-season income variability. The reliable cash flows meant that money commitments could be discharged without depletion of food resources and other assets, and that production outlays for a full *rabi* area could be covered. The short-term benefits of flexibility and security against insufficient or untimely rainfall were cumulative: not only would the credit-worthiness of a cultivator seeking *rabi* seedgrain be considerably enhanced by the fact of his access to canal water, but the risk of a series of poor seasons wearing away the cultivator's resources and making him vulnerable to serious drought was altogether lower. Moreover, the availability of water, grazing, and fodder in canal villages assisted the farmer in bringing his plough bullocks through such periods. This is plainly evident from the established practice whereby cultivators from outside the canal villages brought their cattle to graze near the irrigation channels.[72] From Bah *tehsil* (Agra), for example, there was a 'systematic emigration' in drought years to canal villages in Mainpuri and Etawah, with 'large numbers of families' crossing the Jumna with their cattle. In Etawah, 'a great many districts became deserted' due to the 'general emigration' of people and cattle from the Jumna–Chambal *doab* to canal tracts during famine.[73]

Maintenance of foodgrain stocks

The holding of foodgrain stocks was an important long-standing practice primarily aimed at combating famine, since food stores permitted the regulation of uneven flows of output both within and between seasons. Stocks could be used to cover shortages of food or cash resulting from production shortfalls, and allow production to continue wherever possible through providing wage goods and seedgrain. In famine years they could operate to permit the cultivator and his dependants to achieve some insulation from the effects of scarcity prices.

A series of bad years could wear away stock levels and make the effects of a serious drought catastrophic. The 1860–1

[72] Benett, *Report on Scarcity and Relief Operations*, p. 36.
[73] UP Scarcity, March 1906, 1, Agra Commr. to UP Govt., 14 Jan. 1906.

famine, as it affected the Meerut division, was clearly more severe than that of 1868–9 largely because of the differences in antecedent conditions and the impact these had upon the state of food stocks: despite the 'utmost care' taken in 1860–1 so that proprietors should not be compelled to alienate their shares to meet the revenue demand, the Meerut Collector regretfully recorded, 'still they were frequently compelled to do so to procure food'.[74] In contrast, the favourable seasons prior to the almost rainless 1868–9 season meant that although the 'common field labourers came gladly to the relief works', even in the dry tracts 'the people had stored up food and there was no destitution'.[75]

So important are stocks to the welfare of rural inhabitants during drought periods that one historian has argued that 'the disappearance of domestic stocks which people were accustomed to keep in the past' was one of the fundamental causes of the apparently increasing incidence of famine in the post-1860 period.[76] The abnormal rise in food prices during drought years, the argument goes, was 'itself a measure of the shortage of foodgrains in the country as a whole', reflecting the fact that the foodgrain export trade had siphoned off the food surplus of normal years and that cultivators, both through choice and through the pressures of planters and those making cash advances, opted for commercial rather than food crops.[77]

Changing external conditions thus, according to B.M. Bhatia, profoundly affected stockholding practices. To what degree this occurred is difficult to gauge. The 'invisible' nature of cultivators' stocks made it hard for officials to do more than indicate broadly the level of existing stocks relative to those in a few previous years. Moreover, it is necessary to apply the notion of a rundown of stocks as a concomitant of increasing commercialisation of agriculture with extreme caution. The peasant, in all his decisions, showed great deference to security, and his marketing responses reflect this strongly, though differing according to his holding size, tenure, and family size,

---

[74] UPA(R), Files of Meerut Commr., Dept. xii, File 5/1861, Ser. 125, File on Revenue Administration Report 1860–1, Meerut Collr. to Commr., 1 Oct. 1861.
[75] UPA(R), Files of Meerut Commr., Dept. xii, File 9/1869, Ser. 39, File on Revenue Administration Report 1868–9, Meerut Commr. to BOR, 4 Dec. 1869.
[76] Bhatia, *Famines in India*, p. 9.   [77] *Ibid.*, pp. 9, 27–34.

and the effects of previous harvests and fortunes. The voluminous literature relating to the relationship between marketed surplus and relative prices reflects mainly this complexity.[78] However, from the scraps of evidence which are available, it is possible to identify some of the main processes influencing the level of stocks held at a time of growing commercial penetration and opportunity, and to show how the cultivators on the canal were affected by these processes.

There were indeed, as the marketed surplus debate would suggest, extremely complex links between the village economy and the grain assembly centres, and there is a need, therefore, to treat sceptically the notion that high grain prices implied empty stocks and foodgrain shortages. The holders of grain in the markets exhibited attitudes different from those of their counterparts in the villages. To the merchant, grain was a simple commodity, to be purchased, stored, or sold – locally or at long distance – according to current and projected price levels; and repeated transfers of sealed *khattis* (grain pits), some of which changed hands a dozen times in one week, was entirely normal.[79] In the village, grain was so much more than its money equivalent: it was a consumption good, but it was also a farm input (seed); it served as currency when used to pay for the services of labour and village artisans, and as a convenient form of insurance against unpredictable weather conditions; it could secure credit on the one hand, and yet enhance the power and prestige of those who lent seedgrain within the village on the other. Its many-sided nature within the village made it much less responsive to price movements than that controlled by the traders in the larger markets.

This distinction is important in the context of high prices and scarcity, and it was in an attempt to bridge this behavioural distinction that the agents of substantial trading concerns moved into the districts to pull more grain into the central markets.[80] Hence prices could be lower in the railway towns

[78] Note, for example, the number of articles on this topic included in the bibliography of T.J. Byres, 'Land Reform, Industrialization and the Marketed Surplus in India: An Essay on the Power of Rural Bias', in D. Lehmann (ed.), *Agrarian Reform and Agrarian Reformism* (London, 1974), pp. 221–61. [79] *RAR 1898–9*, p. 7.

[80] UP Sec., Scarcity Dept., File 290/1899, Ser. 56, File on Agricultural Prospects for Kharif Crops, R. Dewhurst (Farrukhabad Collr.) to NWP&O Govt., 16 Sept. 1899. See also Ch. 8 below on this development.

than in the villages, as in Bareilly in March 1914 (due to imports from the Punjab);[81] or the reverse could be the case, as after the successful *rabi* harvest of 1897, when prices were distinctly cheaper in the villages than in the central marts,[82] and in Farrukhabad, where the Collector observed that with the exception of the crops actually being sown – wheat, barley, and *gram* – the prices of nearly all grains were lower in the villages than in the city.[83] It would be wrong, therefore, to equate too closely the recorded (*bazar*) price levels and the actual grain situation outside the towns.

When drought occurred, grain was usually procurable in the towns, but at prices which reflected the value placed upon its storage as security against future uncertainty by a section of the producers, and the expectation of gains from price movements on the part of those more directly connected with trading activity. Stores of grain existed among village merchants and cultivators to an extent basically related to their level of wealth and independence, and to their fortunes in previous years. Drought hit less hard at the cultivators lower down the wealth scale the better they had fared over a series of seasons. It was this, along with their ability to gain credit within the village, which accounts for the fact that the cultivating classes were seldom to be seen purchasing grain from the *bazar* and formed only a small proportion of the relief workers in times of famine; consistently, the overwhelming majority of those claiming relief were Chamars and weavers.[84]

Among holders of grain, differing priorities relating to its uses may have muted the impact of price movements upon the level of foodgrain stocks held in the villages, but the ultimate need for cash made it inevitable that the market price structure would impinge upon current produce and/or the level of reserves at some point. Precisely *how* prices affected the net marketed surplus of foodgrains clearly varied not only

[81] UP Scarcity, April 1914, 14.

[82] *PP* 1898, LXII, *Further Papers Regarding the Famine and Relief Operations*, no. v, T.W. Holderness, 'Narrative and Results of the Measures Adopted for the Relief of Famine during the Years 1896 and 1897', p. 125.

[83] UP Scarcity, Jan. 1908, 115.

[84] UP Sec., Scarcity Dept., File 21/1919, File on Allotment of Funds for Famine Expenditure (Agra), E.A. Molony to R.R. Bahadur (Mainpuri Collr.), 31 Jan. 1919.

according to existing and antecedent conditions in the locality, but according to the interaction of local and non-local market forces. Increasingly, from the 1860s, prices came to reflect Indian and international demand and supply conditions. With 'even the smallest markets in touch with the main wholesale markets',[85] price movements were frequently poorly related – sometimes even unrelated altogether – to local conditions. Given this, it is understandable that the conjunction of price movements and local output levels over several seasons could have an important effect on the level of stores maintained.

The net effect upon the level of grain held in store arising from the operation of this conjunction of forces was made particularly complex by the different types of cultivators' different responses to these forces. Ironically, for the smaller cultivators especially, it was low prices rather than the lure of externally induced buoyant prices which frequently led to the rundown of stocks. This could be serious when, as in the mid-1870s, low prices adversely affected stocks in the wake of a run of indifferent crop seasons.[86] As the Gorakhpur Collector remarked, 'The only complaint regarding the season I have heard from the agricultural classes has been the hardship of having to part with so much grain for so little money, seeing they have, in addition to their ordinary liabilities, to repay the village banker the advances they took from him to tide over the drought of last year.'[87] The same situation occurred in the *kharif* of 1898, when the 'labouring classes', whether paid by harvest share or in cash, had 'profited largely' from the low prices, and were in some places in better shape than the small cultivators who were 'hampered by debts incurred in the 1896–7 famine'.[88]

How the bulk of the peasantry fared in times of high prices depended primarily upon the relative rise in prices received in relation to saleable output, and it most clearly benefited when high prices coincided with good local yields. In 1878, for instance, '*bajra* sold at the usual rate of wheat . . . and for the first time . . . almost alone provided the wherewithal for the

[85] UP Sec., Scarcity Dept., File 144/1914, File on Variations in Price of Foodgrains, Note by W.H. Moreland (Dir. of Agriculture), 13 Sept. 1911.

[86] *RAR 1876–7*, p. 6.

[87] *RAR 1874–5*, p. 3.   [88] *RAR 1897–8*, p. 62.

payment of the revenue *kist*,[89] and for those with grain only a small outflow was necessary to cover cash needs. In Etawah, also, during the spring of 1898, buoyant prices were reported to be allowing many cultivators to replenish their stocks.[90]

Cultivators and *zamindars* with regular and substantial surpluses were clearly in a position to operate with more flexibility, and at times of depressed price levels were not under such pressure to run down grain stocks. In 1899 in Mainpuri, for example, the Collector observed: 'Prices last April and May were very cheap, and zamindars and cultivators who were not compelled to sell to raise money held up more of their stocks than they usually do.'[91] At the same time, when high prices prevailed, the surplus-producing (or -acquiring) individuals could make cash gains, particularly when local output was relatively good, but frequently also when output was down, due to the fact that consumption was only a part of their output.

It is because of the differences in the reactions to prices in the marketing decisions of different categories of cultivators that the net effect on village stores, though hard to determine, was not especially volatile. Larger farmers were likely to be selling when smaller men were able to hold back grain due to good prices; small men were forced by low prices to market a large proportion of their grain when larger men refrained from doing so. Grain stocks could thus remain in the rural areas, even though the control of those stocks could largely rest at times with large cultivators, *zamindars*, and village traders. All of these, by virtue of their connections with the locality, would be restrained from moving grain on to the towns, since to be balanced against the potential cash returns were other opportunities for putting the grain to work locally so as to acquire ongoing advantages, such as medium-term income flows through lending activities, the commitment of future labour services, and political support in village conflicts. When

[89] Benett, *Report on Relief and Scarcity Operations*, p. 23. In times of scarcity, it was the coarser grains which showed the greatest proportional rise in price. For more details, see *PP* 1898, LXII, *Further Papers Regarding the Famine*, no. v, Holderness, 'Narrative', opp. p. 122.

[90] UP Sec., Scarcity Dept., File 290/1899, Ser. 14, File on Agricultural Prospects for Kharif Crops, H.G. Tyler (Etawah Collr.) to NWP&O Govt., 15 Sept. 1899.

[91] UP Sec., Scarcity Dept., File 290/1899, Ser. 39, File on Agricultural Prospects for Kharif Crops, W.R. Partridge (Mainpuri Collr.) to NWP&O Govt., 16 Sept. 1899.

all is said and done, the village trader, also, had to take cognisance of the fact that he would need to earn his living in the village after the hardship was over and the windfall profits were spent.

Village traders, like grain-lending cultivators, did not, however, have limitless resources. Their margins were often pared by competition and by defaults, and they often shared the runs of bad seasons suffered by their clients.[92] The balance between the high prices they could get for the grain in the market and the risks in a dry October of putting it out as a seed loan could frequently tilt so as to cause a flow of grain out of the villages. A trader could also be forced to take credit himself, and thus – while able to continue his operations – lose control over the manner in which he chose to use the grain within his operations. Such a situation was described in the *Report on the Marketing of Wheat in India*: 'If he is financed by an *arhatiya* or large wholesale dealer in a nearby town, or by any other Agency, he may have to dispose of most of his stock in liquidation of his accounts through the *arhatiya* soon after collection.'[93] *Banias* were also exposed to intimidation by villagers who were unhappy at having been refused loans of grain, or at the terms offered. Thus it was reported from Muttra district in 1899: 'There is a feeling of indignation against banias even among educated men, and the chance of getting high cash prices on the one hand and of being looted on the other, has decided most banias to take their stocks to the nearest exporting centre. The stocks are less than 1896 for this reason.'[94] The loss to urban traders of what might be called 'village' control over foodgrain stocks could have damaging implications for the welfare of a locality during difficult seasons. Once the grain was sealed in the *khattis* or stored in the *godowns* of the market towns it could be frozen or moved irrespective of local needs.

[92] E.g., *UPPBEC(E)*, ii, pp. 254–63, Enquiry in Kupa village, Muttra. Three successive harvest failures there had seen the *bohras'* resources diminish to the point where they were unable to 'afford fresh credit'.

[93] GI Directorate of Marketing and Inspection, *Report on the Marketing of Wheat in India* (Simla, 1937), p. 128.

[94] UP Sec., Scarcity Dept., File 290/1899, Ser. 37, File on Agricultural Prospects for Kharif Crops, A.B. fforde (Muttra Collr.) to NWP&O Govt., 16 Sept. 1899.

An illustration of the way in which the conjunction of local and external circumstances led to a position where cultivators and the smaller traders lost control over the grain crops is found in 1877–8 in Agra and Muttra. Prices ranged low in 1876–7, even after the distant demands due to harvest failures in Madras and Bombay and to war in Europe had pulled the level up. The result was that cultivators were obliged to release large quantities of their grain at the low prices prevailing at the 1877 *rabi* harvest; in advance, that is, of the price increases induced by the continued export of grain and the inadequacy of the summer rains. Although prices reached scarcity levels, the 'grain dealers alone benefited from the stimulated export, so that when the failure of crop came . . . there were no stocks to fall back on'.[95]

While rain fell in late October (1877) to allow a larger *rabi* area than usual, the hot breezes and blight caused the outturn to fall below the normal total, something which was aggravated by the fact that exports, once more, were siphoning off supplies required within the district. 'The harvest', the Lieutenant-Governor reported to the Government of India in March 1878, 'is already very much in the hands of the banias, who have given large advances on the security of the standing crops. They will be able therefore to rule the markets, and as there is no reason to suppose there will be much diminution in the demand for grain for exportation, it is improbable that prices will fall to any appreciable extent.'[96] Exports to other provinces did continue and it was not until rain fell at the end of July that the 'resumption of agriculture' provided work to attract people away from the relief works. Here was the substantial beginning of the large-scale movement of grain to distant markets, and its immediate effect had been to confound the officially held theory that no single season's failure could be sufficient to cause such hardship: 'All former famines, deserving the name, seem to

---

[95] *Muttra DG* (1884), p. 51.

[96] Benett, *Report on Scarcity and Relief Operations*, p. 177. Such a situation existed in Agra in late 1899: while the stocks in the hands of grain merchants were said to be ample, the supply available to the population was described by the Commissioner as 'deficient'. UP Sec., Scarcity Dept., File 290/1899, Ser. 67, File on Agricultural Prospects for Kharif Crops, W.H. Impey (Agra Commr.) to NWP&O Govt., 20 Sept. 1899.

have had their origin in the loss of the spring as well as of the autumn harvest.'[97]

In an era of rising prices and widening markets, then, the ability to hold grain – to gain most through price movements, or to lose least from them through insulation – was particularly important in relation to the problems posed by famine to cultivators, and it was their manifest ability to hold up the outflow of grain which helped enable those in the canal tracts to withstand so effectively the trials of a drought year. If the cultivators preferred not to use their canal water to raise *kharif* food crops, this did not mean that they were unable to feed themselves in the rainless intervals between one *rabi* harvest and the next. 'Even before the harvesting of this year's cold-weather crop', stated the report of an enquiry into the 1896–7 famine and relief operations, 'considerable foodstocks were believed to exist with the grain-dealers, and with agriculturists in the fertile canal districts.'[98] This ability on the part of a substantial proportion of the cultivating community to hold onto stocks of foodgrains was clearly a long-standing feature in some districts. Of Meerut, for example, during the 1868–9 famine, it was observed that 'the cultivators had become so independent of the petty traders upon whom reliance is usually placed for advances of seed and money, that they were hoarding accumulations of stores in the hope of more favourable markets'.[99]

Though the Meerut Jats were said in 1896 to have 'prayed openly for the continuance of famine',[100] they were cautious in the manner in which they actually profited from scarcity conditions, and it was the primacy they gave to their own security which limited the canal's effectiveness in assisting other stricken areas when famine was at its height. It is quite clear that substantial stocks of food existed in the canal districts

[97] *Agra DG* (1884), p. 446.

[98] *PP* 1898, LXII, *Further Papers Regarding the Famine*, no. v, Holderness, 'Narrative', p. 214. The same applied even in 1877–8, when, for example, Aligarh's granaries were said to have been so well stocked that the district could have 'held out for a year on its own resources, even had the rabi been lost'. Benett, *Report on Scarcity and Relief Operations*, p. 38.

[99] F. Henvey, *A Narrative of the Drought and Famine which prevailed in the North-Western Provinces during the years 1868, 1869 and the beginning of 1870* (Allahabad, 1871), p. 15.

[100] Gillan, *Meerut SR* (1901), p. 6.

during the 1896–7 period. The 1895–6 *rabi* crops were well up to average where irrigated; elsewhere the outturn was disappointing or poor. 'It might have been expected', commented T.W. Holderness, the officer investigating the famine and relief operations, 'that these villages, which had enjoyed good harvests, would, under the attraction of very high prices, have exported some of the surplus stocks they undoubtedly held.'[101] As it turned out, he found that with the drying up of imports from the Punjab and the rapid stiffening of prices at the end of April 1896, cultivators in canal villages shut away their grain, and the flow of food supplies to the deficient areas was reduced to a slow and painful dribble, mainly to the hard-pressed eastern districts and Bundelkhand. In spite of these stores, and in spite of excellent *rabi* prospects for the 1897 harvest, the canal districts continued to import grain till the 'bountiful yield of the spring crops found the cultivators' granaries still stocked with the produce of a previous harvest'.[102] The Bulandshahr district alone of the Meerut and Agra divisions was an exporter of grain in the first quarter of 1897: its net export was 327 tons, while the net import total for the Meerut division was over 16,000 tons.[103]

The low aggregate supply response to the scarcity prices was clearly a reflection of security preference on the part of the cultivators: the peasantry of Meerut division 'held up their produce till the prospects of the future could be assured'. This was in fact done everywhere, but, as Holderness noted, 'the Meerut people could afford to do it on a larger scale than others'. The produce was therefore held up in the villages so that 'even the usual supply did not therefore come into the local markets and imports were required to provide for the townspeople and for that section of the population which has to purchase its food'.[104] Certainly the advantageous prices of the previous years, when the fortunes of the canal districts were generally more favourable as regards output than other parts of the province, helped the class of cultivators which produced a genuine surplus to hold onto its supplies at a time when there

---

[101] *PP* 1898, LXII, *Further Papers Regarding the Famine*, no. v, Holderness, 'Narrative', p. 120.   [102] *Ibid.*, p. 80.   [103] *Ibid.*, p. 120.
[104] *Ibid.* The same situation arose in 1918, when Meerut cultivators, holding 'fairly adequate stocks', were 'not releasing such stocks to the market'. With rural traders also holding stocks 'in view of the season's character', the stocks in the towns were 'comparatively low'. UP Scarcity, May 1919, 90, UP Govt. to GI, 5 Nov. 1918.

was a widespread belief that the country was on the eve of a terrible famine. Mainly for that reason, aided perhaps by a measure of speculation, the stocks remained with the cultivators and with the 'petty dealers whose transactions do not extend beyond their own villages'. The stocks of the large holders in the central markets were thus abnormally low, with the result that urban prices were high and remained so, despite the addition to the supplies with each successive harvest.[105]

Every district in the Meerut division commenced to send away grain in large and increasing quantities as soon as the *rabi* crops of 1897 were gathered. During the second quarter of the year 73,000 tons was the net export figure. Further down the Doab, Agra division's canal districts – with more unirrigated land within their own boundaries to supply – commenced to send grain outside in July, when the prospects for autumn crops became favourable.[106] Only after the worst of the problem was over did the protection to food supplies embodied in the canal actually begin substantially to assist less fortunate tracts outside. While the 'necessitous people of Rohilkhand, Oudh and Gorackhpur were compelled to part with much of theirs', Meerut's peasantry was able to hold up its grain. 'The Meerut division', summed up Holderness, 'in spite of its prosperous condition and its good harvest, gave little assistance to the distressed parts of the provinces and on the whole showed a net balance of imports.'[107]

It was ironic that the only exporting division was Rohilkhand, whose peasantry was forced to part with a portion of its stores, and that a substantial recipient of those exports was Meerut. This pattern of behaviour on the part of the cultivators accentuated the short-term food shortages and called into question the idea of concentrating public investment expenditure in a region or tract most suitable to secure a supply of food for those areas outside which were subject to periodic shortages.[108]

---

[105] *PP* 1898, LXII, *Further Papers Regarding the Famine*, no. v, Holderness, 'Narrative', p. 120. [106] *Ibid.*, pp. 120–1. [107] *Ibid.*, p. 119.
[108] It is worth noting, in addition, that the relatively unimpaired purchasing power of the canal districts enabled them 'in all famines' to purchase fodder when necessary from the 'very seriously affected districts' such as Budaun in 1918, the trade being organised through regular *banjaras* by grain-dealers. UP Scarcity, Aug. 1921, 7, Fyzabad Commr. to UP Govt., 26 Oct. 1920.

Famine and the market for agricultural labour

A dramatic reduction in work opportunities for day labourers occurred in years of drought as the conditions affected both the demand for and the supply of labour. A straightforward fall in the total demand for labour occurred as a result of the reduced crop areas and shrunken yields. This affected the day labourers disproportionately because they were taken on mainly at the points in the agricultural calendar when family labour, together with that of farm servants, was insufficient to cope with the peaks of farming activity. In terms of labour requirements, these peaks were considerably less pronounced in drought years, and it was naturally the marginal (hired) labour which was chiefly affected. Even the increase in demand in well villages for hands to lift water could not, for reasons to do with the physical limitations of both wells and bullocks, in any real sense counterbalance this fall-off in demand for labour.

The fall in the total labour requirements was, from the day labourers' viewpoint, exacerbated by increases in the labour supply, resulting from a tendency among cultivators to attempt to increase inputs of family labour. Two processes appear to have been at work here. First, the normal point at which it did not matter to the peasant whether he increased the family labour input or whether he hired labour for cash or grain wages occurred at a higher level of total labour input in such years. Smaller producers particularly, responding to conditions of uncertainty, preferred to increase the amount of labour put in by the family to paying out in wages grain which might be crucially important to the family's needs over future months.[109] Secondly, many cultivators without the capacity to retain much produce from normal seasons, and with insufficient work to do on their own parched holdings, joined the day labourers in seeking work on the holdings of other cultivators.

Overall, therefore, with the labour supply curve moving outwards, and the demand curve for hired labour shifting sharply to the left, at a subsistence wage there was a very considerable oversupply of labour, and it was only a matter of time before the scanty stores which labourers had saved from

[109] *PP* 1880, LII, *Report of the Indian Famine Commission*, pt I, p. 41.

the short autumn harvest were exhausted. Without autumn rain to provide employment in sowing and harvesting during the cold season, the plight of the groups scraping a living mainly through hiring out their labour services was to wander in search of employment. It mattered little whether or not stores of grain existed within their locality, since their problem was one of being unable to claim a share of the food which was available. Thus in Muttra, after the winter harvest of 1878, 'there was no deficiency of grain but the want of employment and inability on the part of the landless class to provide themselves with food at the prevailing high rates drove increasing numbers to the relief works'.[110] When large areas of the Meerut division lying outside the embryonic canal system's reach remained unsown in the cold season of 1860–1, 'thousands of poor people left their villages . . . the mortality was very great'.[111]

For the canal tracts, however, there is no evidence to indicate that the normally buoyant labour market was substantially affected during a year of drought. Despite the loss of the *kharif* millets in Bulandshahr in 1868, 'there seemed to have been as much difficulty in procuring labour as in ordinary years'.[112] From neighbouring Meerut, in the same year, F. Henvey reported that there was 'always a demand for labour equal to the supply'.[113] Further north again, according to Alan Cadell in Muzaffarnagar, the labouring class had 'the advantage of continuous employment throughout the year' and was 'not exposed to the loss, distress, and danger caused by droughts'.[114] Even at the height of the 1907–8 famine, when the *kharif* prospects were for only a three-*anna* crop on unirrigated land, the Bulandshahr Collector reported that there was 'not the least chance of attracting labour at subsistence wages', and that contractors were 'finding difficulty in getting labour at 5 annas per day'.[115]

---

[110] *Muttra DG* (1884), p. 53.

[111] UPA(R), Files of Meerut Commr., Dept. xii, File 5/1861, Ser. 125, File on Revenue Administration Report 1860–1, Meerut Collr. to Commr., 1 Oct. 1861.

[112] *Bulandshahr DG* (1876), p. 32.

[113] Henvey, *Narrative of the Drought and Famine*, p. 15.

[114] 'Report on the Ganges Canal Tract', p. 18, in Cadell, *Muzaffarnagar SR* (1882).

[115] UP Scarcity, Jan. 1908, 99. A.B. fforde added that 'in 1896 they earned 7 annas a day'. See a similar observation (of Bulandshahr) in NWP Rev., Dec. 1868, 24.

Clearly the ability of the canal tracts to maintain their production flow with respect.to the labour-intensive cash crops, and thus safeguard also the processing, transport, and marketing activities, meant the absence of significant dislocation in the labour market. There was also a heavy demand for labour to assist with the putting down of an extra-large *rabi* area in drought years. Where the intense demand for labour to assist with the commercial crops of the Upper Doab gave way to the less fertile estates lower down, the increased incidence of lift irrigation made labour vital, since speed was of the essence where diminishing access time for irrigators was the method by which the supplies were economised. Often, as in Etawah, the practice was to take water at night, 'no water being available in the day, and baled up out of deep cuttings, at an expense which, in ordinary seasons, the profits would fail to cover'.[116] In all, over 850,000 acres (25%) of irrigation in 1896–7 involved the water's having to be lifted manually onto fields.

There even seems a possibility that the demand curve for labour shifted to the right in many canal villages during a drought. Certainly, the references to 'the comparatively dear rate' paid for lifting water would support this notion.[117] Other activities encouraged when drought struck were the careful maintenance of the *guls* (or their deepening to increase the flow to the village) and the expansion of the *gul* system and field preparation (involving embankments, *kiaris,* and levelling) to permit the extension of irrigation. Public employment connected with the canals (*rajbaha* maintenance such as silt-clearing), plus the excavation of new canals and channels, provided work outside the village. Famines, in fact, played an important role in increasing the areas able to be irrigated, as the noticeably enhanced average areas of subsequent normal seasons testify (see Figure 4).

The perennial 'exchange entitlements' problem for labourers

[116] Crosthwaite, 'Report on the Average Rent-Rates', p. 4. The overall proportion of lift irrigation increased in drought years as the level in the channels fell and the increased value of irrigation altered the ratio of costs to benefits. The same trend was apparent in 1896–7 in Etawah, when the rates of wages for agricultural labour had 'fallen everywhere in the district, except in villages favoured with [canal] irrigation'. *PP* 1898, LXII, *Further Papers Regarding the Famine*, no. v, Holderness, 'Narrative', p. 111.

[117] *Farrukhabad DG* (1884), p. 121.

was avoided in this region not only because of the demand for their services, but also due to the insulation from price movements given by grain wages drawn from stocks. The majority of farm labourers were relatively unconcerned with prices, since 'the habit of paying in whole or in part in kind'[118] made them more concerned with finding work: 'The higher price of food', the 1908 *Administration of Famine Relief* report made clear, 'does not affect the ordinary labourer as long as he can obtain employment.'[119] There were, moreover, inducements within the system to pay labour in kind which the famine, if anything, enhanced: wages in kind obviously reduced the real value of wage costs by the difference between the realised wholesale rate and retail prices. Where, however, cash payments were made to labour, it does appear that the rates varied with the price of grain.[120]

It is probable that resident labourers benefited first from the high labour requirements in canal villages because, even where the demand for labour fell below usual – or where perhaps incoming migrants offered their services more cheaply – the employer had to bear in mind that he would need to call upon the resident labouring group in subsequent seasons of acute short-term shortage of hands. Nonetheless, the displaced labourers of non-canal tracts clearly did find work in canal villages, as is shown by the migration patterns during drought years. The canal villages, stocked with food and standing out as green belts of secured cultivation clustered along distributaries passing through otherwise bare plains, were the focus of local migratory movements and (before the brake of relief works was applied to stem the movement) of longer-distance migration. Indeed, it was the drought periods which marked lasting shifts in the population once the dust had been dampened and the temporary migrants had returned to their villages. Alan Cadell noticed this in Muzaffarnagar *pargana*, where 'with one trifling exception the population had fallen off in every village in the pergunnah which is not watered from the canal. The single exception is the little village of Sitherali, situated between two

---

[118] Miller, *Muzaffarnagar SR* (1892), p. 45.

[119] *PP* 1909, LXII, *Administration of Famine Relief in the United Provinces of Agra and Oudh, 1907–8*, p. 151.

[120] *Bulandshahr DG* (1903), p. 58.

sandhills but surrounded by canal irrigated estates, which in years of drought afford employment and subsistence.'[121]

This reaction was not confined to the richer Upper Doab districts. Evidence in the form of a village survey conducted in Agra reveals that over a thousand people belonging to Dholpur were said to have come to Punahra village between June and September of 1877. Partly attracted by the indigo factory, most of them left when rain fell in October, but 146 were reported to have remained. In Rehjiwa, another of the three canal villages included in the survey, the population between 1872 and 1879 had increased by almost 30%: 'Sixty-four persons joined the village for the first time during the scarcity. A great many [more] came temporarily for the months preceding October.'[122]

It is doubtful that the shift of population to canal villages – to which drought years significantly contributed – resulted in a decline in their relative security against famine. There is no evidence of increasing susceptibility to famine among canal villages for the period under review. This is partly a reflection of the fact that the underlying increase in population prior to the 1920s was relatively small, thus allowing holding sizes and population–resource ratios to be broadly maintained. The pattern of population change was uneven both over time and between districts, but even the faster-growing Upper Doab districts managed a growth rate of only around 0.3% per annum for the period 1872–1921, and well below 1% even during the general growth period of the three decades up to 1901.[123] Some districts even showed a fall, notably Muttra and, to a lesser extent, Agra. If the evidence from Koil *pargana* is at all representative of the differential in population density (a 23% difference between canal villages and the best well villages in terms of net cultivated area, 7% for gross cultivated area), then it was not hard for the productivity effects of a steadily expanding canal-irrigated area to provide the additional resources for canal tracts to at least keep pace with population increase.

It must also be remembered that most of the migrants were

---

[121] UPA, BOR, 'Different Districts', Muzaffarnagar, Box 8, File 35, A. Cadell, RRR Pergunnah Moozuffurnuggur, 1872, para. 36.

[122] Benett, *Report on Scarcity and Relief Operations*, pp. 387–8.

[123] *Census of India, 1921*, XVI, pt 1, p. 26.

drawn from the labouring classes and thus in most cases did not compete for land resources. They filtered in at the bottom of the social and economic structures of the more commercialised villages, where production was frequently held back by labour constraints. This inflow of manpower, allied to certain technical and institutional developments which also took place in the western districts, played a vital role in enhancing production and maintaining the disparities in wealth, and thus security, favouring the canal tracts. An examination of this 'dynamism', as it has been termed – and the canal's role in it – is the subject of the final chapter.

8

Canal irrigation and the dynamism of the western UP

Patterns of growth and prosperity

Now that the operation and effects of canal irrigation in the Doab have been examined in detail, the scope in this final section is widened to consider the role of the canal as a factor in the differential growth patterns experienced during the period. Certainly from around 1870 onwards, there are frequent references in the writings of officials to the marked prosperity of the canal districts, particularly those in the Upper Doab, in comparison with other UP districts, and especially those in the east.

As early as 1872 living standards were conspicuously higher in the west: 'The labourer in Muzaffarnagar', observed the Settlement Officer, 'dresses better than the average petty proprietor of the eastern districts, and wheat . . . forms a larger proportion of the food of the poorer classes.'[1] The Meerut Collector, F.N. Wright, confirming the impression of widely diffused prosperity, could 'find no signs of distress or poverty anywhere . . . I find a stalwart race of cultivators, well clothed with generally good cattle and comfortable houses.'[2] This was the tract where not only the province's best plough cattle but also 'the best class of milkers' were to be found;[3] where occupancy tenants with medium-sized holdings occupied a 'position far above that of the typical cultivator living from hand to mouth, in a chronic state of debt, and with his growing crops pledged to the nearest Mahajan or the village Bania'.[4] In fact, in this area the *maurusi* tenant could 'meet his landlord more or less on equal terms in a court of justice'.[5] 'The villagers

[1] 'Report on the Ganges Canal Tract', pp. 17–18, in Cadell, *Muzaffarnagar SR* (1882).   [2] *RAR 1882–3*, Appendix I, p. 5.
[3] UP Scarcity, Jan. 1920, 8.   [4] *RAR 1882–3*, Appendix I, p. 5.   [5] *Ibid.*

must be tolerably flourishing', a district officer concluded when recalling a visit to the western parts of Muzaffarnagar and Saharanpur, 'else they could not have built the handsome masonry *chaupals* I noticed in 1874–5.'[6]

Even the landless labourer shared not only the affluence but the new-found independence of the occupancy tenants. By the 1870s, Chamars had 'almost entirely emerged from [their former] state of serfdom', and were able 'to select their own masters, either in their own village or elsewhere'. F.N. Wright reported from Meerut:

In many villages they utterly refuse to do 'begari', and in the villages along the northwest of the district refuse to do their old duties for the zamindars connected with sugar-pressing &c.; the old hakks are not sufficient, they demand a full wage for a day's work and the zamindars are at their wits' end to get the cane pressed and the juice boiled in time.[7]

In the growing towns, too, 'the artizans and manual labourers . . . find themselves in a condition of comfort and independence which must some years ago have been utterly unknown to their class'.[8] Indeed, the lure of urban employment helped create labour shortages and drive up wage levels during the busy seasons in rural areas.[9]

How far down the Doab this prosperity extended is less clear. Impressionistic evidence suggests that the prosperity was less marked in parts of Bulandshahr and Aligarh and south-eastwards from there. Certainly the pattern is far from even. Muttra particularly, but also Agra, did not as districts experience the best of fortunes during this period. Indeed, a recent Ph.D. thesis has suggested that Muttra was gripped in the later 1800s by a 'grave neo-Malthusian crisis', and demographic evidence certainly suggests the presence of severe problems.[10] Local variation existed even here, however. Evidence exists to show that overassessment and climatic disruption may not have had such damaging effects upon the

[6] NWP&O Rev., Sept. 1881, 11, Note by W. Irvine.

[7] *RAR 1882–3*, Appendix 1, p. 6.

[8] *Ibid.*, p. 1.   [9] *S&C 1915–16*, p. 17.

[10] S.J. Commander, 'The Agrarian Economy of Northern India, 1800–1880: Aspects of Growth and Stagnation in the Doab', unpublished Ph.D. dissertation, University of Cambridge, 1980.

canal tracts as elsewhere in the district. Canal areas were not without their difficulties in the wet seasons of the 1880s, but they fared reasonably well in the unusually frequent dry seasons between 1890 and 1915, and population shifts from the 1890s appear to have favoured the canal zones at the expense of the trans-Jumna *tehsils*.[11] These villages, nonetheless, fell well short of the prosperity found in the districts to the north.

Much more typical of the districts in the lower parts of the Doab was Etah, where there had been, according to the District Officer, 'within the last generation . . . a decided improvement in the condition of the agricultural classes'.[12] Forced labour did not prevail in the district to any great extent, and the field labourer apparently received better pay and enjoyed greater independence than in former times. William Crooke also detected changes among the cultivators, including a trend away from using buffaloes or borrowed oxen to the ownership of bullocks; and a noticeable improvement in houses, many of which, instead of being mere 'straw sheds', were built of beams and clay.

Although the level of prosperity in such districts as Etah was not generally comparable with that to be found in the north-western districts, Crooke was in no doubt that 'a decided superiority' was to be observed between Etah and those districts with which he was familiar in the eastern UP. There, as population grew and pressure on the land increased, petty proprietors resorted to attempting to regain cultivating possession of their tenants' tiny holdings. The standard of housing and of dress, the puny nature of what cattle the cultivators did possess,[13] and the fact that in this area cattle were simply 'not kept for milking purposes at all'[14] – all contrasted starkly with conditions in the western districts. Those former cultivators who had lost their cultivating rights, the *harwahas*, and who were now labourers and farm servants[15] (perhaps sustaining their interest in cultivation through small

[11] *Muttra DG* (1911), p. 79; see also NWP&O Rev., Nov. 1883, 96, J.B. Fuller (Asst. Dir. of Agriculture), Notes on Certain Villages in Meerut District, 29 July 1882.

[12] *RAR 1881–2*, pp. 8–10.

[13] A large proportion of Jaunpur's cultivators possessed only one bullock or no bullocks at all. B. Misra, *Overpopulation in Jaunpur* (Allahabad, 1932), Ch. VIII.

[14] UP Scarcity, Jan. 1920, 8.

[15] Misra, *Overpopulation in Jaunpur*, p. 19.

plots granted for their maintenance and ploughed with their master's bullocks), were becoming increasingly numerous, and added to the conditions of labour surplus which made employment often hard to find and drove down wage levels. Wage rates, whether for farm servants or daily employment, were half (or less than half) those prevailing in the western districts around 1900,[16] and while the independence of the Doab Chamars was legend, a large proportion of the labouring class in the eastern districts was, according to Duthie and Fuller, 'literally as well as practically the bondslaves of their employers. The condition of slavery offers a refuge from the pressure of competition, and the certainty of daily food may be held some compensation for the loss of liberty which only manifests itself in an insufficiency of both work and nourishment.'[17]

Many found such refuge in emigration. The UP Census Commissioner remarked in 1911 that there was not a single family in the Benares division which had not at least one member in the neighbouring provinces east and south. Many such movements were temporary or seasonal, but the net outflow from Jaunpur was around 80,000 per decade between 1891 and 1921. The outflow reduced the burden upon the local resources, and the flow of remittances – as much as Rs 80 *lakhs* in the case of Benares district in 1927–8 – gave the region a call upon outside resources in meeting the needs of the local population.

Behind this difference in standards of living lay substantial differences in productivity. This showed up in the greater concentration upon cash crops in the western districts, especially sugar and wheat. The contrast in cropping patterns is shown in Table 8.1. The eastern districts grew largely food crops. Compared with Meerut and Muzaffarnagar they grew less than half the area of sugarcane (with the exception of Gorakhpur), a quarter or less of the area of wheat, and no

---

[16] In the north-west region, W.H. Moreland's wage census indicated that a twelve-*pice* daily rate was 'well established' by the early 1900s, while in the east, only Ghazipur and Azamgarh had crept over the six-*pice* rate. Similarly, farm servants in the west received more than one *anna*, while their considerably more numerous counterparts in the east were frequently on one-half *anna* per day. UP Rev., Oct. 1907, 89, Wage Census of Aug. 1906.

[17] Duthie and Fuller, *Field and Garden Crops*, I, p. iv.

Table 8.1. *Comparison of crop patterns in the eastern and western UP: averages of 1908–9/1909–10 crop areas in representative districts (in percentages of net cultivated area)*

Major crops	Western districts		Eastern districts	
	Muzaffarnagar	Meerut	Benares	Gorakhpur
Rice	6	2	25	42
Wheat	34	30	7	13
Barley	5	7	23	18
Gram	13	13	10	6
Other foodgrains[a]	20	24	35	36
Sugarcane	11	10	4	4
Cotton	4	8	–	–
Oilseed	–	–	–	5

[a] Includes maize, but excludes *juar*, which was frequently a fodder crop in the west (average: Muzaffarnagar/Meerut, 6%; Benares/Gorakhpur, 2%).
Source: Returns of the Agricultural Statistics of British India 1905–6/1909–10.

cotton, and only one or two districts grew a significant area of oilseed. In the highly cultivated *parganas* of the canal districts wheat occupied 60% of the *rabi* area: in the canal circle of Khatauli, for example, such a figure was returned for wheat (equivalent to 34% of the n.c.a.), while at the same time 20% of the n.c.a. was under sugar, 4% was under cotton, and over 27% was double-cropped;[18] in *pargana* Kandhla (also Muzaffarnagar), canal villages put down 36% to unmixed wheat, 21% to cane, and 6% to cotton, and 35% was *dofasli*.[19] Moving down the Doab, as Map 4 shows, the proportion of cash crops undoubtedly diminished. The sugar area fell away, to be replaced by indigo (prior to 1900) and cotton. The wheat area in Aligarh was typically around 20% of the n.c.a. and cotton a percentage point or so below that. Sugarcane in Aligarh amounted to just 1%. Nevertheless, that there was a greater concentration upon cash crops than in the east is still apparent.[20]

[18] MCR, BOR, VIII/II, 14(c), H.A. Lane, RRR Parganas Khautauli and Jauli Jansath (Muzaffarnagar), 1919, pp. 9–10.
[19] MCR, BOR, VIII/II, 14(C), H.A. Lane, RRR Pargana Kandhla (Muzaffarnagar), 1919, pp. 9–10.
[20] F.C.R. Robinson's calculations from railborne trade statistics indicate that, by the quinquennium 1911–16, the western UP and Doab, with less than half the population, accounted for nearly three-quarters of UP trade. F.C.R. Robinson, *Separatism among Indian Muslims, 1860–1923* (Cambridge, 1974), pp. 59–60.

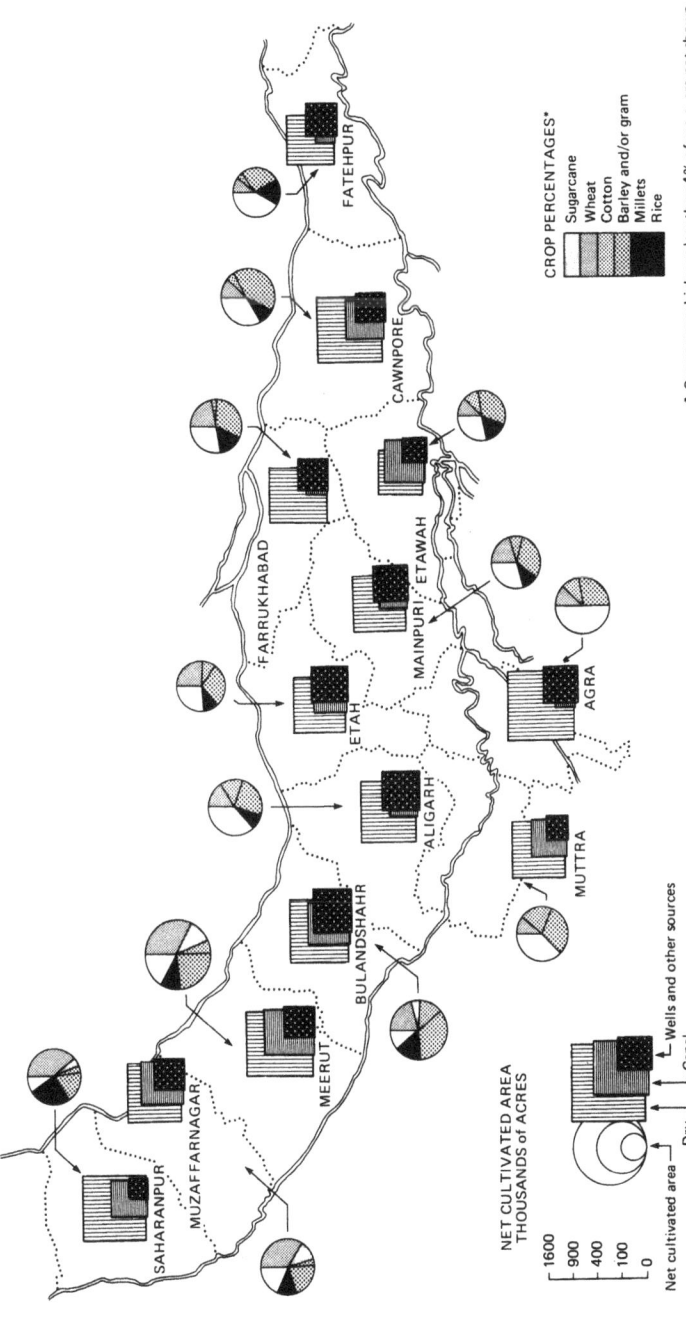

Map 4. Areas under cultivation, irrigation, and major crops, Doab districts, *c.* 1910. (Source: *Returns of the Agricultural Statistics of British India*, 1905–6/1909–10.)

CROP PERCENTAGES*

Sugarcane
Wheat
Cotton
Barley and/or gram
Millets
Rice

* Crop areas which are less than 4% of n.c.a. are not shown

NET CULTIVATED AREA
THOUSANDS of ACRES

1600
900
400
100
0

Net cultivated area

Dry

Canal

Wells and other sources

FATEHPUR
CAWNPORE
FARRUKHABAD
ETAWAH
MAINPURI
ETAH
AGRA
ALIGARH
BULANDSHAHR
MUTTRA
MEERUT
MUZAFFARNAGAR
SAHARANPUR

In addition to the crop pattern differences, the yields of cash crops, according to the detailed work of Duthie and Fuller and the vastly experienced W.H. Moreland, were higher in the west (see Table 8.2). The statistical impressions which emerge from their field measurements broadly concur with the qualitative information dispersed through official reports: with the exception of rice, sugarcane, and peas, crop yields in western districts were consistently one or two *maunds* per acre higher than those in the east.[21] Moreover, according to Moreland, the quality of wheat – a crop vital to the prosperity of the west – 'on the whole decreases as one goes east'.[22] 'Delhi soft white', for which the region was renowned, commanded particularly good prices in Europe, and just prior to the outbreak of World War I earned an increment of Rs 8 or Rs 9 per acre.[23] The same applied to maize. Typically a deep yellow maize was grown throughout the UP, but cultivators in the canal districts took to a more lightly coloured and more productive variety.[24] There seems little to choose between the regions as far as yields of rice or sugarcane are concerned: it was commonly recognised that the fine soil and moist conditions enabled the outturns in the east to compare well with those of the canal districts.

The achievement of higher productivity through concentration upon cash crop production was only really possible through the widespread application of capital to farming operations. State investment in the form of canal irrigation was

[21] Duthie and Fuller, *Field and Garden Crops*, I, p. 6 *et seq.* It is worth noting that their estimated mean outturns for Allahabad and Agra divisions were, in the cases of most crops, below those for the eastern and the westernmost divisions. Indian agricultural statistics must, at best, be regarded as dubious. See, for example, C.J. Dewey, '*Patwari* and *Chaukidar*: Subordinate Officials and the Reliability of India's Agricultural Statistics', in C.J. Dewey and A.G. Hopkins (eds.), *The Imperial Impact: Studies in the Economic History of Africa and India* (London, 1978), pp. 280–314. Such statistics are particularly unreliable when used to detect *trends*; they can, however, be of use in identifying in a general sense the comparative outturns between regions at a point in time. The figures given in Table 8.2 reveal signs of a yield advantage to the west which is also evident in the work of Duthie and Fuller, and which conforms to qualitative impressions scattered in the reports.

[22] Moreland, *Agriculture of the UP*, pp. 204–5.

[23] *RAR 1882–3*, Appendix I, p. 1; UP Rev., March 1915, 65, H.R.C. Hailey, Historical Resume of Efforts Made to Develop Agriculture in UP, 14 Oct. 1914, para. 16.

[24] P. Subbiah, *The Cultivation of Maize* (Allahabad, 1901), p. 8.

Table 8.2. *Comparison of crop yields in representative districts of the eastern and western UP: figures from Moreland's crop experiments conducted between 1896 and 1902 (in lbs. per acre)*

		Muzaffarnagar	Bulandshahr	Benares
Rice	(irrigated)	1,000	1,000	1,000
	(unirrigated)	750	750	750
Wheat	(irrigated)	1,250	1,300	1,100
	(unirrigated)	800	900	600
Barley	(irrigated)	1,500	1,400	1,300
	(unirrigated)	900	900	800
Maize	(unirrigated)	900	1,050	800
Juar	(irrigated)	600	650	650
Gram	(irrigated)	1,000	1,000	800
	(unirrigated)	700	650	600
Cotton	(irrigated)	200	200	–
(cleaned)	(unirrigated)	140	130	100
Sugarcane				
(yield in *gur*)	(irrigated)	2,550	2,050	2,450
Peas	(irrigated)	800	1,000	1,100
	(unirrigated)	550	650	550

Source: Compiled from UP Rev., Aug. 1902, 25, Note by W.H. Moreland.

only one example of this. The powerful bullocks common in the west speeded up ploughing and hauled the especially numerous (and substantial) carts. Just as the canal, efficient draught animals, and carts assisted the farmer in exploiting more fully his land and labour resources, so the increasingly sophisticated sugar mills speeded the processing of his cane crop and improved his extraction rates (and thus the value added). It was common, also, for a costly *pakka* well to lie idle in the wings, ready to fill in for canal supply deficiencies which threatened yields or schedules. Undeniably an indicator of wealth, the capital stock of the canal districts' cultivators contributed to the productivity gap between the west and other UP districts.

Higher productivity is also closely related to the efficiency with which production is organised. A diversified economy which has achieved a significant level of economic specialisation gives rise to cost reductions, and there is evidence of a considerable difference between east and west in this regard. Around 15% of Meerut division's population was classified as urban in the 1881 census, as opposed to 7% in Benares. Moreover, the occupation profiles compiled for the Commission on Agriculture showed that while 76% of Jaunpur's population relied on cultivation for its livelihood, and only 10% and 5% respectively on industry and trade, Meerut's profile showed 42% relying on cultivation, 21% on industry, and 6% on trade.[25]

The west's superiority in terms of living standards was not a long-standing phenomenon, however. Indeed, in the early nineteenth century it was the eastern region that was relatively monetised, and Azamgarh in particular was heavily committed to commercial crops and manufactures, receiving its raw materials and food from adjoining areas. The great commercial centres were strung along the major river routes, through which local trade and the movement of indigo, sugar, opium, and cotton to distant markets took place. Mirzapur was the main trading centre, its merchants, along with those of Benares, linking Bundelkhand cotton to the great cloth-manufacturing district of Azamgarh, which grew very little grain in the 1830s but which concentrated instead upon sugarcane, indigo, and opium and which drew its grain supplies from Gorakhpur, Bihar, and some up-river districts.[26] Bundelkhand, parts of which – like Kalpi in Jalaun district – had one-third of its area under cotton, was the agricultural boom area around 1840, with the town of Kalpi itself allegedly the largest mart in northern India.[27] The thriving centres in the Doab at that time were to be found in the central and lower regions, their importance founded upon the cotton and indigo crops clustered within reach of the main riverways. For a time Farrukhabad, where indigo production was centred, was the most prominent commercial and financial city in the Doab, though Cawnpore was already growing in significance.

While the districts further up the Doab shared in the expansion of trade and the more settled conditions of this period, the degree of commercialisation was comparatively limited. All Central and Upper Doab districts experienced an increase in population and cultivation; indeed, by 1840 little further increase in tillage was possible in Aligarh or in the Doab sections of Agra district. There was still room for expansion in

[25] R. Mukherjee in 'Introduction' to Misra, *Overpopulation in Jaunpur*, p. 7; see also *Report on Agriculture in the United Provinces* (Naini Tal, 1926), p. 121.

[26] A. Siddiqi, *Agrarian Change in a North Indian State: Uttar Pradesh, 1819-33* (Oxford, 1973), pp. 159–61.   [27] *Jalaun DG* (1909), pp. 21–5, 45.

the districts to the north, however, most of which were subsequently to expand their cultivation (n.c.a.) by 25–30%.[28] This was a generally drier, less secure tract, with considerable areas of relatively light population and extensive rather than intensive cultivation, which found an outlet for its surplus grain down river through such riverside towns as Anupshahr. There were, in addition, pockets of undoubtedly high cultivation, where sugar crops were grown and converted into *gur* for sale across in the Punjab. Considerable population increase and intensification of cultivation had taken place in the EJC tracts by mid-century, and a degree of prosperity had been identified for these and some well tracts, but overall the tract was far from a rich one. In the view of the Saharanpur Settlement Officer of 1920, Muzaffarnagar in 1840 'was from the agricultural point of view very similar to Bundelkhand of the present day'. Certainly the revenue rates on cultivated area for Saharanpur and Muzaffarnagar were less than half the level of those prevailing in districts like Etawah and Cawnpore, confirming the impression that the Upper Doab was in many parts noticeably less prosperous than the Lower Doab around 1840.[29]

A striking reversal of fortunes took place in the post-Mutiny period. The eastern region went through a process similar to Geertz's 'involution'.[30] As population growth far outpaced the expansion in net cultivated area, the ecological setting of these eastern districts – fertile soil, generally reliable and adequate rainfall, favourable conditions for the construction of cheap wells – enabled productivity per unit area to be increased in line with the population. The process was 'involutional' rather than developmental due to the fact that no basic reorganisation took place: traditional inputs were simply increased within the same overall framework. The basic ingredient of this effort was the more thorough application of labour, which was directed at improving water control, land preparation, and weeding operations, and the juggling of crop combinations to achieve

---

[28] *Aligarh DG* (1911), p. 32; *Agra DG* (1906), p. 27. Of the Upper Doab districts, Meerut, for instance, increased its cultivation by a quarter between 1836 and 1900. *Meerut DG* (1904), pp. 35–7.

[29] *Report on Agriculture in the UP*, p. 119.

[30] C. Geertz, *Agricultural Involution: The Process of Ecological Change in Indonesia* (Berkeley, 1963).

the required levels of food intake and cash-earning produce. Under such a strategy, output per unit of labour did not increase. As population pressed on land resources and relative factor scarcities shifted accordingly, the response was a general substitution of heavy- for light-yielding crops.

Since enhanced water control was necessary to facilitate this process, it is not surprising to find the 1927 *Report on the Seasons and Crops* noting that new well construction in the provinces was habitually headed by Gorakhpur, followed by Azamgarh and Basti.[31] The spread of *boro* (dry-season) rice crops reflected this extension in irrigation and facilitated double-cropping, as did the expansion in the cultivation of maize, which was preferred to other *kharif* millets due to its heavier outturn and early maturity. The *rabi* area under *gram* and peas expanded – peas being a nitrogenous crop (important to double-cropping) whose stalks were used for fodder. Barley was increasingly preferred to the lighter-yielding *kharif* millet *juar*, and often also to wheat due to its lower cost, higher yield, and less precise growing requirements. The small wheat area in comparison with that of the west was taken by officials to be another indicator of the poverty of the region; rice, the main staple, they considered a less nutritious and less valuable crop than wheat. There was a tendency, too, for the heavier-yielding *dofasli* practices to squeeze out costly sugarcane cultivation: Ghazipur and Benares both experienced falling sugarcane acreages over the second half of the century; Jaunpur saw its area fall from 14% of n.c.a. in 1841 to 8% in 1891 and 5% by 1908–10; and Gorakhpur's sugar area fell to 4% in 1908–10 and only hovered around the 7% mark in the 1920s. Apart from arable crops the area of groves was progressively eaten into by the pressure of population,[32] while diminishing grazing areas precipitated a decline in the numbers of draught cattle in the most crowded subdivisions.[33]

---

[31] J.K. Mathur, *The Pressure of Population: Its Effects on Rural Economy in Gorakhpur District* (Allahabad, 1931), p. 38. The number of wells in Gorakhpur was recorded to have risen from 84,487 in 1913–14 to 103,552 in 1932–3; in Azamgarh, from 41,559 to 48,495.

[32] These trends are outlined in Ganguli, *Trends of Agriculture and Population*, pp. 102–7. For specific district cases, see Mathur, *Pressure of Population*, Ch. vii; and Misra, *Overpopulation in Jaunpur*, Ch. viii.

[33] See Misra, *Overpopulation in Jaunpur*, Ch. vi.

While 'involution' processes were causing reduced specialisation to take place in the east, the western districts were undeniably dynamic. Commercial crop production expanded dramatically, particularly so in the decades leading up to 1900, but even afterwards it showed continued improvement: on a comparison of five-year averages, Meerut increased both its valuable sugarcane area (from 9% to 13% of n.c.a.) and the extent of *dofasli* (from 21% to 28%) between 1901 and 1940. Moreover, according to recorded statistics the proportion of pure wheat had improved at the expense of mixed wheat.[34] The signs of growth and expansion were everywhere. A European estate-owner named Michel, with extensive property in south-central Meerut, commented upon the increase in the number of carts from a total of four to 100 in the few years since his arrival in the 1860s; 'the same tendency is everywhere apparent', added the 1876 *Gazetteer*.[35] The Director of Agriculture, referring to Muzaffarnagar, was similarly struck by the 'market places crowded with carts and a railway station to which additions are yearly made'.[36] The diffusion of progressively more sophisticated sugar mills and the manufacture of these – as well as the fabrication of carts – in the principal towns were part of this process of expansion. The continuous improvement in productivity through the application of technical inputs and innovations (both technical and institutional), lowered costs and relieved constraints and thus created new opportunities for growth. It is this dynamic process which brought about the contrasting pattern of prosperity which separates the west from other areas, especially the eastern districts.

What factors account for the relative dynamism of the western districts, and for the rate and character of the growth process which occurred there? The divergence in growth experiences within the UP may be simply a reflection of differential resource endowments. In other words, the early advantage of the eastern districts might have been due to their relative proximity to ports, which was subsequently rendered less decisive when railways opened up a more productive zone

[34] Cooke, *Meerut SR* (1940), Appendix VI, p. 52.
[35] *Meerut DG* (1876), p. 313.
[36] NWP&O Rev., Aug. 1886, 52, Note by D.M. Smeaton (Dir. of Agriculture).

made up of fertile soil, a relatively sparse population, and a body of agriculturalists more 'entrepreneurial' or more skilled than its counterpart to the east. There are other possible explanations of the dynamism of the canal districts, and part of this chapter will focus upon the technical aspects of growth: the extent to which production conditions – especially factor proportions and existing technological conditions – interacted to create incentives for investment, innovation, and organisational improvements. Technical possibilities, of course, do not exist in a vacuum. The willingness and ability on the part of individuals and groups to respond to opportunities are affected by the institutional setting. Land tenure is perhaps the principal institution affecting the peasants in this regard, but another vital institution affecting growth relates to the whole marketing process. The role of each of these influences will be examined in detail. Case studies of two major commercial crops associated with the canal, sugarcane and indigo, will then be presented with the aim of illustrating the way in which technical and institutional conditions affected the level and nature of benefits arising from expanded commercial crop production, and of showing how differences in these conditions affected growth and living standards in different regions. The final section of this chapter will pull together many of the strands of this discussion and will argue that canal irrigation, through its role in conditioning the very economic structure of the western districts, was the fundamental ingredient of the growth process which occurred there, and thus of the distinct level of prosperity.

Natural resources and railways: a discarded hypothesis

The proposition that differential resource endowments, in conjunction with improved communications through railway construction, are sufficient to explain the divergent growth paths and wealth differences substantially evaporates upon examination. The difference in man–land ratio between the eastern and western UP, for example, was certainly too small to account for the contrasts. Duthie and Fuller show that while the 1881 population per cultivated square mile in Meerut division amounted to 741, that in the trans-Gogra districts

(Gorakhpur and Basti) amounted to 915 and in the Benares division to 978. Although in 1921 Meerut's population per cultivated square mile (858) was still below that of the easternmost districts (Gorakhpur 970, Ballia 997, Azamgarh 1,083, and crowded Jaunpur 1,138),[37] the thriving canal circles with their large areas of sugarcane, wheat, and cotton could exceed all of these in density – as, for example, in Muzaffarnagar's Kandhla *pargana*, where there were 1,162 heads per cultivated square mile.[38] Meerut's rich Baraut *pargana* had 940 inhabitants per square mile as early as 1866.[39]

The point is underscored by the size of holdings. Professor Stokes has outlined the erroneous assumptions which caused the student–authors of the Gorakhpur and Jaunpur investigations to compute figures for average holdings of 3.5 acres in Jaunpur and 0.65 acres for tenant holdings in Deoria subdivision in Gorakhpur as against 7.8 acres for Meerut. In fact the differential was nothing like as dramatic. Professor Stokes pointed to estimates made by the Settlement Officer in 1919, based on plough areas, which suggested that only 16% of Deoria was in holdings of four acres and below. This latter impression was confirmed by the *All India Rural Credit Survey* of 1951–2, which estimated that, even then, 70% of holdings area was in units averaging 8.5 acres and that 40% of the cultivated area was in units larger than that average figure. In Meerut, according to C.H. Cooke's 1940 settlement estimates, around 12% of the cultivated area was held in units of less than five acres, while 36% of the cultivators held more than ten acres, which, in aggregate, accounted for 60% of the sown area.[40]

Comparisons of the availability of land appear wholly

[37] Land not cultivated is not necessarily economically useless, and differences in density per square mile are even less striking: 1921 figures show Meerut with 652, Azamgarh 691, Ballia 679, Gorakhpur 721, and Jaunpur 745. Misra, *Overpopulation in Jaunpur*, p. 4. By 1958, in fact, with 984 heads to the square mile, Meerut had become the most thickly populated district in the UP. W.C. Neale, *Economic Change in Rural India: Land Tenure and Reform in Uttar Pradesh, 1800–1955* (New Haven and London, 1962), p. 11.

[38] MCR, BOR, viii/ii, 14(c), H.A. Lane, RRR Pargana Kandhla (Muzaffarnagar), 1919, pp. 1–2, 9.

[39] W.A. Forbes, Baraut Pargana Report, 1866, p. 9, in Buck, *Meerut SR* (1874).

[40] E.T. Stokes, 'Dynamism and Enervation in North Indian Agriculture: The Historical Dimension', in E.T. Stokes, *The Peasant and the Raj: Studies in Agrarian Society and Peasant Rebellion in Colonial India* (Cambridge, 1978), pp. 228–42.

inadequate to account for the east–west growth discrepancy, but differences in other natural resources are not sufficient either. There is no doubting the fertility of the eastern region, founded upon rich soils, reliable rainfall, and easily utilised groundwater. With a reservation in the case of wheat,[41] natural conditions did not put the east at a relative disadvantage in a physical sense in the production of commercial crops, particularly sugar and indigo. If Basti and Gorakhpur possessed rain-reliant villages in northern areas towards the hills, Meerut itself possessed 230 villages, principally in *parganas* Loni and Barnawa, which were reported to 'suffer from drought as severely as any in poorer districts'.[42] Rainfall in the western districts was substantially below that of the eastern area: while the former could look for 28 inches per year, the Benares–Jaunpur region received an average of more than forty inches. A deviation from the norm, in either total or seasonal distribution, was generally more critical for agriculture in the less humid western region and irrigation consequently much more important. Even so, a comparison of irrigated areas suggests that the west was at a relative disadvantage in terms of water supply. The districts from Muzaffarnagar down to Aligarh irrigated on average 45% of the n.c.a. between 1905–6 and 1909–10 (see Map 4) – a proportion exceeded by Jaunpur and Azamgarh (both 56%), which relied principally on wells, though exceeding that of Gorakhpur (33%), which received a higher rainfall, particularly in the northern areas.[43] The significant comparison here related perhaps to the dry areas, which constituted over 50% in the western districts and which, in consequence of the generally lower and less reliable rainfall, were more susceptible to crop disruption than were the eastern

---

[41] Outbreaks of rust, due to dampness, were more serious in the central and eastern UP and could reduce outturn by 25–30%. The drier west was seldom seriously affected. UP Rev., July 1906, 173, Note by W.H. Moreland. Apart from this, however, it was the differences in care over seed selection and in the adoption of superior varieties which accounted for the yield and quality advantages of the west in wheat production. *Ibid.*; NWP&O Rev., March 1880, 22, Note by E.C. Buck.

[42] UP Rev., March 1906, 246, BOR to UP Govt., 1 Feb. 1906.

[43] Compiled from GI Commercial Intelligence and Statistics Department, *Returns of the Agricultural Statistics of British India 1905-6/1909-10* (Calcutta, 1911), Table 2-F, pp. 76–83. This comparison tends to exaggerate the western irrigation figures, since the years 1905–6 and 1907–8 were particularly dry.

divisions. B.N. Ganguli, a geographer writing in the late 1930s, pointed out the very much larger deficiencies of crop areas – particularly the *rabi*, which was in many districts more than one-third down on normal – in the western region in the dry years of 1913–14 and 1918–19.[44] In Ganguli's words, the 'arid districts of the western Gangetic plain . . . badly require the full amount of the scanty rainfall they usually receive'.

The 1926 *Report on Agriculture in the UP* agreed that the differences between the Upper Doab and districts in the east were not due to differences in fertility, arguing instead that it was due to 'the difference between the men who form the bulk of the cultivators in the two areas'.[45] They conceded that the climate was possibly a factor – Meerut was 'dry and bracing', while Gorakhpur was 'damp and enervating', but it was to the Jats, 'comparative latecomers to Hindustan', that the 'prosperity of the Upper Doab must largely be attributed'. As a general explanation of the performance of western districts, this argument – a hardy standby for successive generations of officials – has been seriously questioned by Professor Stokes, who has presented evidence to show that Jat standards, like those of other castes, varied according to circumstances and that it was 'the nature of the land rather than innate caste characteristics that appeared to determine the character of agriculture and the mental attitudes of the people',[46] thus following Edmund Leach's view that the constraints of economics were prior to those of morality and law. The purpose of this chapter is to show how the fundamental conditions of production in the western districts were conducive to peasant enterprise, and that *all* castes seemed to respond to them. The Jats, it must be remembered, even in Meerut only accounted for 30% of the total population.

Although railways revolutionised trade lines and altered the relative competitiveness of trading centres and specialised

---

[44] Ganguli, *Trends of Agriculture and Population*, pp. 80–4. As an indication of the general security of cultivation conditions generally enjoyed in the east, Jaunpur's net cultivated areas for 1901–2/1927–8 varied within the narrow range of 628–657,000. The three lowest figures fell short of the average of the other 24 years by 3.8%, 4.5%, and 4.6% respectively. Calculated from annual areas tabulated in Misra, *Overpopulation in Jaunpur*, p. 31.

[45] *Report on Agriculture in the UP*, p. 120.

[46] Stokes, 'Dynamism and Enervation', p. 233.

localities, it is clear from what has been said above that this is simply not a case of the railways coming along after 1860 and unlocking the natural potential of the western UP at the expense of the hitherto more accessible eastern districts. The railways obviously widened the markets for *gur*, cotton, and wheat, and improved prices overall; they must therefore have benefited an area like the Upper Doab, and possibly to a greater extent than the eastern districts, but this is not sufficient to explain the conditions of dynamism in the west. The fact is that many areas came to be served by rail communications, yet few responded in so dynamic a fashion once the initial gains from making use of idle resources had been realised. This points to the importance of cultivating communities having the technical and institutional capabilities to respond to new demand conditions. The evidence from *pargana* Koil, for example, illustrates clearly the technical limitations imposed by well irrigation upon cropping responses, though the non-canal villages were for the most part better located in relation to the railway than the canal villages. Moreover, it is worth noting that the significant growth of sugarcane cultivation and *gur* manufacture on the EJC and the Ganges Canal during the pre-rail era was adequately sustained by land and river transport links. Even much later, when the railways had captured much of the trade, 'a considerable amount of sugar' exported from Muzaffarnagar was 'still carried on camels which came down in large numbers from the Punjab for the purpose'.[47] Naturally, railways provided the favourable conditions for continued expansion, growing markets, and rising incomes. But the supply side is of particular importance in explaining differential responses. The western districts were reducing their costs through increased efficiency – as well as improving their quality in many cases – and the canal was central to this process.

The technical basis of higher productivity: canal irrigation, the supply of labour, and technical innovation

From a technical point of view the critical input affecting growth in the western UP was canal irrigation. While the

[47] *Muzaffarnagar DG* (1926), pp. 60–1.

eastern districts had virtually no canal irrigation, the extensive systems in the western tract allowed an expansion in cash crop production as peasants redirected the labour and bullock resources freed from wells and benefited from the sheer speed with which canal water could be spread over fields. The details of the canal's impact have already been examined (Chapters 3 and 4), but while its very substantial effect upon productivity was closely related to the large labour economies it permitted in the actual process of irrigation, it was also related to the levelling out of the labour-intensive peaks in the agricultural calendar. With the work spread more evenly through the year, and with cropping intensities less constrained by labour and bullock supply bottlenecks, more labour could be applied to the land.

If, therefore, on the one hand the canal proved a labour-saving technology, in its encouragement of more intensive cultivation it was undoubtedly *labour-using* overall. It thus merely aggravated the traditional labour shortages of the west. Peasant decision-making therefore took place in a very different environment from that of the labour-surplus east. The high costs of hired labour and the opportunity costs associated with using family labour encouraged productivity-raising innovations. In the north-western districts this was particularly the case with sugar, the area of which expanded greatly in canal areas. Shortages of labour, and also of capital (such as draught animals), created serious bottlenecks at harvest period.

Technical innovations – new cane presses – saved both labour and capital. Traditionally, cane was cut into short strips and crushed in the *kolhu*, a large mortar-and-pestle contraption turned by two bullocks.[48] The *kolhu*'s grinding only extracted a limited proportion of the cane's juice, and left wooden and grit impurities behind, which lowered the value of the produce. In addition, it took three men, two boys, and a pair of bullocks to work the press. And even then, 'working day and night a *kolhu*

---

[48] The mortar was normally made of stone in the western areas, though it was frequently wooden in Oudh and the eastern districts where, as in Gorakhpur for example, there was difficulty in obtaining stone. G. Watt, *A Dictionary of the Economic Products of India*, vol. VI, pt 2 (Calcutta, 1893), p. 283.

will not press more than $1\frac{1}{2}$ acres of cane in a month'.[49] These limitations came into prominence in the western districts in the 1860s and 70s, when canal irrigation encouraged cultivators to expand sugarcane cultivation. In 1861–2 so much cane was grown that 'there were hardly hands enough to take it in and manufacture it'.[50]

The invitation to inventiveness was not left unanswered for long. A pair of Bihar estate-owners, Thomson and Mylne, patented the Behea sugarmill. Two rollers set in a wooden frame squeezed rather than ground the juice from the cane, producing *ras* of better quality. In addition, it could be worked by a single bullock and, due to the fact that the cane did not have to be pre-cut, two men.[51] Its adoption was rapid, particularly in the western districts, where, by the mid-1880s, it was reported by F.V. Corbett that not one *kolhu* was at work in the entire Doab.[52] Corbett, in fact, attributed the largely increased cane area on the EJC to the introduction of this mill. While, over time, the Behea was adopted in parts of the eastern districts, in the west it was quickly superseded by the Nahan mill, consisting of three iron rollers set in a cast-iron frame.[53] Shortly afterwards the Babu mill, patented by Babu Bhawani Dass of Bijnor – the only mill with four rollers – was reported to be gaining ground in parts of Meerut and Rohilkhand.[54] While the chief output advantage of the Behea over the *kolhu* was in quality of output, the three- and four-roller mills undoubtedly extracted more than their two-roller counterpart, though relying on the use of strong bullocks to be able to do so. By 1902, therefore, 'the mills commonly used in Meerut division' were markedly superior to 'those employed in the east, which include the inferior two-roller iron mills and the old-fashioned stone and wooden mills'.[55]

[49] Duthie and Fuller, *Field and Garden Crops*, i, p. 58; see also S.M. Hadi, *The Sugar Industry in the United Provinces* (Allahabad, 1902), p. 61.

[50] *PRRC 1861–2*, p. 25.

[51] For a detailed description, see Hadi, *Sugar Industry*, pp. 54–61; and Duthie and Fuller, *Field and Garden Crops*, i, p. 58.

[52] *IRR 1885–6*, p. 42; *IRR 1887–8*, p. 144.

[53] Hadi, *Sugar Industry*, p. 54.   [54] *Ibid.*

[55] *Ibid.*, p. 23. By the turn of the century, two-roller mills were comparatively scarce in Muzaffarnagar district, numbering 236; there were 1,687 of the larger type (two large rollers and a small one) and 1,809 large sugar presses with three large rollers. *Muzaffarnagar DG* (1903), pp. 34–5.

Apart from the difference in value of the extracted produce, the great attraction of the new mills was the saving of labour and bullock power at a critical point in the agricultural calendar. While the traditional *kolhu* technique took around twenty days to crush an acre of sugarcane, my estimates – drawn up from three independent and indirect sources – suggest that a liberal crushing time for one acre in Meerut would be eleven to fourteen days, or from one-half to two-thirds the *kolhu* rate. Overall, therefore, ignoring the employment of child labour required in the use of the *kolhu*, the latter would utilise sixty man-days and forty bullock-days, as against 22–8 man-days and eleven to fourteen bullock-days for an acre of cane in the western districts.[56]

It was also capital-saving as far as draught power was concerned. The relative capital costs of the equipment itself are not so clear, however. When demand exceeded supply, in the 1880s hire rates for the two-roller mills appear to have been upwards of Rs 60–70 per year; but the price came down thereafter, to reach around Rs 12.5 by 1901, while the larger types could then be hired at Rs 20 and Rs 33 respectively.[57] In fact, it was the practice for a few families to co-operate in hiring the crusher, working it by their bullocks in turn, the cost per week amounting to Rs 2.25–4.0, depending on the type of mill.[58] This is a small charge in comparison with the total cash value of the crop and the factors freed by the new techniques.

Not only the sugar presses, but the cattle which worked them, were of superior quality in the west[59] – largely because fully one-half of the agricultural cattle of the districts of the Meerut division were purchased from outside, principally from

[56] Hailey and the *Meerut DG* furnish estimates of crushing times which, when considered alongside the average outturn of sugarcane (360–380 *maunds* per acre) and *gur* (36 *maunds* per acre), and allowing the extraction rate suggested by Hailey of 9–10%, yield the estimates given in the text. H.R.C. Hailey, 'Prices of *Gur* and Cane in the United Provinces', *AJI*, ix (1914), pp. 217–26; *Meerut DG* (1876), p. 229. F.N. Wright corroborates these figures in his *Memorandum on Agriculture*, p. 81, where he estimates the time taken to express the juice from one *bigha* of cane at eight days. Taking one *bigha* as equivalent to 5/8 acre, this corresponds to the estimate given in the text.

[57] *Muzaffarnagar DG* (1903), pp. 34–5.

[58] Hailey, 'Prices of *Gur* and Cane', p. 220.

[59] Duthie and Fuller, *Field and Garden Crops*, i, p. xi.

Mewat in Rajputana and from the Punjab districts of Rohtak and Hissar.[60] The highest proportion was found in Meerut itself, and that proportion declined steadily towards the east, falling to virtually nil by Fatehpur and Allahabad.[61] Such cattle had to be stall-fed. The area of *chari* was more than ten times as large in Meerut as in any other division.[62] It was in Meerut that concern for high-quality animal stock was most noticeable: in the early 1870s it was observed that 'the better class of zamindars who take an interest in the breeding of cattle have been of late years importing bulls from Hissar, and in some villages, the sharers have subscribed among themselves and purchased Hissar bulls, the expenses of which are borne by the village'.[63] The preference for stronger draught animals obviously entailed a substantial outlay on the cultivator's part: while locally bred oxen and buffaloes cost around Rs 20 and Rs 25 respectively in the 1890s, 'good plough cattle' cost Rs 50–70, and even more.[64] Because they were stall-fed, such cattle were also more costly to maintain. The Imperial Council of Agricultural Research survey put the figures for Meerut, Rohilkhand, and Gorakhpur at Rs 129, Rs 62, and Rs 47 respectively. Even allowing for the much greater number of working days attributed to Meerut bullocks (153 days, as against Rohilkhand's 113 and Gorakhpur's 70), the cost per working day was 13 *annas* 6 *pice* in Meerut as against 10 *annas* 10 *pice* in Gorakhpur.[65]

Despite the greater bullock costs per working day incurred by Meerut cultivators, however, acre for acre the superior bullocks of Meerut were actually cheaper to maintain than those of Gorakhpur. Here we have yet another example of the tendency to substitute capital for labour, raising its productivity. The number of working days per acre amounted to only 10.6 in the Meerut sample, as opposed to 16.1 in Rohilkhand and 13.3 in Gorakhpur; moreover, a pair of bullocks was found to command 15.8 acres in Meerut, more than double that in Rohilkhand (7.25) and three times the figure in Gorakhpur

[60] *Ibid.*　　　　　　　　　[61] *Ibid.*, p. 25.　　　　　　　　　[62] *Ibid.*

[63] *Meerut DG* (1876), p. 223.

[64] *Muzaffarnagar DG* (1903), p. 78; Miller, *Muzaffarnagar SR* (1892), p. 45.

[65] GI Imperial Council of Agricultural Research, *Report on the Cost of Production of Crops in the Principal Sugarcane and Cotton Tracts in India*, vol. III (New Delhi, 1939), p. 12.

(5.2).[66] Their chief advantage, perhaps, related to field preparation. In Meerut, noted Hadi, 'an acre of land can be conveniently ploughed up in a day', whereas 'in the eastern districts, where bullocks are weak and ploughs small, two days are commonly devoted to an acre'.[67] Duthie and Fuller had noted also how, in eastern districts, 'the general weakness of the cattle only permits of land to be ploughed, so to speak, by instalments'.[68] Weak cattle clearly precluded the introduction of heavier ploughs. In the rice districts of Oudh and Benares division the typical plough was of 'ludicrously small size', often weighing only seventeen or eighteen pounds. 'Speaking generally', Duthie and Fuller observed,

the efficiency of the plough may be said to increase as we go westwards, the ordinary plough of the central doab weighing about 28 lbs., while that of the Western Districts . . . weighs nearly 50 lbs., is bound with iron round the edges of the sole, and instead of a short spike for a share has a long iron bar which projects behind, and can be thrust forward from time to time as its point wears down.[69]

Thus improvements in one area of activity had repercussions in others. The stronger bullocks not only were a prerequisite to the effective use of the Nahan sugar press – with its higher juice extraction rate – but permitted the faster preparation of the ground for sugarcane cultivation by speeding up ploughing and by the use of the *lakkar*, or roller fashioned from a tree trunk, across the field.[70] Other improvements – whether the seed drill attached to the plough, which was in widespread use for wheat-sowing in the west around 1900,[71] or organisational change, such as the extensive use of carts in canal tracts to substitute for labour in such tasks as taking manure to outlying fields[72] – had the effect of speeding up processes and making labour available for redeployment. Here is an important part of the explanation of the growing contrast in cropping patterns

[66] *Ibid.*   [67] Hadi, *Sugar Industry*, p. 43.

[68] Duthie and Fuller, *Field and Garden Crops*, I, p. x.

[69] *Ibid.* The *Meerut DG* (1876), p. 313, noted an increased demand for bar-iron for ploughshares.

[70] Duthie and Fuller, *Field and Garden Crops*, I, p. x.

[71] Moreland, *Agriculture of the UP*, p. 201. Outside the western districts the seed was sown, as before, behind the plough.

[72] E.g. UPBORL, IV(c), Ser. 5, L. Parshad, Pargana Handbook of Tahsil Deoband, Saharanpur, *c.* 1891, p. 8.

between the western and eastern UP. Labour-saving innovations in the west did not mean that less labour was applied to the land; on the contrary, they permitted *more* labour to be applied to each acre and allowed more intensive cultivation. This is reflected in the Imperial Council of Agricultural Research survey finding that adult males spent 145 days each year on crop production in Meerut as against 113 in Rohilkhand and 80 in Gorakhpur.[73]

In the eastern districts, emphasis was increasingly placed on *land*-saving rather than labour-saving strategies. As with all cases where one input – in this case labour – increases while other inputs remain relatively fixed, diminishing marginal physical returns set in. But whereas the increased application of capital in the west, embodying small but significant technical innovations, might be labour-saving in intent, there were land-saving developments as well. The adoption of new varieties accompanied canal irrigation. The use of better strains of wheat and maize in the north-west has already been mentioned. Higher-yielding sugar varieties also were quick to take hold. Soon after the Ganges Canal opened, *agaul* cane was imported for the first time from Bareilly and acclimatised in a village near Meerut city. From there this thick cane spread into Muzaffarnagar, where it quickly ousted the thin *desi* (indigenous) strain called *daulhu*, which had been 'almost universal' in 1841.[74] The spectacular success of *agaul* was later to be dampened by the repeated appearance of rind fungus, which caused such damage to the crop during the 1880s wet cycle that it was replaced by local forms of *ukh* (sugarcane). Even so, the Upper Doab was still noted for 'the superior quality of the canes grown . . . especially the *dhaur* cane', which, noted one investigator, 'is famous for its high outturn'.[75] *Desi* was only to be found in the *khadirs*: 'Everywhere on the upland improved varieties are now grown whose yield is at least 50 per cent

[73] *Report on the Cost of Production of Crops*, III, p. 15. This particular statistic is not the most reliable measure of the extent of year-round employment, though it is indicative of the basic differences existing between the two regions. It is worth noting also that there were almost twice as many hired man-days per family-day recorded in Meerut (10/16) as in Gorakhpur (10/30).

[74] Hadi, *Sugar Industry*, p. 52. See also *Meerut DG* (1876), p. 226.

[75] *Meerut DG* (1876), p. 110.

higher than that of *desi*.'[76] Heavier presses, extracting more of the juice of higher-yielding cane, were also effectively land-saving innovations. Commercialisation gave rise to the expansion of productivity-raising specialised services within the west. Villages with large areas under canal irrigation had little time left to cart their produce to market, and many came to rely on villages nearby with less canal irrigation to move their produce.[77] The new carts, frequently remarked upon, were thus often the acquisitions of peasants from non-canal villages. Dr Peter Musgrave has identified a Jat village, Sikandrapur, in Meerut, which switched from cultivation to carting the produce of the canal tracts;[78] similar villages could be found in *pargana* Khatauli[79] and *pargana* Jalalabad (Meerut), where the Baluchi proprietors preferred to 'employ themselves almost entirely with their camels in the carrying trade, letting out their lands to their neighbours and the lower classes of tenants'.[80] Competitive by their very nature, such services were of particular value to the smaller cultivators without carts.

The supply of bullocks also became more specialised. Rearing costs in Meerut division were high and the cultivators preferred to purchase mature bullocks which could commence work immediately. The system which evolved consisted of *banjaras* making purchases at the large cattle fairs in the Hariana cattle-breeding tract. Frequently financed by *banias*, the *banjaras* purchased two types of animal: mature bullocks for immediate sale in the north-western districts, and young stock which went to districts – principally in the Agra division, but

---

[76] Cooke, *Meerut SR* (1940), p. 8. By 1914 around 5,000 acres were reported by the Director of Agriculture to be under *saretha* cane, which had a yield advantage of around 20% in *gur* over all other varieties. UP Rev., March 1915, 65, H.R.C. Hailey, Historical Resume of Efforts Made to Develop Agriculture in UP.

[77] The distribution of carts among the villages of *pargana* Koil is suggestive of this pattern: the number of carts per 100 acres of n.c.a. was on average greater in the villages with less than half the land under canal irrigation than those with more than half.

[78] P.J. Musgrave, 'Rural Credit and Rural Society in the United Provinces, 1860–1920', in Dewey and Hopkins, *Imperial Impact*, pp. 216–32.

[79] According to the Settlement Officer's Notes in Muzaffarnagar Collectorate Records, Muzaffarnagar, Combined Assessment Statements, vol. v, Pargana Khatauli, 1325F (1917–18).

[80] Buck, *Meerut SR* (1874), p. 39.

also to parts of Muzaffarnagar, Aligarh, and Hardoi – where grazing was available and rearing less costly. The pattern was for the animals to be given light work there until reaching maturity, when they were resold to the Upper Doab districts at enhanced prices.[81]

The local economy, therefore, underwent a qualitative change as specialised services emerged in response to new demands and opportunities. Agencies for the hire and sale of sugar mills were scattered all over Muzaffarnagar; indeed, some of the 'more pushing traders' took to 'manufacturing them on their own account'.[82] A Khatauli carpenter had even applied for a patent for an improved sugar mill, apparently 'receiving his inspiration from some of the thriving sugar-planters in the neighbourhood'.[83] The demand of petty operators for simple agricultural machinery led to the establishment of fifteen flourishing factories in Meerut producing carts of various sizes in the years before 1908;[84] while their commercial requirements accounted for the remarkably 'large number of small towns' Muzaffarnagar contained.[85]

The institutional dimension

Land tenures

While relative factor proportions affected the adoption of specific innovations, so did institutions, especially tenures. Some institutional forms will not pose obstacles to the introduction of a technical advancement which generates new income streams through efficiency gains; moreover, the

---

[81] UP Rev., April 1914, 2, W.H. Moreland, Note on Cattle Supply of UP, p. 10.

[82] NWP&O Rev., Aug. 1886, 53, Note by C.J. Connell. J.O. Miller reported in 1880 that the demand for iron sugar mills had 'led to the establishment of large workshops in the principal towns for their manufacture and repair'. MCR, BOR, VIII/II, 14(c), Ser. 5, J.O. Miller, AR Tahsil Jansath (Muzaffarnagar), 1890 (?), p. 12.

[83] *Ibid.* Mention is also made of the 'fine iron-turning lathes in the Meerut district between the Hindan and the Jumna'. *IRR 1884–5*, p. 7C. In the context of the current discussion of 'appropriate technology', the ability to manufacture and repair locally is recognised as an important component of the successful diffusion of technology.

[84] A.C. Chatterjee, *Notes on the Industries of the United Provinces* (Allahabad, 1908), p. 137.

[85] UP Rev., Sept. 1916, 42, Note by W.J. Lupton.

potential for further such streams can trigger off a sequence of reinforcing technical and institutional adjustments. Equally, institutional factors may well prove an enduring obstacle to technical change, particularly where the gains from innovation are of insufficient benefit to groups with the power to perpetuate existing institutions. Further, institutions may favour specific types of technical change: examples are commonplace in the present-day underdeveloped world of the incongruence between relative factor scarcities and the type of technology in use – as in the deployment, for example, of capital-intensive and labour-saving appliances to produce, on scarce land, semi-luxury products for an elite market despite severe un/underemployment, capital scarcity, and widespread malnutrition. In short, the type of technology adopted – along with the product mix and the distribution of returns – reflects the structure of agrarian society.

Eric Stokes' hypothesis – which holds that peasant proprietorship was the key to growth in the west[86] – is supported by contemporary observers, like Auckland Colvin:

In the pergunnahs of this district which I have assessed the balance is altogether in favour of the village communities . . . [where] the villages are full and substantial. Cultivation is incessant, careful and of the best crops. The landholders are for the most part intelligent, pleasant men . . . their cattle numerous, and well-to-do . . . But in the villages owned by landlords of other castes – i.e. zemindaree villages – the site is covered by a few huts of malees or chumars. The cultivation

---

[86] Professor Stokes suggests that the dynamism and enervation exhibited by the respective regions are to be explained chiefly in terms of institutional constraints relating to tenure. The main distinction he identifies in this respect is the predominance in the west of the *bhaiachara* tenure, which accounted for over half the proprietary holdings in Meerut in 1940. This tenure was closely associated with owner–cultivator operations, and contrasted with the widespread landlord–tenant forms of production associated with the joint *zamindari* and imperfect *pattidari* tenures of the more secure eastern districts, where land scarcity made possible a quasi-rental surplus for landowners. The inherited tenure system was thus the critical difference between the two regions as far as their growth paths were concerned. Both felt the pressure of population and subdivision, but the owner–cultivator tenures adapted most readily to the need to increase production under the Indian system of *petite culture*, while in the east, 'landlord structures that had outlived their role . . . survived because even a minute fractional share in the joint patrimony validated caste status for other purposes'. Stokes, 'Dynamism and Enervation', pp. 232–5.

is in greater proportion of poorer crops. There is less manure, and hence less cane, cotton and maize.[87]

Landlordism in the east – characterised by subleasing, often at a rack-rent to low castes – when combined with the relative factor scarcities, constitutes an entirely plausible explanation of 'involution'. The picture in the west needs to be filled out a little, however, to take account of the fact that the *bhaiachara* tenurial form was not the only one in the Upper Doab, and that dynamism was not confined within the boundaries of the *bhaiachara* villages. Kandhla *pargana* may have had nearly two-thirds of its area recorded as *bhaiachara*, but another of the most fertile and thickly populated Muzaffarnagar *parganas*, Khatauli – which was predominantly owned by landlords – was described as an area where 'villages are rich and well-to-do . . . and good substantial houses, frequent groves and well grown cattle, attest as well as the fine crops of cane and rice and wheat the prosperity of the people'.[88] Indeed, figures for Khatauli in the 1880s show that neither the distinction between *bhaiachara* and *zamindari* tenures nor that between Jats and other castes explains the differences in the relative prosperity of the villages, measured by conventional standards such as the proportionate area under sugarcane, wheat, or *dofasli*. Details from the unambiguously single-*zamindari* and *bhaiachara* villages are summarised in Table 8.3. There are no significant biases in the sample introduced by way of soil quality, irrigation, or location; the tenurial specifications are distinct and, though the sample is not especially large, the inference might well be drawn (from the crude attempt at ranking the villages in Table 8.3) that it is the single-*zamindari* villages which display superior cropping features. Clearly the tenurial element in the explanation of dynamism in this area needs to be extended to take account of these findings.

A key element in this prosperity appears to be related to the possession of secure cultivating rights. Where cultivators possessed secure occupancy rights 'the characteristics of the

[87] 'Settlement Report of the District of Muzaffarnagar', p. 106, in Cadell, *Muzaffarnagar SR* (1882).
[88] UPA, BOR, 'Different Districts', Muzaffarnagar, Box 8, File 35, A. Cadell, RRR Pergunnah Khataoli, 1872, p. 6.

Table 8.3. *Cropping features in relation to tenure, irrigation, and caste: village statistics from* Pargana Khatauli, Muzaffarnagar, *c. 1890*

Rank (based on intensity of cash crop cultivation)[a]	Village name	(B)/(Z)[b]	Tenure Percentage of total cultivated area under:			Cash crops and *dofasli* (average for 1877–88) Percentage of total cultivated area under:			Principal caste of cultivators	Irrigation (percentage of total cultivated area) (C) = Canal (W) = Well
			Sir + khudkasht	Occupancy + expro-prietary	Tenants-at-will	Sugar-cane	Wheat alone	Dofasli		
1	Rukanpur	(Z)	0	100	0	32	36	21	Rawa	98 (C)
2	Gangdhara	(Z)	4	92	4	24	52	13	Rawa	88 (C)
3	Mohammadpur	(B)	30	22	48	28	30	23	Shaikh	93 (C)
4	Kheri Rangran	(Z)	2	92	6	24	35	22	Rajput	87 (C)
5	Madhkarimpur	(Z)	1	79	20	25	34	26	Rajput	99 (C)
6	Bainsi	(Z)	0	92	8	20	36	17	Jat	80 (C)
7	Barsu	(Z)	1	82	17	15	41	15	Sani	70 (C)
8	Sardhan	(Z)	0	86	14	23	33	14	Rawa	67 (C)
9	Sherpur	(Z)	0	82	18	19	41	8	Jat	82 (C)
10	Matheri	(B)	69	11	20	18	33	17	Jat	91 (C)
11	Chandsamand	(B)	3	36	61	10	49	8	Gujar	30 (W)
12	Dahaur	(Z)	0	91	9	18	36	11	Rajput	78 (C)
13	Munawarpur Khurd	(B)	38	11	51	15	33	14	Taga	67 (C)
14	Rustampur	(B)	71	3	26	12	38	12	Taga	95 (C)
15	Sonta	(B)	82	13	5	17	34	10	Jat	66 (C)
16	Naola	(B)	63	9	28	9	39	13	Taga	52 (C)
17	Sikandarpur Khurd	(B)	31	7	62	9	33	17	Pathan	19 (W)
18	Akbarpur	(B)	78	11	11	13	31	13	Taga	73 (C)
19	Kakrala	(Z)	0	67	33	10	37	8	Gujar	38 (C)
20	Anchauli	(B)	70	10	20	5	41	6	Rajput	6 (W)
21	Majahidpur	(B)	68	10	22	10	25	12	Rajput	40 (C)
22	Ambarpur	(Z)	0	84	16	7	37	3	Taga	25 (C)

[a] The ranking used here is based simply upon the combined total for each village of the sugarcane, pure wheat, and *dofasli* percentages, arranged in descending order. This is admittedly a crude measure of cropping intensity, but the order is not too dissimilar to a ranking based solely upon sugarcane percentages.

[b] (B) = *Bhaiachara*. (Z) = Single-*zamindari*.

village are the same as in the village communities', Colvin noted of Muzaffarnagar.[89] Secure occupancy rights were often synonymous with moderate rents, and settlement officers were struck by the moderation (and stability) of occupancy rents in the north-west.[90] In 1889, the prevailing occupancy rates in Jansath *tehsil* were found to be those fixed at the previous settlement: 'Practically all occupancy tenants already hold at these rates.'[91] Moreover, in the western *parganas* of Muzaffarnagar – long the recipients of EJC water – 'numbers of occupancy tenants' were found to 'still pay the old dry rates for land that had long been irrigated, and for which their landlords are paying owner's rates'.[92] In eastern Shikarpur there were villages alongside the canal in which occupancy tenants were paying Rs 3 or a little over for irrigated land for which a tenant-at-will would pay Rs 9 or Rs 12.[93] Indeed in the discussions of the early 1870s over the application of permanent settlements to the area, one powerful reason raised against it was the low level of assessments current in Bulandshahr and Meerut because rents were low. Bulandshahr, due to the large numbers of occupancy tenants, displayed an 'artificially low' rent level; while in Meerut, despite the spread of canal irrigation and increased prices, rents had hardly risen at all: 'All over the country, and especially in this pergunnah [Baghpat], we find the low rents of older days, when landlords were glad to entice tenants at almost nominal rents in order to lighten their own burden.'[94] And yet we know that competition for land became increasingly keen in the western districts. The provision of canal facilities frequently ran ahead of local capacity to utilise the water, especially in previously sparsely populated tracts such as Jansath and Bhukarheri, but natural population growth, migration, and the accumulation of labour-saving capital eroded the scarcity value of tenants. How is it possible to account for this striking situation, whereby market forces failed even over the long run to equate the price

[89] 'Settlement Report of the District of Muzaffarnagar', p. 106, in Cadell, *Muzaffarnagar SR* (1882).

[90] NWP&O Rev., Aug. 1886, 53.

[91] NWP&O Rev., Jan. 1891, 15, J.O. Miller (Settlement Officer) to Meerut Commr., 4 Oct. 1889.

[92] *Ibid.*   [93] *Ibid.*

[94] NWP Rev., Feb. 1872, 58, GI (Home Dept.) to NWP Govt., 26 May 1871.

paid for the use of land with its relative scarcity? The question is important because of the role of secure tenure and moderate rents in facilitating capital accumulation and property improvement.

Market forces were, of course, often not the key factor in the determination of rent levels. 'Rents here', wrote the Etawah Collector, 'as in other places, depend much more on local competition and on the relative position and strength of the zemindars and tenants than on any accurate calculation of the cost of cultivation and profits.'[95] Frequently rent levels reflected traditional social norms: 'Many of the old fashioned yeoman proprietors raise their rent very slowly', observed one official of Aligarh. 'They cultivate a large seer, and from this they make their chief profit, not from their rents. Custom, too, and public opinion, have as yet prevented any great increase in rents in these villages.'[96] The same tendency was noticeable in *bhaiachara* villages, where the rentals were 'largely made up of payments that are merely nominal or revenue quota'.[97]

In *zamindari* estates, low occupancy rates were often the result of lax estate management. In 'weak' estates – like the Skinner estate in Bulandshahr – proprietors abstained from raising rents to anything approaching the value of land because, 'from historical reasons, the tenants are too powerful for them'.[98] In Saharanpur, 'landlord and tenant are more on an equality than even in Muzaffarnagar. The landlord cannot always appropriate the best land and sometimes has to be content with an inferior holding.'[99] Moreover, the Upper Doab cultivators were apparently more able to protect, and even extend, their occupancy rights than was the case elsewhere. While over most of the UP occupancy status was especially

---

[95] NWP&O Rev., Jan. 1892, 53, Note by E.B. Alexander.

[96] A.B. Patterson, 'Report on the Assumed Rates to be Used in the Settlement of Tehseel Eglas', 27 Aug. 1870, *Revenue Reporter*, v, n.d., p. 43.

[97] MCR, BOR, viii/ii, 14(c), Ser. 2, T. Stoker, AR Tahsil Sikandarabad (Bulandshahr), 1886, p. 14.

[98] Patterson, 'Report on the Assumed Rates', p. 43. Low rent rates sometimes concealed the fines and *douceurs* extracted by the agents, *karindas*, *mukaddams*, *patwaris*, and suchlike. UP Rev., May 1888, 31, Settlement Officer to Meerut Commr., 31 Oct. 1887.

[99] UP Rev., March 1887, 22, Note by D.M. Smeaton (Dir. of Agriculture).

endangered when famine disrupted production,[100] the security and wealth of Meerut division gave landlords few opportunities to obtain ejectments on the grounds of failure to pay rent.[101] When the landlord resorted to the courts, either to seek an ejectment on grounds other than failure to pay rent or to obtain a rent enhancement, the Meerut *maurusi* tenant was well able to withstand such assaults on his rights: 'He meets his landlord more or less upon equal terms in a court of justice. He asserts his rights with extreme persistency and is quite as ready as his landlord to indulge in the luxury of an appeal.'[102] The smaller landlords were reluctant to go to court for enhancements, and where applications were at all numerous – usually in areas which had just received canal water – close examination frequently revealed that the applications were confined to a few villages. Three villages accounted for 109 out of 115 cases in Shikarpur (Muzaffarnagar) brought during the currency of Cadell's settlement; two villages in Bagra accounted for 122 out of 148, and one in Charthawal (a village of wealthy landlords benefiting from the canal) accounted for 439 out of 734.[103] Even where the larger landlords did apply for enhancements, the occupancy tenants were treated with 'much leniency' by courts which refused to increase rents or awarded only inadequate enhancements after the introduction of canal irrigation.[104]

Far from being deprived of his rights, the Meerut tenant was frequently extending the area over which he had secure tenure. The area under occupancy rights increased by 35,000 acres in Muzaffarnagar between 1892 and 1920, prompting the Settlement Officer to comment: 'This bespeaks a prosperous peasantry, for such valuable rights are normally allowed to accrue only with a substantial payment to the proprietor.'[105]

The courts were merely one element in the distribution of power in the villages, where the dictates of the market were circumscribed by the cost of enforcing the terms of any transaction. Gujar communities in *tehsil* Budhana (Muzaffar-

---

[100] UP Rev., Nov. 1885, 59–70.   [101] *RAR 1882–3*, p. 11.

[102] *Ibid.*, Appendix I, pp. 1–2.

[103] NWP&O Rev., Jan. 1891, 22, J.O. Miller (Settlement Officer) to Meerut Commr., 22 May 1890.

[104] See e.g. Miller, *Muzaffarnagar SR* (1892), pp. 60–1.

[105] H.A. Lane, *Muzaffarnagar SR* (1921), p. 8. The price paid in the form of *nazrana* for perpetual leases came to Rs 180 per acre in Jansath, or 23 years' rent. *Ibid.*, p. 9.

nagar) – not the most careful cultivators – were, nonetheless, 'prosperous and more successful than the Jats in avoiding transfers to outsiders; they withstand, frequently with success, the attempts of purchasing Bohras or Mahajans to raise prevailing rents'.[106] In Jawan village, *pargana* Khatauli, according to the Settlement Officer's handbook notes, the Rajput tenants restricted competition for village lands by the effective expedient of not allowing tenants of other villages to cultivate there, 'causing damage to their fields if they attempt to do so'.[107] Rajput proprietors in some Budhana villages, especially Itali and Nagwa, found it impossible to wrest the best village lands – held at low occupancy rates – away from the Jat cultivators whose homes were outside the village.[108] Not only were the occupancy tenants often substantial cultivators with large holdings, but they often possessed proprietary rights in land in their own villages, especially in Jat areas, where the Jats' 'own estates will not . . . contain them; they have burst forth and travel far to cultivate lands assessed at lighter rates, yet they hold to their villages, and will not leave the infinitesimal ancestral shares in order to better themselves in other tracts less populated than their own'.[109] Instead, they extended into new estates developed near their ancestral villages. In *pargana* Bhukarheri, recently provided with canal water, Cadell noted that out of 48 townships, 'Jats and Jhojhas predominate largely in 35, and Gujars in only 9 and in all of these Jats and Jhojhas, residents of other villages, cultivate largely'.[110] Jat communities bargained collectively, strengthening the bargaining power of individual cultivators, and preserving rights gained in periods when tenants were comparatively scarce. 'Solidarity of caste feeling', after all, was

[106] MCR, BOR, viii/ii, 14(c), Ser. 5, H.C. Ferard, AR Tahsil Budhana (Muzaffar-nagar), 1890, p. 3.

[107] Muzaffarnagar Collectorate Records, Muzaffarnagar, Combined Assessment Statements, vol. v, Pargana Khatauli, 1325F (1917–18).

[108] NWP&O Rev., Jan. 1891, 25, H.C. Ferard, Note on Tahsil Budhana (Muzaffar-nagar), 15 May 1890.

[109] Forbes, Baraut Pargana Report, p. 10, in Buck, *Meerut SR* (1874).

[110] UPA, BOR, 'Different Districts', Muzaffarnagar, Box 8, File 35, A. Cadell, RRR Pergunnah Bhukarheri, p. 14. In Chaprauli and Baraut, also, Jats cultivated 'the greater part of other villages where they have no proprietary rights'. Forbes, Chaprauli Pargana Report, p. 8, in Buck, *Meerut SR* (1874).

'more pronounced in the Jat community than in others',[111] and W.A. Forbes notes the instance of an estate being sold on the Ganges bank in Meerut, where 'a large number of Jats from scattered villages, in no less than three pergunnahs on the Jumna side of their district, clubbed their resources, bought the village, and sent forth a colony to inhabit and cultivate it'.[112]

Collective action of a different kind was encountered in parts of Bulandshahr, the district which marked the transition from the predominance of owner–occupier tenure to that of landlord and tenant.[113] Here powerful *zamindars* made 'persevering and only too successful efforts . . . to eject their hereditary tenants', particularly in the south-eastern *parganas*.[114] They harassed *maurusi* tenants until they lost their occupancy rights, largely through the enforced exchange of fields.[115] The tenants normally continued to cultivate as tenants-at-will. 'As a result, in all the villages owned by the different members of the wealthy and influential Lalkhani family', wrote the Buland-shahr Collector, 'it is exceedingly rare to find a tenant who ventures to claim a right of occupancy.'[116]

The pressure which large *zamindars* were able to exert upon their tenants in this area of south-east Bulandshahr was partly due to the reliance on renting of the local cultivators. They did not possess the considerable advantage of secure holdings in their home villages like the Jats in the northern districts. Their weakness was exacerbated by oligopolistic collusion on the part

---

[111] Lane, *Muzaffarnagar SR* (1921), p. 9.

[112] W.A. Forbes, 'Memorandum upon the Mode of Ascertaining and Fixing Rates of Assessment', 17 Nov. 1866, p. 4, in Buck, *Meerut SR* (1874). In such a way Jats frequently competed 'with bunnias to buy encumbered estates'. Forbes, Baraut Pargana Report, p. 10, in Buck, *Meerut SR* (1874).

[113] According to Currie's settlement figures, 1,206 estates were classified as *zamindari*, 324 as *pattidari*, and 273 as *bhaiachara*. Currie, *Bulandshahr SR* (1877), p. 73.

[114] NWP&O Rev., March 1880, 23, Note by E.B. Alexander.

[115] See NWP&O Rev., Nov. 1885, 69, D.F. Addis (Joint Magistrate) to Bulandshahr Collr., Memorandum on Loss of Occupancy Rights, 5 March 1885, which describes the process in detail, with special reference to Addis' investigations into Ransi village following its 1879 purchase by the *Rais* of Danpur.

[116] NWP&O Rev., Sept. 1881, 10. The *zamindars*' agents also welcomed the destruction of occupancy rights. It increased their influence within the village and enabled them to benefit from the customary practice whereby the receiver of a fresh *patta* paid one rupee to the *karinda* or *patwari*. There were, of course, no such renewals where the tenants were *maurusi*. NWP&O Rev., Nov. 1885, 59, Note by R.L. Singh (Deputy Collr.).

of large *zamindars*. Ex-*maurusi* tenants stayed on as tenants-at-will because 'in addition to the common "love of home" . . . among several of the large taluqdars in these parganas there is a private but well-known understanding that no tenant deserting a village of one of them without his permission should find a house to live in or a field to cultivate in any village of another'.[117] To prevent the accrual of occupancy rights, tenants were made to take different fields each year, rendering all effort at improving or maintaining soil fertility pointless. The contrast between the lightly managed Paikpara, Malkena, and Skinner estates and those of Danpur, Dharampur, Pindrawal, and Chhitari further south was marked: 'Cultivation has invariably increased', noted Deputy Collector Singh, 'in the estates where the right of occupancy has not been largely extinguished and its growth not much interfered with, [while] in estates where the right of occupancy is differently treated' the reverse was the case.[118] The varying balance of strength produced a patchwork of contrasting effects in this district. Where occupancy rights were numerous, tenants felt sufficiently secure to sink masonry wells and plant mango groves.[119] Nowhere was the patchwork pattern more noticeable than in the north-eastern *parganas*, Agauta, Siyana, and Shikarpur, where,

in the highest rented circles, the pressure of rents was felt. Undoubtedly the outward signs of diffused prosperity which appeared to me so noticeable in this district as compared with Basti are much better marked in the east of Agauta and the west and centre of Siyana than in the west of Agauta and more especially than in Shikarpur. The houses are larger and better, and the stock of cattle round the homestead is on the whole distinctly better than in the higher rented parganas.[120]

Tenurial realities, then – which are distinct from legalistic revenue categories – were critical in determining the cultivators' security, independence, and incentives, and were reflected in crop patterns, yields, and capital accumulation. The pattern of use – even the very adoption – of so fundamental an

[117] NWP&O Rev., Nov. 1885, 59, Note by R.L. Singh (Deputy Collr.).
[118] *Ibid.*
[119] Currie, *Bulandshahr SR* (1877), p. 63.
[120] MCR, BOR, VIII, 14(c), Ser. 2, D.C. Baillie, AR Parganas Agauta, Siyana, and Shikarpur (Bulandshahr), 1889, p. 35.

agricultural input as canal irrigation was conditioned by tenures. One canal engineer remarked upon the 'rapid extension of irrigation in the villages owned by Bhaiachara communities [while] where the land belongs to a large zemindar the increase, if any, is slight' – apparently cultivators in *zamindari* villages were afraid that landlords would use the advent of canal irrigation to enhance their rents and deprive them of *maurusi* rights.[121]

The north-western districts were by no means without their unprotected statutory tenancies – insecure and liable to rack-rents – which amounted to just below one-third of the cultivated area in Muzaffarnagar and Meerut.[122] In fact, rents paid by Meerut division tenants-at-will were the highest in the UP, since with so little land coming onto the 'market', competition was fierce.[123] Cultivators from the canal villages were often forced to pay wet rates for a few *bighas* of unirrigated land for their dry crops. This was especially the case, apparently, in Meerut's Baraut *pargana*, where

proprietors of canal irrigated land often wish to keep this for growing cane or some other valuable crop, but they also want a little land to grow kharif on, either for their own food or for that of their cattle. This land they seek in villages held by zamindars with a large ghair-maurusi area, and to get it they offer to pay as high as Rs 6, 7 or even 8 an acre.[124]

But this situation did not in any way hamper the incentive to growth and development in the Upper Doab because of the way in which it was incorporated into the production system. The same cultivator who (as an occupancy tenant or proprietor in a

[121] NWP&O PWD(I), Feb. 1878, Irrigation Revenue Report 1876–7, IOR pagulation 120.

[122] Lane, *Muzaffarnagar SR* (1921), p. 8. The combined area under *sir, khudkasht,* exproprietary, and occupancy tenures in Muzaffarnagar increased in proportion to the total area during the period between the 1892 and 1921 settlements. By 1921 the 'secure' tenures covered more than two-thirds of the total area. *Ibid.,* p. 23.

[123] The gap between the average rent paid by occupancy tenants and that paid by tenants-at-will (Rs 6½ as against Rs 15) was noticeably greater in the Upper Doab than elsewhere in the UP. *United Provinces Provincial Congress Committee: Agricultural Distress in the United Provinces* (Allahabad, 1932), p. 25.

[124] NWP&O Rev., Nov. 1879, 21. For a similar example from Bulandshahr, see NWP&O Rev., Aug. 1888, 39, T. Stoker (Settlement Officer) to Meerut Commr., 29 March 1887.

bhaiachara village) improved his irrigated land was willing (as a tenant-at-will) to pay high rents for unirrigated land (which was not susceptible to much improvement) on which dry crops could be grown, freeing his irrigated land for more valuable crops – and for investment in improvements.

Cultivators in Meerut's Dasna *pargana*, where, as in south-east Bulandshahr, there were no large areas of secure proprietary holdings and occupancy rights, were less independent in their crop choices, and the pattern of utilisation of canal water supplied to the tract reflects this. There was no agricultural reason why this Meerut *pargana*, which was right alongside sugar-growing estates, should not produce sugar.[125] Yet it was 'little grown', and its cultivation showed no signs of increasing, because cultivators gave 'greater attention to indigo'. This was due to the presence of a huge indigo factory at Masuri, owned by a European, Michel, with a large estate in the canal circle, and to the several smaller *kothis*, whose *zamindar*–owners were reported to 'naturally take pains to keep up the cultivation of indigo'.[126] Indeed, to E.B. Alexander Dasna exemplified

the point of great change . . . between the upper portion of the Meerut and the lower portion of the Bulandshahr districts. We here begin to pass out of the country of bhaiachara villages with high cultivation and large cane areas into that of large zamindars with ghair-maurusi tenants and indigo cultivation unduly forced up for the benefit of the owners of the estates who mostly keep indigo factories.[127]

Up to the turn of the century, when chemical dyes dramatically affected the demand for indigo and caused agriculturalists to turn to cotton, canal villages in the central and lower sections of the Doab were marked by the widespread cultivation of the dye plant. Landlords, by the assertion of their control over access to land, by the use of their credit resources, and effectively by the provision of marketing services, were frequently able to determine a significant part of local production, bringing into existence patterns of income dis-

---

[125] Indeed, sugar replaced indigo after the closure of the Masuri factory. *Meerut DG* (1904), p. 43.
[126] UPBORL, IV(c), Ser. 25, Handbook of Tahsil Ghaziabad, Meerut, *c.* 1900, p. 5.
[127] NWP&O Rev., Nov. 1879, 21.

tribution, consumption, and employment, forms of capital accumulation, and local multipliers which differed from those of the sugarcane areas. The whole nature of the local economy, including its capacity for growth, was thus affected.

It would, of course, be difficult to argue that tenure alone held the explanation for the lower level of prosperity and growth achieved in the districts further down the Doab. It may have been a key factor in Dasna, where conditions in the surrounding area were comparable; but the central and lower parts of the Doab, were *reh* was common and the soils were largely unsuitable for sugarcane, were on the whole less fertile than either the Upper Doab or the eastern districts. Overall, their population was denser; their infrastructure was less developed; they had less canal irrigation and more canal water was given 'lift'. But even if these districts were less dynamic and wealthy than Meerut division, they were still better off than the east, where the norm was 'not one of peasant proprietors, but of a petty landlord class driven close to the soil but still continuing to exploit it largely through lower caste sub-tenants and predial labourers'.[128] Such a situation favoured maximum use of *variable* inputs (rather than fixed ones), a tendency encouraged by the abundance of labour. It produced an output mix and technical regime which reflected the pressure on land in the context of static technology, and a system of carefully layered consumer entitlements as an alternative to capital accumulation and increased productivity.

Competition and efficiency in marketing facilities

The striking contrast between the marketing systems of the eastern and western UP contributes significantly to the divergent growth patterns. In the east coarse rice was raised in numerous scattered villages; and – partly because rice was retained in the village to pay wages – only a small proportion of the crop was destined for the marketplace. Indeed, rice-marketing (as compared to wheat- and *juar*-marketing) in India is generally unorganised and local.[129] Poor communica-

[128] Stokes, 'Dynamism and Enervation', p. 239.
[129] See U. Lele, *Food Grain Marketing in India: Private Performance and Public Policy* (New York, 1971), p. 44.

tions favoured the periodic markets (*painths*), in which itinerant traders linked the peasant to his market. The rice trade was 'in the hands of small men and concentrated in village markets or small mandis'.[130]

The market system in the western UP corresponded more closely to that existing in the Punjab districts of Jullundar, Ludhiana, and Montgomery, where intensive wheat crops were produced in large concentrated settlements. *Painths* were plentiful – *tehsil* Koil had 33 around 1900 – but they operated as retail centres only.[131] This network of weekly and bi-weekly gatherings, linked together in a complex and overlapping manner, provided a limited range of frequently required consumer goods and services.[132] Apart from petty quantities of grain, mostly used for exchange purposes, *painths* were not part of the assembling market network, which mainly consisted of a network of substantial *mandis* served by their satellites, or secondary *mandis*, which did not possess storage facilities and operated as collecting centres for the main ones – sometimes for the harvest period only.[133] Although specialisation naturally reflected local production – Baghpat, Khekara, and Dankaur,

---

[130] *United Provinces Provincial Banking Enquiry Committee, 1929–31, Report* (Allahabad, 1931) (hereafter *UPPBEC(R)*), p. 152. For example, it was reported of Jaunpur: 'There is not much trade. The two marts of Shahganj and Mungra Badshapur are the only places which show any activity of business.' *RAR 1884–5*, p. 80. On the contribution of the transport difficulties to the local marketing situation the comments of the Collector of Gorakhpur made in *RAR 1882–3*, p. 6, are particularly graphic: 'At least one-half of the grain which arrives in Gorakhpur arrives by river. The only metalled roads in the district are the military roads to Benares and Fyzabad. The kutcha roads are everywhere difficult and some of the most important of them are quite impossible for carts during the rains. The race of bullocks is unusually small, and with such roads the carts must be unusually small also. The consequence is that much the greater part of the harvest never comes in to the market at all, but is stored and ultimately consumed where it is grown, and grain, instead of money, is the usual medium of payment in the villages. The amount of grain in store at present must be very large, probably equal to a whole year's consumption of the district.' Perhaps the most frequent complaint of the eastern districts in this regard related to the lack of feeder roads to link up with the railways. Alan Cadell complained in the 1880s that although the East India Railway had been opened in Mirzapur in 1863, 'not a single mile of metalled road has been constructed to facilitate the approach to it of traffic from the south'. *Ibid.*, p. 5.

[131] *Report on the Marketing of Wheat*, p. 136.

[132] See G.W. Skinner, 'Marketing and Social Structure in Rural China', 2 pts, *Journal of Asian Studies*, XXIV (1964–5), pp. 3–43, 195–228.

[133] Kandhla, in Muzaffarnagar, was such a case.

for instance, being concerned more with *gur* and *shakar* than with wheat – the major centres, Ghaziabad, Hapur, Meerut, Shamli, Deoband, Muzaffarnagar, Saharanpur, Hathras, and Sikandarabad, were major wheat *mandis*. The grain stocks of Ghaziabad and Hapur were never less than 85% wheat.[134] These centres possessed substantial storage facilities in the shape of *khattis*. There were 3–4,000 *khattis* – normally with a capacity of around 700–800 *maunds* – in Hapur by the late 1920s.[135] These bulking facilities were owned or hired (at Rs 20–50 per year) by the moneylenders who financed the movement of produce from the rural hinterland. Unlike the sack storage in the *kothis*, or *godowns*, of the large consumer markets such as Cawnpore and Lucknow, where wheat had usually to be disposed of before November to minimise losses from weevils, *khatti* storage was safer and cheaper.[136]

While, as in the eastern markets, the smaller *mandis* were the preserve of local traders and *banias*, activity in the larger markets was more directly responsive to the international wheat market; it was here that the exporting firms made their purchases and the *banias* gave way to adequately financed commission agents (*pakka arhatias*) frequently operating both on their own account and on behalf of large wholesalers and exporting firms.[137]

In line with the change in the character of the market network in the western districts came distinct developments in the organisation and practices of the larger *mandis*. One such development was the practice of dealing on a 'clean wheat basis', whereby dirt and impurities had to be removed. The canal colony markets of the Punjab operated such a system, and the larger centres – among them notably Ghaziabad,

---

[134] For details of the market network, see T. Prasad, *The Organisation of the Wheat Trade in the North-West Region of the United Provinces* (Allahabad, 1932), p. 1 *et seq.*; and J.K. Pande and R.B. Gupta, *Prices of Cereals in the United Provinces* (Allahabad, 1938), p. 10.

[135] *Khattis* were to be found at Ghaziabad, Hapur, Deoband, Shamli, Meerut, and Muzaffarnagar. Muzaffarnagar possessed 1,500 such pits and Meerut possessed 100, though evidence suggests that the *khattis* in Meerut were of a smaller size. Saharanpur possessed none, owing to the high water level, although Muzaffarnagar had earlier solved this problem by converting its *kachha khattis* into *pakka*. *UPPBEC(E)*, II, p. 125.

[136] *Report on the Marketing of Wheat*, pp. 205–16.

[137] For fuller details, see Prasad, *Organisation of the Wheat Trade*, pp. 2–6.

Hapur, and Muzaffarnagar – adopted the same practice: a move connected with the improvement in the quality of wheat varieties.[138] The physical–spatial organisation of *mandis* also underwent rationalisation during the last two decades of the century. The older *mandis* – Cawnpore, for example – were organised in an inefficient and costly way, particularly from the seller's viewpoint:

The efficiency of the market lies mainly in the trading conveniences and facilities which it provides for the speedy determination of prices – which should be uniform for the same quality of produce – and for the easy and economical handling of the goods brought into it. In those markets where the shops and godowns of both sellers and buyers are situated in the heart of the city or town, or are far away from the railway stations or docks, the costs of assembling and distribution are relatively high. In addition, the inconvenience caused to the trading public may be considerable and the conditions are such as to foster malpractices on the part of the less reputable traders. Further, the weighman and other labourers, who would, if the market were centralized, be able to cope with many carts daily, are unable to handle more than a few.[139]

Marketing charges in such markets were higher; where the marketing practices were standardised, charges were significantly lower – particularly as they included fewer miscellaneous levies. In the western districts, the average total merchandising charge for a sample of ten markets came to Rs 3.5 per Rs 100 (with Hapur the lowest at just below Rs 2.5); in the central UP charges ranged from just under Rs 5 in Banda to Rs 8 in Bareilly; and further east, in Benares, the charge was Rs 7.5.[140]

The dominant pattern of marketing in the less advanced areas involved significant reliance on itinerant local traders (*beoparis*). Operating either on their own resources or with working capital provided by *banias* in nearby *painths* or small towns, they moved a large part of the marketed surplus, usually

[138] *Report on the Marketing of Wheat*, p. 151.
[139] *Ibid.*, p. 138. In the case of Cawnpore, there was insufficient room at the shops to clean and weigh the grain, with the result that this took place three to five miles from the point of sale. The *kachha arhatia* negotiated the sale before sending the weighman and labourers to the buyer. Should the latter not approve of the deal, it was frequently too late in the day for the process to be repeated. *Ibid.*, pp. 136–8.
[140] *Ibid.*, pp. 161–3.

by camel, to local market centres. In the more commercial areas the process significantly departed from this pattern. In the north-west it was estimated that about one-third of the marketed wheat was taken to the assembling markets by the cultivators themselves.[141] As the Collector of Meerut observed in 1891:

In this district, I have been much struck by the fact that the village bania is not what he used to be; he no longer exclusively collects agricultural produce and brings it to market; it is the agriculturalist himself who loans his own cart or hires mules and donkeys and brings in his produce to sell in the open market.[142]

In this area the village *bania* was estimated by Prasad to account for only 30% of the wheat produced and disposed of (see Figure 6). Much of this was obtained through the *kachha hisab* loan system, and a significant proportion was retained by him for village consumption and seed requirements. Of the wheat coming into the *bania*'s hands, typically around one-quarter was dispersed within the village, or by timely releases to nearby markets, prior to the time when weevil damage threatened the stocks.[143] Village buyers constituted the other major marketing agency, the category being made up of small-town grocer–*banias*, village *kumhars* (potters), and the *beopari*, who was in fact usually a cultivator possessing a cart. These village buyers supposedly accounted for a third of all the wheat marketed, taking the produce straight from the threshing floor, for cash, and moving it immediately to the market. Of this proportion, around two-thirds was handled by the grocer–*bania*, who often kept an account with a local *arhatia* in a nearby *mandi*;[144] while the *kumhars* – who owned mules and donkeys to carry clay – went into trade when their hereditary occupation was undermined by the replacement of traditional earthenware by metalwares.[145] The clientele of the village

---

[141] *Ibid.*, p. 127.

[142] *RAR 1890–1*, p. 74; see also *UPPBEC(E)*, II, p. 339, relating to Kandhla *mandi* in Muzaffarnagar district, to which, it was reported, 'the principal crops of the area are brought . . . by the cultivator'.

[143] Prasad, *Organisation of the Wheat Trade*, pp. 28, 133. Very occasionally in this area the *bania* would be financed by an *arhatia* in a nearby town, thus obliging him to release a greater proportion at harvest time.

[144] *Ibid.*, pp. 28–9. [145] *Ibid.*, p. 13.

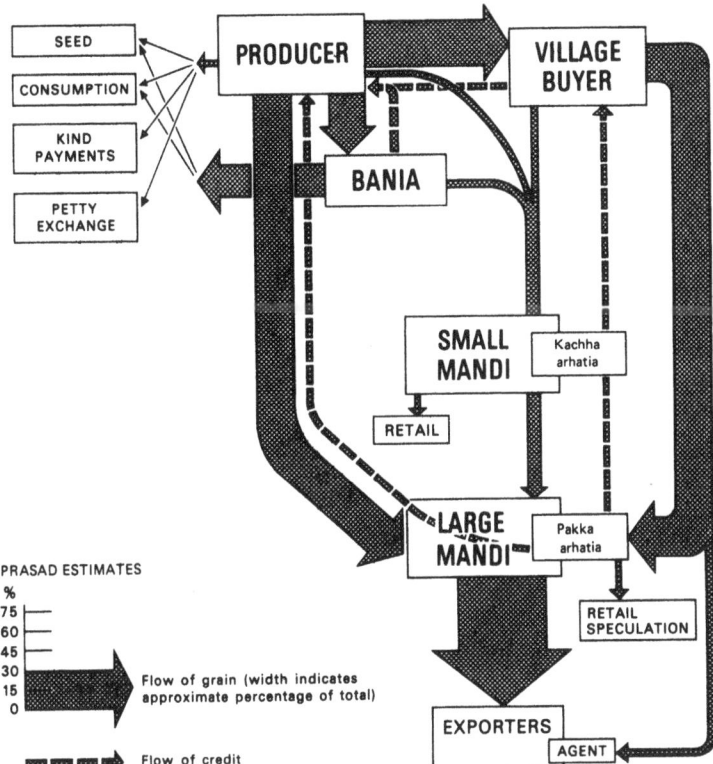

Figure 6. Meerut grain flows, 1920s. (Source: Percentage estimates from Prasad, *Organisation of the Wheat Trade*.)

buyers consisted of the smaller men requiring cash sales, as well as those cultivators with no time to go to the market. In some areas, yet another type of village buyer was to be found. In the late 1920s it was reported of Kandhla that 'sometimes the agent of the wholesale grain merchant goes to the villages and buys the grain there and then he brings the produce to market'.[146]

Finance was required to integrate the system and keep the grain flowing upwards. The tentacles emanated from the *pakka*

---

[146] *UPPBEC(E)*, II, p. 339. During the 1930s depression landlords became a more important assembling agency than previously, due to the growth in the practice of collecting rents – as well as repayments of seed and maintenance loans – in kind and thus marketing their tenants' produce as well as their own. *Report on the Marketing of Wheat*, p. 129.

arhatias in the large markets, dealing either as agency houses for large exporter–millers, or on their own account. Being owners of *khatti* stocks, these large traders were able to employ these as collateral in borrowing operating capital from *shroffs* and, increasingly, joint stock banks and the Imperial Bank. In Hapur alone, the banks advanced an estimated Rs 50 *lakhs* to *arhatias* during the wheat season.[147] Such funds would often be used to finance the operations of the *kachha arhatias* in the smaller *mandis*, and the system was sometimes extended to include *beoparis* – even cultivators – in order to attract custom.[148] This practice of financing the producer himself developed as an inevitable corollary to the growth in producer-marketing: the richer cultivators bringing their produce to the market themselves were 'in some cases looking to the *arhatia* rather than the village *bania* for finance'. In the same manner, 'large buyers who themselves are shippers overseas are now beginning to deal direct with large cultivators'.[149]

What did this marketing structure mean for the prices received by the cultivator? Within the confines of internationally determined price levels adjusted for long-distance transport costs, prices in local markets were determined by local transport costs and the marketing charges involved in the successive stages of bulking – the levels of which reflected the degree of technical and organisational efficiency as well as the relative bargaining power of those involved. In both transport and marketing the structure in the western UP was relatively competitive, with efficiency levels high by Indian standards.

The market for transport services was generally free from serious imperfections. This was ensured by the atomistic character of transporting activity: entry was free, substitution of pack animals for carts was feasible, and there were many cultivators in possession of such facilities who could be drawn out by seasonal prices.[150] If no significant element of monopoly

---

[147] *UPPBEC(R)*, p. 62.

[148] A case is cited of one *arhatia* who financed *beoparis* and cultivators to the tune of Rs 3,500; indeed, it was not unknown for cultivators to 'bank' with their *arhatia*. *Ibid.*, p. 266.

[149] *Ibid.*, p. 166.

[150] Monopoly power could potentially be exerted by railway companies, but this did not seem to be the case in the local context since the advantage of rail transport was in long-haul transport. Pande and Gupta noted that *thela* charges were invariably

is noticeable with respect to transport, a similar situation existed with regard to itinerant traders. Again, entry into *beopari* activity was fairly unrestricted: it was frequently carried on by cultivators who owned a cart and were able to take a break from actual cultivation. As a result the Banking Enquiry found that 'the margin . . . allowed to the *beopari* is little more than the cultivators would have to spend if they themselves went to the market. If the *beopari* can make a profit out of this margin it is because he has specialised in the business and can effect various economies which the ordinary cultivator would fail to secure.'[151] One calculation put the allowed margin at around 0.75% less than the current wholesale rate, taking into account the incidental charges confronting the cultivator,[152] and in the light of opportunity costs and – particularly important to smaller peasants – the risks involved, the terms offered by the *beopari* (usually involving a fairly immediate cash settlement) were attractive to the less substantial cultivator.

The very advantages of the *beopari* – possession of transport, knowledge of the *mandi* procedures, and acquaintance with the *arhatia* – were qualifications not difficult to acquire in the conditions of the north-west, particularly given the competition for patronage between *arhatias* in the major marts.[153] Remoteness clearly interfered with the acquisition of such contacts and knowledge by the operating peasants. Even so, it would be hard to press the case that the *beopari* operating with his camels in the outlying *nagla* would be able to exact monopoly profits on a regular basis; higher unit movement costs and the uncertain rate he could obtain through his quick-turnover system (using the local *painth*) would have an effect on his margins. Nor did monopoly access to knowledge of current prices unduly shift

below those of the railways and that grain was transported by road between *mandis* 'at considerable distance from each other'. Pande and Gupta, *Prices of Cereals in the UP*, p. 71.

[151] *UPPBEC(R)*, pp. 265–6.

[152] *Report on the Marketing of Wheat*, p. 133. *Beoparis* were able to effect small economies in allowance for refraction and other charges in kind and in weighment.

[153] Frequently, *arhatias* despatched agents two or three miles out along approach roads to tout for trade; the advances sometimes given to *beoparis* and cultivators were for the same object. 'Competition between *arhatias*, at Hathras especially, is so keen that it is not easy to secure or retain the custom of a client without treating him specially well' – such as by providing accommodation, good drinking water, and tobacco. *UPPBEC(R)*, p. 266.

the distribution of returns from cash production in favour of the trader. Whenever a cultivator did part with his grain at an unfavourable rate it was not because he lacked information: 'The prices of grain are known in all rural areas', the Banking Enquiry found. 'Even in the remotest villages . . . the prices of grain are a popular topic, so the dealer's chance of profiting by the cultivator's ignorance is greatly reduced.'[154]

Commonly, marketing was closely tied in with credit. Here again it seems that both credit availability and the range of sources for loans expanded considerably in the north-west. Higher and more certain returns to agriculture attracted credit as the change of the balance in portfolios was accentuated by the rising supply of loanable funds resulting from trade: there was a noticeable preference for placing capital in trading activity rather than tying it up in the complications of landownership.[155] The canal, particularly, contributed to a fall in interest rates, due to its effects on 'increased prosperity and certainty of agricultural result': Alan Cadell noted that eastern Muzaffarnagar proprietors were able to borrow at 12–18% per annum, and tenants of 'good character' at 15–24%.[156] It was particularly profitable for larger cultivators – those who took their own grain to the market – to acquire additional supplies of marketable produce through moneylending, which, apart from providing them with direct profits, frequently yielded social and economic benefits to them in other spheres of village life.

If relative freedom of entry into these related aspects of marketing produced a broadly competitive structure, the benefits were tiny or non-existent to the small, hard-pressed operator who found little competition for his small surpluses and dubious repayment capacities.[157] The plain fact is, though, that increased credit competed for clients among a larger proportion of the cultivating community than was the case either previously or elsewhere in the UP.

[154] *Ibid.*, p. 265.

[155] See Musgrave, 'Rural Credit and Rural Society', p. 230.

[156] 'Report on the Ganges Canal Tract', pp. 4, 18, in Cadell, *Muzaffarnagar SR* (1882).

[157] A fact not fully appreciated when the government stepped up the marketing and credit co-operatives programmes in the 1950s. As J.W. Mellor points out, 'co-ops found themselves competing in a business where the margins over real interest were relatively low and competition substantial'. J.W. Mellor *et al.*, *Developing Rural India: Plan and Practice* (New York, 1968), p. 66.

Though it may have been the general case that in many aspects of distribution free entry meant the absence of monopoly profits, this situation can only exist within wider parameters. Imperfections at *any stage* of the marketing process can affect the distribution of returns as between production and marketing. A 1911 enquiry by Moreland revealed that prices frequently differed between markets by more than the cost of carriage between them. The incidence was greatest among small markets, which were the most susceptible to collusion:

No district officer admits the existence of a ring in the trade, but some of the replies indicate either that there are informal understandings or that the influence of an individual predominates. Thus, in Partab-garh, where the market is very restricted, the weighmen come from the larger periodical markets in the district and inform the dealers in the headquarters market of the prices that have been paid, and the dealers then fix the headquarters rates after consulting the Chaudhri. This is very like a ring.[158]

The small periodic markets, common throughout most of the UP and especially in the eastern districts, were frequently blemished by corruption, collusion, and monopsony. Perhaps the most striking example of this involved the practice of dumping by large trading firms:

There is some evidence to show that exporting firms are occasionally guilty of dumping agricultural produce to facilitate purchase at lower prices. Thus in Rae Bareli, the charge levelled against the leading exporting firm is that it shatters prices with a travelling load of 10,000 maunds of grain, which is dumped in an area in which prices are steady to lower them *ad lib*, and then, after affecting purchases, the prices are forced up again.[159]

While such action by a trading unit large enough to affect the market price effectively procured for it a larger share of the local surplus, within the purely local context distribution between producer and intermediary clearly favoured the latter in eastern UP markets in comparison with those of the west. The cultivator who took his produce directly to an eastern market paid one-and-a-half to two times as much in market deductions

[158] UP Sec., Scarcity Dept., File 144/1914, File on variations in Price of Foodgrains, Note by W.H. Moreland (Dir. of Agriculture), 13 Sept. 1911.
[159] *UPPBEC(R)*, p. 160.

as did the average merchant, who in turn paid more than his western counterpart. The situation was summed up in the *Report on the Marketing of Wheat*: 'Apart from the gradually increasing charges as one proceeds from west to east, the share borne by the seller rises. Payments by kind are also more common in the eastern tracts.'[160]

Such observations reflect the qualitative difference between typical rural market structures and those of the western districts, where a greater degree of competition existed between markets. In the west, the important feature was that cultivators and itinerant traders, for various reasons, were able to choose between local (small) *mandis* and the large ones, and, making a simple deduction based on Prasad's estimates, it would appear that around 85% of the wheat deposited in the main *mandis* between April and mid-July was taken there directly – only around 15% being passed on through smaller *mandis* (see Figure 6).[161] For distances of up to fifteen miles, where the route was not tortuous, those with grain to market generally preferred the central *mandis*:[162] it was apparently a frequent sight to see cultivators and *beoparis* passing secondary *mandis* on their way to larger markets. The reasons were outlined by Pande and Gupta:

A direct sale in the principal *mandi* always proves more advantageous to the seller than one in the secondary *mandi*: not only are the middleman's profits eliminated to the seller's advantage, but the *mandi* expenses charged to the seller are, on average, less in the principal *mandis*; other marketing expenses are also lower.[163]

Several factors contributed to the more efficient marketing complex in the west; and once again the importance of competition at various levels in the bulking process is paramount. Intermarket competition was clearly fostered by the practice of traders in terminal markets gauging local price variations before placing orders with wholesalers in the main assembling markets. Stocks flowed relatively smoothly between the major markets to equalise prices net of transport and

---

[160] *Report on the Marketing of Wheat*, p. 137. The significance of the reference to kind payments relates to the fact that it was easier to procure an additional margin from the seller than when the charges were on a cash basis.

[161] Prasad, *Organisation of the Wheat Trade*, p. 30.   [162] *Ibid.*

[163] Pande and Gupta, *Prices of Cereals in the UP*, p. 96.

incidental expenditures.[164] In addition, Chambers of Commerce in both the major markets, Hapur and Ghaziabad, were actively involved in improving the competitive circumstances of their markets in relation to the different levels of competition. Hathras reduced *octroi* charges on wheat and sugar in 1873, and from 1887, to encourage its expansion as a collecting and distribution centre, levied *octroi* from traders only on any excess of imports over exports in the commodities handled within the municipality.[165] Although Ghaziabad abolished grain and sugar *octroi* in 1904, and abolished the levy completely in 1913,[166] Hapur appears to have made the most progress overall in improving market organisation and regulating standards and charges. Regulation at the level of the primary markets encouraged direct marketing and widened the range of influence of these markets, thus putting pressure on the smaller centres. This pressure was accentuated by other developments than the growing practice whereby the large trading concerns sent agents to deal directly with the localities:

With the expansion of railway communication, traders have left the big markets, such as Hathras and Cawnpore, and established themselves at smaller markets where they can deal more directly with the cultivator. The result has been the springing up of a great number of small markets in these provinces. These markets have been beneficial to the cultivator in so far as they tend to reduce the market difference that formerly existed between the village and bazar rates.[167]

A typical case involved Ralli and Company of Calcutta, who deputed a representative to Muzaffarnagar district for several months of the year: his appearance in the *bazars* caused prices to increase 'by leaps and bounds'.[168] The increasing trend, identified by collectors, for *zamindars* and large cultivators to store wheat within the villages with a view to taking advantage of off-season prices only encouraged such traders to enter the villages from where they frequently bought and packed the grain, and even directly despatched it by train.[169] Competition and improving accessibility encouraged signi-

[164] *UPPBEC(R)*, p. 267.   [165] *Aligarh DG* (1909), p. 247.
[166] *Meerut DG, Supplementary Notes and Statistics*, vol. IV (Allahabad, 1917), pp. 16–17.
[167] UP Rev., Aug. 1917, 45, H.R.C. Hailey (Dir. of Agriculture), Note on Wheat Stocks.
[168] *RAR 1886–7*, p. 67.
[169] UP Rev., Aug. 1917, 45, H.R.C. Hailey (Dir. of Agriculture), Note on Wheat Stocks.

ficant developments in the pattern of wheat-marketing in the western UP. Instead of the commodity moving up in stages through the different market levels – from *painth* to intermediate *mandi*, and from there to primary *mandi* (a pattern substantially intact in the eastern districts) – the wheat of the Doab flowed increasingly through specialised channels outside this traditional structure. The intermediate stages, with their gross (and costly) inefficiencies and their lack of storage facilities, were bypassed and the bulking process more and more reduced to a single stage.[170] It is not surprising to find that, under pressure from this practice, the price differentials between the different levels of markets narrowed significantly.[171]

Once more, evidence testifies as to the superior organisation and efficiency of the western districts. The share of the produce absorbed by marketing costs – a key element in the distribution of returns from cultivation – was smaller than in the east. When a cultivator took his wheat crop to Hapur, for instance, he paid a cash sum of Rs 2.5 per Rs 100 of grain in levies and charges. This was thirteen *annas* more than the average experienced *beopari* would forfeit for the same load, but less than half the Rs 7 he would pay in a small *mandi*. Even with the more competitive structure identified for the western markets, a Rs 100 load finding its way via a village buyer and small *mandi* to the primary market would incur around Rs 12 in cumulative market charges, with transport costs in addition.[172] On market charges alone, a considerably larger proportion was soaked away in the eastern districts, where even direct marketing in Benares involved an average charge of Rs 7.5 – more than a small western *mandi* – virtually all of which was borne by the seller. The implications for the distribution of returns from production and for the restriction of the market are obvious.

The above figures exclude transport costs. Here again, the

---

[170] A similar pattern is described for Indonesia in A.G. Anderson, 'The Rural Market in W. Java', paper presented at the Second New Zealand Conference on Asian Studies, University of Canterbury, Christchurch, May 1977.

[171] In Aligarh, for example, while Aligarh/Koil and Hathras were the main markets, cultivators were reported to be able to get 'a full price for their agricultural produce' at four other markets in the district. Ahmed Ali, *Aligarh SR* (1943), p. 2.

[172] Prasad, *Organisation of the Wheat Trade*, pp. 34–5. Included in the figures for small *mandis* are labour charges, formal weighment fees (*tulai*), false weighing, and purchaser's margin.

western districts benefited from growing economies of scale. Even at the turn of the century, Meerut district was 'exceptionally well provided with railways and metalled roads'; thereafter they 'improved considerably'.[173] Improvements were always required on the bulk of rural roads and lanes but in Meerut, 'as far as the inter-*mandi* trade' was concerned, 'nothing more [was] needed' in the 1930s, and once the Hapur–Mawana–Lakhsar line was completed, there was not 'a single village greater than fifteen miles from a railway station'.[174] Such developments fostered growth: in 1935 the Canal Department itself built a steam tramway 'intended solely for sugarcane traffic direct from the rich canal tract around Mawana to the large new sugar mill at Daurala' nine miles north of Meerut city [175]

Road improvements led to significant economies of scale in transport, reducing journey times and increasing the loads carried. Carts such as the two-wheeled *chakra* and the four-wheeled *chaupai* could, when drawn by three bullocks, carry loads of 45 and 60 *maunds* respectively on metalled roads. *Unt garis*, camel carts with large rectangular cages, were capable of carrying 25 *maunds* 40 miles in a day. During the harvest they carried grain and pulses from rural areas to *mandis*, and in the off-season they were deployed between distant Doab *mandis* carrying vegetables, salt, iron sheets, and cotton bales and seeds.[176] Prasad estimated that inter-*mandi* transport rates were around one *anna* per *maund*-mile, which translates into about 2% of the value of grain for a twenty-mile journey. He also estimated that 70% of the wheat carried along *kachha* roads was conveyed by bullock cart. These figures indicate that transport costs were too low to narrow the market and hinder the growth of Meerut trade, as officials argued was the case in some eastern areas.

[173] Cooke, *Meerut SR* (1940), p. 4.
[174] Prasad, *Organisation of the Wheat Trade*, pp. 14, 16.
[175] Cooke, *Meerut SR* (1940), p. 4.
[176] For details see *UPPBEC(R)*, pp. 219, 273–4; and Pande and Gupta, *Prices of Cereals in the UP*, p. 72.

Case studies of major crops

Impressive commercial crop growth rates are of course not synonymous with evenly diffused prosperity throughout rural society. As is suggested by the link which has been established between the system of tenure and the way water was used, the degree and character of the growth and the distribution of its benefits are bound up with the way the production and marketing of the various crops are organised. The following sections examine two major cash crops, the cultivation of which in the Doab was largely due to the availability of canal irrigation. The study of these crops – covering markets, profitability, the degree and character of local processing, and the relative bargaining strengths of those involved – assists significantly in understanding and explaining the patterns of growth and prosperity observed both within the Doab and between the western and eastern parts of the UP.

Regional differences in the sugar industry

There were substantial differences in the ways in which the cultivation, processing, and marketing of sugarcane were carried on respectively in the Upper Doab, Rohilkhand, and the eastern districts. The level and distribution of returns varied markedly as a result of technical and institutional differences in the arrangements for sugar production. This section illustrates how conditions were such in the north-west districts as to allow them to gain substantially more than the other areas, *and* in ways which benefited a larger proportion of the population.

Although immediately prior to the First World War over one-fifth of UP sugarcane was converted into sugar, the bulk was made into *gur*;[177] and virtually all the cane exported from Meerut division – amounting in 1912–13 to twenty *lakhs* of *maunds*, or around half the UP exports – was in the form of *gur* or *shakar*.[178] Meerut *gur*, in fact, commanded the highest prices, as a result of the varieties of cane grown and the attention given to

---

[177] *Pounda* (chewing cane) occupied a negligible acreage.
[178] *Gur*: cakes of unrefined sugar made by boiling *ras*. *Shakar*: the product resulting from slightly more prolonged boiling and from being stirred while cooling.

quality in manufacturing. A broad shallow boiling pan was in general use by the turn of the century, which facilitated quality control; moreover, great care was taken in clarifying the juice. The finest *gur* fetched around Rs 5 per *maund* in the 1910–14 period; and the more typical 'good-quality' produce earned Rs 4 to Rs 4.5.[179] Little wonder, then, with an acre of cane yielding 35 *maunds* of *gur*, that the 'prosperity of the farmer' in Muzaffarnagar was held to 'depend upon this crop'.[180]

Because production in the north-western districts was largely financed from the cultivator's own resources, he had the advantage of being free to choose how, when, and in what form he disposed of his cane. With bullocks freed from *rabi* irrigation duty in canal villages, cultivators normally found it more profitable to convert their cane into *gur* than to sell it as a standing crop, as cut cane delivered, or as *ras* (juice) for conversion into sugar. Even in the parts of Bulandshahr where the varieties grown were inferior to those in Meerut, Muzaffarnagar, and Saharanpur, and where *gur* typically fetched Rs 3.75 per *maund*, sugar-making survived because the cultivator could convert his product to *gur* if he was dissatisfied with the price offered by the factories.[181] In the Upper Doab, factories – competing with imported sugar from high-productivity producers in Java and the West Indies – could not pay prices as high as those obtained through *gur* manufacture.[182]

Although W.H. Moreland[183] and George Watt[184] drew attention to the inefficiency of individual processing by Upper Doab cultivators in comparison with the *khandsaris* over the Ganges in Rohilkhand, who achieved economies of scale, the latter system was renowned for its exploitation of the peasant. In contrast to Meerut, only 44% of the Rohilkhand sugar

---

179 Hailey, 'Prices of *Gur* and Cane', p. 224.
180 MCR, BOR, viii/ii, 14(c), H.A. Lane, RRR Pargana Muzaffarnagar (Muzaffarnagar), 1919, p. 15.
181 Hailey, 'Prices of *Gur* and Cane', p. 222. Bulandshahr did, however, enjoy the advantages of powerful bullocks and improved crushing equipment, which enabled its cultivators to obtain around 6% more *gur* per unit weight of cane than the average achieved outside the Meerut division. *Ibid.*
182 This was the conclusion of both Hailey (*ibid.*, pp. 219–23) and the *Report of the Indian Sugar Committee* (Simla, 1920), p. 238.
183 See UP Scarcity, Sept. 1905, 2.
184 Watt, *Dictionary of Economic Products*, VI, pt 2, p. 282.

product exports in the 1890s consisted of *gur*; the rest was derived from the juice boiled briefly to form *ras*, from which *kachhi chini* or *khand* – a whitish floury sugar – was formed.[185] The *khandsaris*, who controlled this processing, sited a dozen or so bullock mills in a cluster around a *bel*, or boiler unit, which converted the *ras* into *khand*.[186] Of course, the *khandsaris* were still vastly inefficient in comparison with modern factory methods, but although the factories extracted more than twice as much sugar from the cane, the experimental Nawabganj factory found it difficult to survive.[187] The price of *khand* was higher than that of factory-made sugar: *khand* had a 'small artificial market of its own' in which prices were determined 'almost entirely by strength of caste prejudice or sentimental considerations'.[188] In addition, there was the nature of the relationship between *khandsari* and cultivator. In contrast to other areas, where the cultivator generally manufactured the final product or part-processed the cane for further refining, the *khandsari* purchased the product of the cane field in the form of *ras*, for which there were two rates: that reached under the *khushkharid* system, where agreement was made at harvest time, and that procured under the 'advance' or *badni* system. The discrepancy between the two made all the difference. At the *khushkharid* rate of Rs 64–72 per 100 *maunds*, cane converted into *khand* produced little profit at prices current in the early 1910s. But the bulk of the *khandsaris'* *ras* was obtained at the rate of Rs 48 per 100 *maunds* through the *badni* system. The indigenous industry relied for its very survival upon this system of obtaining raw material.[189]

The *gur* made in Rohilkhand – 5.75 *lakhs* of *maunds* were exported around 1910, mainly to Rajputana – was of lower quality than that of Meerut, being darker and inferior in consistency. The cane varieties were inferior; the crop was frequently cut before it was ripe (an action forced upon the

---

[185] *Ibid.*   [186] *IRR 1899–1900/1901–2*, p. 31.

[187] *Report of the Indian Sugar Committee*, p. 206. Sugarcane mills extracted a greater proportion of the sucrose from the cane (average net yield being 9.5% as against the 4% by traditional methods), and were more effective than indigenous processes in preventing the inversion of sucrose into glucose. See A.K. Bagchi, *Private Investment in India, 1900–39* (Cambridge, 1972), pp. 363–5.

[188] *Report of the Indian Sugar Committee*, p. 237.

[189] Hailey, 'Prices of *Gur* and Cane', p. 222.

cultivator 'by the poverty of his cattle and the claims of other crops'); and a narrow deep pan was used in boiling, which made it difficult to avoid overheating and caramelisation.[190] Good-quality Rohilkhand *gur* brought Rs 3.5, although much of the crop fetched far less. But many cultivators were unable to turn their cane into *gur*, since their crops were pledged to the *khandsari*, who combined the roles of landlord, moneylender, and trader.[191] The importance of this lack of choice was starkly demonstrated by the varying fortunes of 'independent' and 'dependent' cultivators in the dry year 1907–8, when, due to low yields, *gur* prices were very high, frequently around Rs 5–6:

Independent growers of cane . . . recouped themselves in part for the loss due to the bad season, but refiners who had not secured stocks in advance suffered from the effects of high prices. In parts of Rohilkhand the unusual phenomenon was observed of cultivators breaking their contracts to deliver juice to the manufacturers and making *gur* for their own profit instead. Where, however, manufacturers were able to enforce their contracts they have undoubtedly strengthened their hold over the growers, who were unable to deliver juice in full satisfaction of their advances.[192]

Independence of decision-making and the existence of alternatives were critical determinants of the distribution of the profits of sugarcane cultivation.[193]

In the eastern districts the price of *gur* was low, reflecting its inferior quality and the distance from large markets. Here, as in northern Bihar, once local consumption was met *gur* was mainly sold to refineries; indeed, up to the 1920s these were the only tracts in north India where the factory industry was established on any significant scale.[194] *Gur* in this area was badly made, 'often a sticky mass containing an appreciable percentage of fibre and mud. That made from the wooden

---

[190] *Ibid.*, p. 224.
[191] While the *khandsari* frequently combined all three functions, sometimes he operated in collusion with the landlord. An enquiry conducted in 1882 by the Director of Agriculture revealed that 43% of Pilibhit indebtedness was to *khandsaris*. NWP&O Rev., Oct. 1882, 52.
[192] *RAR 1907–8*, p. 7.
[193] Watt, *Dictionary of Economic Products*, VI, pt 2, p. 282.
[194] *Report of the Indian Sugar Committee*, p. 268. Misra lists six sugar mills in the district, employing in total around 1,700 workers for the winter months. Misra, *Overpopulation in Jaunpur*, p. 22.

crushing mills, either from the bacterial action set up or the extraneous matter which finds its way into the mill, is particularly poor.'[195] The price, in the early 1910s, rarely rose above Rs 3 a *maund*, and was frequently Rs 2.5. Although at that time 6.5 *lakhs* of *maunds* were exported from Gorakhpur division to other districts for the bottom end of the market, 3.5 *lakhs* were shipped to refineries at Cawnpore.

Here again, the *badni* system played a crucial role. According to the Gorakhpur *Gazetteer*, 'very few villages' were reported to 'grow cane altogether without advances'.[196] Advances in this area were chiefly provided by local factories, and it was through this method that the expansion of sugarcane areas in non-canal tracts mainly took place. Cultivators would receive advances for *ras* if they were close to the factory; otherwise, they would receive advances for juice already boiled in *karahis* lent or hired out by the factory-owner. Where there was no significant competition between the factories, either because their numbers and operational areas were limited or because factory agreements concerning zones were in existence, the bargaining position of the cultivator was generally weak. Moreover, in contrast to the situation in the north-west, where *gur* quality and its value in the nearby expanding Punjab markets gave it a significant premium over sales to the factories, the quality and market value of eastern UP *gur* was typically low, and the value of the option for the independent cultivator consequently limited. In addition, much of the eastern districts' *gur* sold for factory processing did not generate incomes and employment locally, but rather benefited the inhabitants of Cawnpore, where much of it was refined.

Indigo production in the Doab: growing limits to exploitation

Indigo cultivation and processing made an appreciable contribution to the commercialisation of the Doab, especially below Meerut district. In Bulandshahr, Stoker estimated that between 1881 and 1885 the district manufactured an annual amount of 6,400 *maunds* of indigo 'cake', valued at Rs 1,080,000, which gave employment to 82,000 people for six

[195] Hailey, 'Prices of *Gur* and Cane', p. 225.
[196] Cited in Watt, *Dictionary of Economic Products*, VI, pt 2, p. 283.

weeks of the year.[197] Another settlement officer calculated that Aligarh produced over 4,500 *maunds* worth Rs 630,000 in 1880.[198] Indigo was mainly grown outside the sugarcane areas, in *parganas* with a large share of canal irrigation, such as Khurja (Bulandshahr), where indigo consistently took up around 20% of the *kharif* area (roughly 10% of the total cultivated area between 1878 and 1885).[199] With canal water available, cultivators could sow indigo in the spring (when it was called *jamowa*) and cut it in August, which gave them time to sow a *rabi* crop in the same field.[200]

Because *jamowa* raised the nitrogen content of the soil, and its growing season was so short, the peasant's husbandry was hardly disrupted.[201] But there was much criticism of the manner in which the indigo industry was organised.[202] The disappearance of indigo, in the face of competition from synthetic dyes, according to one contemporary, was

not to be regretted. The way in which this industry is carried on here is very baneful to the country. The cultivation of the plant is forced and the produce taken at very low rates, so that the profits, instead of being shared in any reasonable proportion among the producer, the manufacturer and the capitalist, who is usually the proprietor of the land, go almost entirely to the latter.[203]

[197] Uttar Pradesh Board of Revenue Library, Allahabad, Tahsil Handbook of Bulandshahr District, n.d., p. 5.

[198] Smith, *Aligarh SR* (1882), pp. 36–7.

[199] *Returns of the Agricultural Statistics of British India 1890–1*, Table A.

[200] Stoker, *Bulandshahr SR* (1891), vol. II, p. 80. Shiurajpur *pargana*, in Cawnpore, put down 10.5% of the *bangar* area to indigo. *Cawnpore DG* (1881), p. 28. Crop statements have a tendency to understate the extent of the crop, since 'being cut in the rains, and replaced in the cold weather by a rabi crop, it has very frequently escaped entry in the measurement papers prepared during the latter season'. Ridsdale, *Etah SR* (1874), p. 6.

[201] See Ch. 4 above for details. *Jamowa* was commonly preferred in the Doab and yielded, according to various estimates, eighty to 120 *maunds* of plant per acre. Alternatively, the cultivator could plant an *asarhi* crop, planting with the rains, cutting in September, and leaving the stumps in the ground till the following year, when a *khunti* or *ratoon* crop would be derived from the same plant. The yield in the first year would be about three-quarters of that of *jamowa*, but about the same the following year, though the land was not available for use in the *rabi*. Watt, *Pamphlet on Indigo*, pp. 25–6.

[202] One canal officer sceptical as to the agricultural, social, and psychological value of the crop was W. Good, who raised most of the criticisms made against indigo cultivation in *IRR 1877–8*, p. 19A.

[203] *Bulandshahr DG* (1903), p. 55.

Indigo undoubtedly made a lot of money for landlord–capitalists through their factories, and the very way in which the industry was organised meant that the structure of benefits, and the secondary effects, were different from those emanating from sugar production in the districts to the north. But there is much evidence to suggest that the crop impinged upon the cultivator far less than critics implied: indigo cultivation did not interfere unduly with wheat production; nor was the grower, it appears, without alternatives – occasionally very lucrative – in disposing of his produce. Such factors are important in explaining the greater prosperity of the indigo districts of the Doab compared with the poverty-stricken east.

Although for a large part of the nineteenth century indigo was mainly produced by Europeans – particularly in Bihar and Lower Bengal[204] – the balance shifted as it spread westwards into less productive regions. While ample canal water encouraged plant growth, much of this growth was useless stem formed at the expense of the dye-producing leaf, which tended to be of an inferior colour.[205] The quantity and quality of the dye were inversely related, with the Doab producing a large volume of low-priced indigo.[206] High-cost European ventures faced a precarious existence in this marginal area, often operating outlying factories only when high prices made it profitable to do so. While, around 1890, Europeans contributed slightly more of the indigo leaving Benares division than did the Indian factories (6,515 *maunds* against 5,750), in the Doab the latter accounted for 24,000 *maunds* alongside a European total of 3,120.[207] Some ventures, such as that of Michel at Dasna (Meerut), whose modern factory had a capacity of 500 *maunds* per season, were well established;[208] but many hopeful European schemes went the way of Messrs Mercer and Company, who purchased 32 villages in Bulandshahr but

---

[204] Watt, *Pamphlet on Indigo*, p. 43.　　　[205] *Etah DG* (1884), Appendix I, p. 11.

[206] Watt, *Pamphlet on Indigo*, p. 37. This is reflected in the relative prices fetched by the finished product: taking the figures for 1879–81, Tirhut indigo appears to have been worth 25–30% more in Calcutta than Doab cake, with that of the Benares division enjoying a 15–20% margin also. Only 'very seldom', reported Duthie and Fuller, did the Doab price come to within 20% of the Tirhut level. Duthie and Fuller, *Field and Garden Crops*, I, p. 47. It appears also from the figures that when demand fell, the eastern producers tended to hold their prices better than those in the Doab.

[207] Watt, *Pamphlet on Indigo*, p. 43.　　　　[208] *Bulandshahr DG* (1903), p. 54.

found the operation unviable and sold out to the local large landlords, many of whom had already set up indigo factories.[209]

The main problem encountered by Europeans was the margin between their costs (estimated at Rs 125 per *maund*) and the costs of Indian manufacturers (around Rs 85 per *maund*). European costs were higher, William Crooke explained, because indigo could

> only be grown profitably for any length of time by proprietors. Besides the great influence which a landlord has over his cultivators, he is able to utilise his collecting establishment in the work, and can use them for other purposes when the work is over. There is none of the home farm cultivation which is the rule in Bengal.[210]

Admittedly the European producer, with better equipment and closer supervision, could obtain a better price for his product than his Indian counterpart, who looked 'more to quantity than quality';[211] but he could never escape the fact that he was operating in a marginal zone, vulnerable to market fluctuations, and Europeans were reluctant to commit themselves to the Doab in the manner in which their counterparts involved themselves further east. In Bihar and Bengal, indigo was cultivated directly by capitalists ('planters') on land which they owned or leased from *zamindars*; whereas in the Doab, factories purchased the plant from the cultivator at a price fixed at the time of sowing or delivery. The *badni* system initially suited both cultivator and factory – particularly the cultivator, who got an assured buyer for an otherwise useless crop. Duthie and Fuller estimated that, around 1880, most indigo was grown under the *badni* system. Normally the factory bound itself in March–April to purchase the plant at a price fixed upon, and paid an advance to the cultivator to seal the bargain. As the cultivator was frequently hard-pressed to find the cash to pay the spring instalment of his land revenue, such an advance was particularly tempting to him. The price was obviously below the likely competition price, but so long as it was not below Rs

---

209 *Meerut DG* (1876), pp. 308–9.
210 *Etah DG* (1884), Appendix I, p. 11. See Watt, *Pamphlet on Indigo*, pp. 42–3, for a description of the organisation of production in Bengal; and C.M. Fisher, 'Planters and Peasant: The Ecological Context of Agrarian Unrest on the Indigo Plantations of North Bihar', in Dewey and Hopkins, *Imperial Impact*, pp. 114–41.
211 *Aligarh DG* (1875), p. 473.

16–18 per 100 *maunds*, Duthie and Fuller considered, the system was 'not more objectional than that followed by the government in furthering opium cultivation'. There was, however, a more disturbing side to this system:

> One of the principal objects of the factory in making advances is often not so much to arrange for a crop in the present as to gain such power over the cultivator as will enable it to compel him to grow indigo on its own terms in the future. Very frequently, therefore, in the first agreement made with the cultivator the plant is priced at a favourable rate, but a stipulation is entered binding him down to deliver not less than a certain amount of it which is often knowingly fixed at an impossible figure on penalty of forfeiting $2\frac{1}{2}$ to 3 times what balance there might be against him. With the chance of obtaining cash down, the cultivator pays but little heed to the stipulations of a more contingent nature, and hence a single bad season may involve him in obligations to the factory ... The power thus acquired over a cultivator may be used either to compel him to grow plant at the factory's will, or to sell it at a price much lower than it would otherwise command. That these are solid advantages may be judged from the fact that the value of a factory is often estimated by the amount of outstanding debts it has, or in other words, by the degree to which surrounding cultivators are under obligations to it.[212]

The *badni* system could be seen at its worst in the south-eastern *parganas* of Bulandshahr, and especially in the Chhitari estate during the 1880s. When the owner, Kunwar Abdul Ali, 'fell under the influence of his advisers', notably Chaudhri Ashfak Husain, rents were forced up and an indigo factory established. Factories were commonplace among the Bulandshahr *zamindars*: 'It is not considered that the owner of a canal village gets the full benefit of his property unless he works an indigo factory in it', observed the Settlement Officer.[213] But cultivators on the Chhitari estate were pushed to an 'unknown extreme', and forced to cultivate indigo on

> all their outlying fields except the swampy or dry sandy lands where the crop would not grow ... which disturbed the whole conditions of agriculture, unduly restricted the area of maize and other coarse foodgrains on which the people subsist, and threw such strain on the

---

[212] Duthie and Fuller, *Field and Garden Crops*, I, pp. 48–9.
[213] This example is taken from MCR, BOR, VIII/II, 14(c), Ser. 2, T. Stoker, AR Tahsil Khurja (Bulandshahr), 1889, pp. 33–4.

cultivators and the canal, that the tillage of indigo itself became imperfect and unprofitable.[214]

Even the low price paid for the plant (around Rs 17 per 100 *maunds*) was not credited to the producer's rent until some months after its delivery.

On the face of it, then, the system operated in a way which enabled the *zamindar* to channel a larger portion of the agricultural surplus into his pocket, and to enhance the value of this surplus through manufacturing processes. On this interpretation, the canal was a useful instrument for increasing the exploitation of the peasantry by the indigo-planters. Yet this conventional picture is almost certainly overdrawn. The balance of forces was by no means everywhere so decidedly in favour of the *zamindar*–capitalist as the stereotype of unyielding exploitation would suggest. Two related conditions were necessary for the system to function in favour of the *zamindar*–capitalist: the factory-owner had to be able to maintain his monopsony purchasing position over the suppliers of plant; and profitability had to be maintained. While the price was buoyant and the seasons good the producing arrangement worked well for the Factor, although even in the best years he had to write off some bad debts and face losses where *ryots* sold some of their contracted produce on the open market.[215] Being near the geographical margin of indigo production, the Doab was especially sensitive to bad seasons and low prices. In George Watt's words, 'If the Calcutta rate for indigo drops and bad seasons intervene, the weakness of the system manifests itself. The *ryots* abscond or are sold up, and the Factor is left with a ledger full of debts which there is no hope of ever clearing.'[216] The dangers inherent in the advance system became acute as the factory's monopoly was undermined:

With the increase in the number of factories in the district the market for plant becomes of course much wider, and it then becomes possible

---

[214] It was not always the case that the proprietor managed the indigo factories himself. One of the Chhitari factories, a large one at Gangagarh, was let in 1887 to a *bania*. The *Rais* credited the cultivators for their plant at the rate of Rs 19 per 100 *maunds* delivered at the factory, obtaining Rs 42 himself from the *bania* – a figure which included rent of factory and 'other conveniences'. *Ibid.*

[215] Watt, *Pamphlet on Indigo*, p. 43.   [216] *Ibid.*

for a cultivator to grow indigo unfettered by agreements and to rely on obtaining a good price on the competition of one factory against another. This is the system known as *khushkharid* or 'good bargain', so named of course from the cultivator's point of view. In a district where factories are numerous, the difference between the price paid for *badni* and for *khushkharid* indigo is very great; when the former is contracted at Rs 18, the latter will often sell for as much as Rs 26 per 100 *maunds*.[217]

Another authority confirmed that while Rs 20 per 100 *maunds* was a common but low rate in cases where advances were given (and carriage undertaken by the manufacturer), 'Rs 25 to 27 is got when the *ryot* carries for himself, or when he sells at his own option at harvest. In one instance, so great was the competition between two rival factories for plant that Rs 32 and even Rs 40 were given for 100 maunds.'[218] Within the confines of distance constraints and beyond the attempts at extra-market influence and collusion by powerful local *zamindars*, competition was inevitable, given the relative freedom of entry into this kind of manufacturing activity. Further, independent cultivators were always in a position to benefit from *khushkharid* arrangements, since a manufacturer would generally wish to run his capital at full capacity to minimise the fixed cost element in unit costs. Even where competition between factories was absent, therefore, the factory would run a dual price system on plant supplies so long as the higher-tier price allowed a positive net revenue effect.

Faced with occasional heavy losses as indebted cultivators left their villages and settled elsewhere or sold their indigo to the growing number of competing factories, the *badni* system fell out of favour in many areas, giving way in the 1880s to cash deals at harvest.[219] Duthie and Fuller, in fact, traced 'the gradual increase in the prosperity of a village, or its gradual recovery from the effects of a series of disastrous seasons, [to] the increase of the area under *khush-kharid* at the expense of that under *badni* plant'.[220] The higher-cost European ventures were particularly affected by the rapid increase in the number of factories which was undermining the *badni* system. By 1890, for

---

[217] *Ibid.*, p. 49.   [218] Wright, *Memorandum on Agriculture*, p. 29.
[219] Watt, *Pamphlet on Indigo*, p. 43.
[220] Duthie and Fuller, *Field and Garden Crops*, I, p. 48.

example, Sikandarabad *tehsil* (Bulandshahr) had 48 factories, Anupshahr 44, and Khurja 28.[221] The planter became, in F. N. Wright's words, 'but one bidder against many', so that 'indigo no longer pays European factories in the North-Western Provinces, and the place of the old squire-like planter has been taken by the native capitalist'.[222]

But there was more to the decline of *badni* than an increase in the number of factories. There was considerable freedom of entry (at different levels) in the actual processing of the plant, and the independent cultivator was free to choose among a number of alternatives when he decided how to cultivate indigo and how to dispose of his produce. It is this flexibility, in addition to the competitive situation, which explains why the crop was 'largely cultivated voluntarily'.[223]

Many cultivators built vats in which they steeped their indigo, producing a coarse dye (*kachha nil*) which they then sold to the factories for final processing. In *pargana* Marehra (Etah), 'almost every village [had] its factory'.[224] Where conditions permitted, small-scale capital accumulation evidently took place in the indigo industry: 'Even the tenants spent money in building pairs of miniature vats on the borders of their fields.'[225] For cultivators not involved in small-scale processing, there were other ways of increasing returns from the indigo fields. Frequently a cultivator cutting for his own rough manufacture by vats would only cut half as much plant as he would have done for the factory, leaving the stalks for seed. Doab seed was highly prized by Bengal planters, and an extensive trade grew up, mostly despatched from Cawnpore. The trade was in the hands of Calcutta brokers under contracts to supply a certain number of factories with the seed they required; and the fairly inflexible nature of the season's demand made the price subject to violent variations, ranging from a normal level of around Rs 6 per *maund* to a very profitable Rs 42, which gives some indication of the inelastic nature of the demand involved.[226] The

[221] *Bulandshahr DG* (1903), p. 54. The total figure for Aligarh was put at 171, *tehsil* Sikandra Rao possessing 91. *Aligarh DG* (1875), p. 473.

[222] *Cawnpore DG* (1881), p. 109.

[223] NWP&O Rev., April 1885, 7, F.N. Wright (Meerut Collr.) to Dir. of Agriculture.

[224] Ridsdale, *Etah SR* (1874), p. 159.   [225] *Cawnpore DG* (1909), p. 43.

[226] Watt, *Pamphlet on Indigo*, p. 43. Wright's estimates, based on fairly low prices and average yields and on the cultivator's opting for the sale of fifty *maunds* of *jamowa* and

balance of bargaining power was not, therefore, as unevenly tilted as has been thought – which makes the reference in a Cawnpore revenue document to the 'deplorable loss the peasantry has suffered from the failure of the dye industry' more comprehensible.[227]

The canal as the principal source of dynamism

Through enhancing the peasant's ability to utilise more fully his available resources (by saving labour and bullocks, extending the range of cropping options, and reducing some of the risks his production was constrained by), the external input of canal irrigation had an immediate impact upon productivity. The very fact of the spread of canal irrigation, at the rate of 50,000 acres per year from 1860 to 1920, provided an impetus to growth through allowing the redeployment each year of millions of labour-hours away from well irrigation into more intensive cultivation and related processing and trading activities. In addition, however, because the canal was (as was demonstrated in Chapter 4) ultimately a labour-using technology, it perpetuated the historic conditions of labour scarcity in the tract – conditions which in turn constituted a bottleneck to exploiting more fully the potential of the canal. The dynamism imparted by the spread of canal irrigation was thus enhanced by the encouragement it gave to the development and diffusion of a chain of small-scale labour-saving complementary innovations, of a technical, biological, and organisational nature, which enhanced each unit's capacity to work its land and successively pushed back the production frontier. But the canal did more than simply create the appropriate technical conditions for expansion: it played a central role in the moulding and gradual adjustment of institutional structures in the west into forms broadly attuned to growth and to coping with change.

---

five *maunds* of seed, put the price received at Rs 47. With the costs of production – including imputed labour costs and rent – being Rs 23 (canal 'flush') and Rs 27 (canal 'lift'), the estimated profit would be Rs 24 and Rs 20 respectively. Wright, *Memorandum on Agriculture*, pp. 29–30.

[227] UP Rev., Jan. 1905, 51, Revenue Report and Assessment Report of Tahsil Bilhaur (Cawnpore).

It is certainly true that the canals were initially routed through a region of especially favourable conditions for adapting to canals and to extending commercial farming. It was pointed out in Chapters 5 and 6 that the canals were only minimally under the control of the state – neither the pricing mechanism nor the canal bureaucracy effectively directed the use and distribution of canal water – and, with pricing levels fixed so as to allow even the poorest regions to use water, very large surpluses were reaped by tracts with appropriate physical, economic, and social characteristics. The canal was thus a boon to the fertile districts of the Upper Doab, where, as Professor Stokes has pointed out, abundant land and scarce labour had given rise historically to the predominance of direct cultivation by owner–cultivators as exemplified in the Jat *bhaiachara* villages of Meerut division. Substantial peasant proprietor units proved ideal vehicles for the kind of petty accumulation which was suited to the technical circumstances the canals brought with them – particularly when set against the backdrop during this period of upward price movements, expanding communications, and a decline in the share of output expansion taken in land revenue.

It is the role of the canal in consolidating and enlarging a class of medium-to-large peasants which links technical factors with the development of institutions favourable to growth in the western UP. The canal, in reproducing conditions of persistent labour scarcity (and thus the highest returns to direct cultivation using family labour), also permitted the cultivation of relatively large holdings. *Bhaiachara* villages were among the first to receive water, and these populous communities were quick to spill over into sparsely populated surrounding (*zamindar*) estates opened up for cultivation by the introduction of the canal. Viable tenants were initially scarce; holdings were often in excess of fifteen acres and rents were low. Conditions of production, for the reasons given, so favoured these peasant units that they were able to resist (either individually or collectively) subsequent encroachment upon their cultivating rights (and attempts at rent enhancement) when land became relatively less plentiful and when their wealth became obvious. The realities of tenure were thus conditioned over the period by economic and technical circumstances favouring the medium-to-large peasant cultivator.

Similarly, the marketing structure increasingly came to reflect the distribution of economic power within rural society. The competitiveness of the marketing system in the western UP was a reflection of the autonomy of a large proportion of the surplus-producing peasantry, and of the fact that wheat cultivation was especially facilitated by canal irrigation. Generally independent of creditors and traders – and often with his own means of transporting his produce – the substantial Meerut peasant could exercise choice in his marketing decisions and thus exert pressure not merely for the modernisation of marketing processes but for a more competitive environment overall. In the context of a technological regime favourable to growth and to individual peasant units, therefore, the degree of peasant autonomy and the competitiveness of marketing structures were mutually enhancing. Markets, credit, and transport all came to reflect the underlying economic structure – of which the canals were so crucial a part – of productive resources; rather than being concentrated, they were spread among a relatively well-off and independent peasantry.

If the canal was instrumental in bringing about the new conditions in the western tract, it was not only the canal villages – nor the renowned agriculturalist castes – which benefited. The non-canal villages could not insulate themselves from the changes taking place in their locality. They could, for example, benefit considerably from the external economies arising out of the expansion of commercial activity, and in ways other than by directly tapping the wealth of the canal villages through offering services such as carting. *Pakka* roads conveying surpluses from canal areas also reduced marketing costs for many dry villages, thus increasing returns to cash-cropping. The reorganisation of marketing facilities to encourage self-marketing also benefited cultivators in the dry villages. But there were external diseconomies as well. Factors of production were increasingly priced on a regional basis as market relations spread, and these affected the dry farmer, who had to pay high rates to labourers to prevent them from moving to the canal villages or urban centres; whose stock-rearing and maintenance costs had risen as the canal caused cultivation to expand and grazing areas to shrink;[228] and whose rented-in fields adjacent to canal areas now involved a higher rental due to the increased cultivating and rent-paying capacities of canal villages.

It is this *dual* effect – of potential benefit and actual cost – which helps explain why members of castes hitherto undistinguished as agriculturalists should have responded energetically in one locality, while their caste brethren in other areas showed little inclination to alter old habits and practices. Instances of improved economic performance are not, as many settlement officers believed, to be explained merely in terms of the example provided by the high farming found in the villages of the Jats. This should not be dismissed as a contributory factor, but there were more tangible forces at work.

The Jats were indeed legendary for their skill as cultivators. This skill, and a complementary value system which tended to mobilise family labour resources more effectively than in most other castes, accounted for their terraced fields in the ravineland of *pargana* Puth, where they produced 'the richest crops on what the Pathans would call "barren waste" '.[229] While the less numerous Jhojha and Rawa castes possessed similar characteristics, others – such as Gujars and Rajputs – inherited a different set of social traditions, skills, and ways of making a living. These traditions had been formed in an earlier time, and were consistent (i.e. so as to be viable) with the conditions of that time. They now confronted the contradictions between their social organisation and their objective circumstances in an age of profound change. If the potential returns to adjustment were significant, so were the costs of retaining their former patterns of activity. Comparing the *khadir* zone of Meerut's *pargana* Loni with the canal-irrigated upland, W.A. Forbes pointed to the effect which changed

---

[228] A note by BOR member C. Thornhill, relating to *parganas* Chaprauli, Baraut, Puth, and Sarawa, illustrates one other important force towards this: 'Sir H. Elliot's expectation that a "nominal" jumma would act as an incentive to cultivation has been disappointed, because the land yields as much as the indolent zemindars care for in grazing rents. It is desirable therefore that such a fair rent on cultivable area should be assessed, as will compel zamindars to bring it under cultivation or transfer it to those who will do so.' Included in Buck, *Meerut SR* (1874), p. 71.

[229] W.A. Forbes, Puth Pargana Report, 1867, p. 13, in Buck, *Meerut SR* (1874).

circumstances had had upon the pastoral Gujars: 'The Goojurs are the chief owners of the land on the lower river tracts, and nothing can be more backward than their present condition as far as agriculture goes, but on the high lands we find them quite a different race: here they vie with and are certainly not surpassed by their Jat neighbours.'[230] The Gujars' response, however, was clearly not irrespective of tenurial circumstances. Forbes again: 'The spirit of industry is working rapidly amongst even the unsettled Goojurs, wherever the cultivating communities hold proprietary rights in land.'[231]

If the new structure of returns also encouraged service castes to take to cultivation themselves,[232] it appears to have had a similar effect even upon *fakirs* in the locality: in *pargana* Baghpat, two whole estates and a part of a third were 'owned and cultivated by Goshaens, who, having dropped the habits of their religious order, drive the plough and farm as highly as their Jat neighbours'. The latter informed the Settlement Officer that the 'ex-facqueers' had thriven ever since they 'filled their houses – that is, taken wives to them'.[233]

The cost of labour, growing competition for land, and growing population (and thus subdivision) all put pressure on the 'non-agricultural' castes to adjust to the new conditions, as is clearly suggested by Table 8.3. Even where the canal *rajbahas* did not trace their way into the tracts containing the non-agriculturalist castes – as in the Rajput areas of Muzaffarnagar's Thana Dhawan and parts of *pargana* Jhinjhana – signs of such pressures were clearly apparent to settlement officials.[234] Frequently, those caste groups whose social systems made them slow to respond to the new circumstances found themselves surrendering the initiative to the more adaptable Jats and their ilk, with the result that they were

---

[230] *Ibid.*, p. 32.   [231] *Ibid.*

[232] 'I observed through the district [that] the mechanics and lowest castes, who in the olden days never thought of holding land, are found as landlords . . . and we have a carpenter, a blacksmith, a butcher and a presser of oil in our list.' *Ibid.*, p. 62. J.O. Miller reports the same tendency in *pargana* Jansath. MCR, BOR, viii/ii, 14(c), Ser. 5, J.O. Miller, AR Tahsil Jansath (Upland Portion) (Muzaffarnagar), 1880, p. 30. See also NWP Rev., Feb. 1872, 84.

[233] W.A. Forbes, Baghpat Pargana Report, 1870, p. 62, in Buck, *Meerut SR* (1874).

[234] MCR, BOR, viii/ii, 14(c), J. Singh, RRR Tahsil Kairana (Muzaffarnagar), 1920, p. 4.

increasingly squeezed out of the prime areas.[235] The 'Jat's readiness to pay a higher rent than the average cultivator', according to one official, had the effect of improving the standards of cultivation 'even among castes which [were] usually indolent and unskilful'; even the Rajput showed 'less pronounced views than elsewhere on the indignity of labour'.[236] That astute observer W.H. Moreland was himself convinced that the increased cost of labour was 'reacting to a certain extent on the social customs' and that, 'speaking generally, the higher castes [were in the 1910s] probably doing more work than was formerly the case'.[237] Again, the cultivation standards of the Chamars appeared to vary according to the circumstances in which they operated.[238] What qualitative evidence there is does tend to testify as to the relative behavioural flexibility of different castes when faced with new circumstances.

Such conditions contrasted starkly with the experience of the eastern districts, where technology was comparatively static. There was no significant modern input capable of sufficiently affecting economic potential so as to prise open the social and economic structure. As population growth ate into the margins for adjustment, attention became fixed upon the ever more meticulous layering of consumer entitlements as an alternative to the investment of resources in capital accumulation and increased productivity. The institutions were a bar to progress because the advantages of change for those controlling the key assets of land and capital were never significant enough. In a static environment – and one displaying a marked absence

---

[235] Land was certainly under cultivation by the best cultivators in Khatauli and Jansath, for example. Jats, Rawas, and Sanis were to be found in Khatauli's richest tracts, while Rajputs were 'chiefly found in inferior villages on the banks of the river'. MCR, BOR, iv(c), Ser. 7, L.P. Barma, Handbook of Pargana Khatauli, *c.* 1890, p. 6. In Jansath, Gujars, Rajputs, and Shaikhs were observed 'mostly [to] occupy the poor estates to the south-west'. MCR, BOR, iv(c), Ser. 7, L.P. Barma, Handbook of Pargana Jansath, *c.* 1890, p. 5.

[236] Lane, *Muzaffarnagar SR* (1921), p. 7. Forbes noted fifty years earlier the 'altered character' of Rajputs in Dasna, which had shortly before received canal water. Forbes, 'Memorandum upon Ascertaining and Fixing Rates', p. 4, in Buck, *Meerut SR* (1874).

[237] *Agriculture of the UP*, p. 142.

[238] See e.g. W.J. Lupton, *Handbook of Pargana Kishni, Tahsil Bhongaon, Mainpuri District* (Allahabad, 1904), p. 5.

of external economies – those in a position to influence the pattern and rate of growth responded rationally to a slowly deteriorating population/resource situation by trying to wring more out of the existing system.

The western districts therefore constituted an appropriate technical and institutional environment for maximising peasant production. It was important that the nature of the technology, in preventing possibilities for scale economies conducive to polarisation, reinforced the landownership pattern which favoured peasant cultivators. The unit of operation remained limited in size and this, plus the ultimately labour-using consequences of labour-saving technology (in that the canal and the other innovations permitted more intensive cultivation), guaranteed a widely spread distribution of income. There is more to this issue, though, than a straightforward question of welfare. It is quite clear that the very *nature* of growth is important to the achievement of growth itself. In the case of the canal districts, independent owner–cultivator units, able to capture the increasing returns to production over the period under review and to apply the small-scale technical innovations which came forward to meet their needs, were the ideal vehicles for achieving high productivity levels and growth. The stimulation to local expansion and diversification in general was related closely to the particular multiplier effects emanating from an agricultural sector characterised by small units.[239]

Ironically, when today's growth models are looking more and more at the nature of the growth process, it seems that the historical experience of the western UP contains useful lessons which become even more relevant as the nature of the Green Revolution's growth – with its polarisation of holding size, its impact on employment and output structure, and so on – becomes more apparent. Moreover, it is ironic also that the

---

[239] A good latter-day comparison is Taiwan, whose agricultural development (even under the Japanese, but certainly since the early 1950s) has been structured around an owner–cultivator peasant system with policies designed to improve the returns to the agricultural producer, not so much through improving the agricultural terms of trade as by encouraging the application of small-scale technological advances. See G. Ranis, 'Equity with Growth in Taiwan: How "Special" is the "Special Case"?', *World Development*, VI: 3 (1978), pp. 403–08.

technological input which had been dismissed as a 'costly experiment' should have been so central not only as a driving force behind the prosperity of the western UP, but to the very notable diffusion of those benefits.

APPENDIX 1

Career outlines of engineers important in the formative phase of canal irrigation development

Baird Smith, Richard (1818–61). Madras Engineers, 1836–9; Bengal Engineers, 1839–61. Colonel. Assistant to Cautley (q.v.) on EJC, 1840–3; Superintendent of EJC, 1843–50; Sikh Wars, 1845–6, 1848–9; furlough in Europe, during which commissioned by Court of Directors to write report on *Italian Irrigation*, 1850–3; Deputy Superintendent of Canals, NWP, 1853–4; Director of Ganges Canal and Superintendent of Canals, NWP, 1854–9; defence of Roorkee and Chief Engineer in Delhi assault, 1857–8 (brevet Lieutenant-Colonel, 1858); Superintendent-General of Irrigation and Officiating Secretary to GI PWD, 1859–61; investigation of 1860–1 famine, 1861.

Baker, William Erskine (1808–81). Bengal (later renamed Royal) Engineers, 1826–77. General; KCB, 1870. Assistant on Delhi Canal, 1829–36; Superintendent of Delhi Canals and Sind Canals and Forests, 1836–45; Sikh Wars, 1845–6; Director of Ganges Canal Works, 1845–8; Secretary to GI PWD, 1854–5; Military Secretary to India Office, 1859–61; member of Council of India, 1861–75.

Brownlow, Henry Alexander (1831–1914). Bengal Engineers, 1849–86. Lieutenant-General. Indian Mutiny, 1857–8 (brevet Major); Executive and Superintending Engineer, including remodelling of Ganges Canal, 1860–9; furlough, 1869–71; Chief Engineer, NWP&O Irrigation Dept., 1872–82; Inspector-General of Irrigation and Deputy Secretary to GI PWD, 1882–6.

Cautley, Proby Thomas (1802–71). Bengal Artillery, 1819–54. Colonel; KCB, 1855. Assistant on reconstruction of Doab Canal (EJC), 1825–30; siege of Bharatpur, 1825–6; Superintendent of EJC, 1831–43; projection and construction of Ganges Canal, 1836–44; furlough and study of hydraulic works in Britain, Italy, and Egypt, 1845–8; election to Fellowship of Royal Society, 1846; Director of Canals, NWP, 1848–54; preparation of Report on the Ganges Canal Works, 1855–9; member of Council of India, 1858–68.

Cotton, Arthur Thomas (1803–99). Madras Engineers, 1819–77. General; KCSI, 1866. Assistant Engineer, north coast of Ceylon, 1821–3; expeditionary force to Burma, 1824–6; Irrigation Engineer, Madras PWD, including planning and execution of Cauvery works (1828–36), Godavari *anicuts* (1840–52), and projected *anicut* on Krishna River; retirement from official service, 1862; project consultant to private canal companies, 1862–5; involvement in controversial exchange with Cautley (q.v.), 1863–5; active in intermittent campaigns for irrigation extension in India, 1865–78.

Crofton, James (1826–1908). Bengal Engineers, 1844–82. Lieutenant-General. Executive Engineer, including assistant to Dyas (q.v.) in construction of (Upper) Bari Doab Canal, 1846–59; furlough, 1860–2; Officiating Superintending Engineer, EJC,

1863–4; preparation of *Report on the Ganges Canal*, 1864; Chief Engineer, Punjab Irrigation Dept., including design of Sirhind Canal, 1865–74; drafting of North India Canal and Drainage Bill (Act VII of 1873), 1872–3; Inspector-General of Irrigation and Deputy Secretary to GI PWD, 1874–82.

Dyas, Joseph Henry (1824–68). Bengal Engineers, 1843–68. Lieutenant-Colonel. Construction of (Upper) Bari Doab Canal, 1850–7; Officer-in-Charge, Punjab Canals, 1857–9; furlough, 1859–63; Chief Engineer, Punjab Irrigation Dept., 1864; Chief Engineer, NWP Irrigation Dept., 1864–7.

Greathed, William Wilberforce (1826–78). Bengal Engineers, 1844–76. Major-General; CB, 1859. NWP Irrigation Dept., 1847–52; Sikh Wars, 1848–9; furlough, 1852–4; Consulting Engineer, East India Railway extensions, Allahabad, 1855–7; distinguished military role during Indian Mutiny and in China campaigns, 1857–60 (brevet Lieutenant-Colonel, 1861); Assistant Military Secretary to Horse Guards, 1861–6; Chief Engineer, NWP Irrigation Dept., 1867–72; furlough, 1872; major role in construction of Agra Canal and LGC, 1873–5.

Moncrieff, Colin Campbell Scott (1836–1916). Bengal Engineers, 1856–83. Colonel; KCSI, 1903. Indian Mutiny, 1857–8; Executive and Superintending Engineer, NWP Irrigation Dept., 1859–7?; Chief Engineer, Burma, 187?–83; Under-Secretary for Public Works, Cairo, 1883–92; Under-Secretary of State for Scotland, 1892–1902; President of Indian Irrigation Commission, 1901–3.

Strachey, Richard (1817–1908). Bombay Engineers, 1836; Bengal Engineers, 1836–75. Lieutenant-General; KCB. Assistant to Baker (q.v.) on Delhi Canals, 1839; Executive Engineer, Ganges Canal reconstruction, 1841–7; Sikh Wars, 1845–6 (mentioned in despatches, brevet Major); study of botany and geology, Naini Tal, 1847–50; furlough in England, 1850–5; election to Fellowship of Royal Society, 1854; Director of Irrigation Works, Bundelkhand, 1855; Officiating Under-Secretary to NWP PWD, 1856–7; Consulting Engineer to GI Railway Dept. and active in reorganisation of GI PWD, 1858–62; Secretary to GI PWD, 1862–5; Inspector-General of Irrigation, 1866–9; Acting Secretary to GI PWD with seat on Legislative Council, 1869–71; Inspector of Railway Stores, India Office, 1871–5; member of Council of India, 1875–7, 1879–89; President of Famine Commission, 1878–80; Chairman of Meteorological Council, 1882–1905; President of Royal Geographical Society, 1888–90; Chairman of East India Railway Company, 1889–1907; member of Herschel Committee on Indian Currency, 1892.

Yule, Henry (1820–89). Bengal Engineers, 1840–62. Colonel; KCSI. Assistant on WJC and restoration of other indigenous works, 1843–5; Executive Engineer, Ganges Canal, 1846–54; Sikh Wars, 1846–7, 1848–9; Under-Secretary to newly formed GI PWD, 1855; Secretary and Chronicler to Sir Arthur Phayre's embassy to Burma, 1855–8; Secretary to GI PWD, 1858–61; retirement in Italy, 1862–75; member of Council of India, 1875–88.

APPENDIX 2

Quantitative analysis of the impact of canal irrigation upon crop patterns

Stepwise multiple regression was employed upon Koil village data to examine the effect on individual crops and combinations of crops of the following independent variables: the respective proportions of n.c.a. under canal irrigation, well irrigation, and tenants-at-will, and four indices of transport costs involved in marketing. The transport cost variables ranged from simple road distance measures to (a) the main market in Aligarh and (b) the nearest railway station, to more sophisticated cost estimates of transport to these points which took account of the proportion of each journey over four different road quality classifications based on the *maund*-mile costs prevailing for each type of road. During the exercise in stepwise multiple regression (in which the computer systematically selects from a batch of variables those which add most to the explanatory power of the equation and brings them in one by one) the canal variable was consistently brought in first, and in most instances the inclusion of a second variable caused the F statistic for the equation to fall. The following tabulated results thus represent the linear relationships between the crop and canal variables.

Regression analysis of a sample of 74 canal and well villages: summary of output[a]

Dependent variable (crop area)	Constant	Variable X_1 coefficient	R^2	F statistic
Commercial crops[b]	40.02	Canal = 0.35	0.71	176.3
Indigo	0.46	,, = 0.22	0.86	434.3
Wheat	22.36	,, = 0.17	0.51	73.6
Maize	10.15	,, = 0.15	0.54	165.6
Barley (alone and mixed)	20.48	,, = 0.11	0.31	32.2
Dofasli	22.46	,, = 0.41	0.76	223.6
Rabi	63.73	,, = 0.21	0.64	126.1
Kharif	57.72	,, = 0.20	0.50	71.5

[a] Significant at the 99.9% level.
[b] Aggregate of wheat, indigo, sugarcane, and cotton.

Cotton, the chief *kharif* cash crop in well villages, showed no systematic relationship (even a negative one) either with the canal-irrigated area or with the well-irrigated area. The well variable, in fact, was not significantly related to any of the dependent variables.

Given the integrated nature of the crop patterns, it did seem likely that multiple regression would produce restricted and fragmentary results, and possibly miss some more complex relationships involving groups of variables. To test this a programme was chosen which analyses the effect of a *set* of independent variables upon another set of dependent variables, producing a measure of the overall correlation between the two sets and indicating the relative importance of each of the components within the sets. Canonical correlation analysis is designed to achieve this by deriving a linear combination from each of the two sets of variables in such a way that the correlation between the two linear combinations is maximised. When the dependent (crop) variables were placed against the set of independent variables, the canal variable was overwhelmingly dominant. Just over 93% of the variance in crop acreages was accounted for by this variable alone in the sample of 74 villages.[1] This result should not be surprising, given the integrated nature of the crop patterns. The variations in individual crop coefficients, in fact, confirmed the essential pattern shown in the above table in terms of the crops most affected by the canal. When a non-canal sample of 112 villages was examined in the same way, the eigenvalue, or measure of the variance in crop acreages accounted for by the well area, was a low 0.308.[2] Partly, of course, this reflects the greater inaccuracy of the well irrigation statistics, but it mostly mirrors the fact that well irrigation was of limited importance in determining crop patterns.[3]

---

[1] Chi-square: 181.99.   [2] Chi-square: 38.56.

[3] The method of analysis, its results, and its implications are dealt with in more detail in I. Stone, 'Peasant Agriculture: An Analysis of Production Determinants in Pargana Koil (Aligarh, U.P.) 1900–01', paper presented at the Second New Zealand Conference on Asian Studies, University of Canterbury, Christchurch, May 1977.

APPENDIX 3

Irrigation of major crops as a percentage of total area irrigated by each canal
(average for 1902–3/1904–5)

	Upper Ganges Canal	Eastern Jumna Canal	Lower Ganges Canal	Agra Canal
Wheat	40.1	39.4	40.3	30.2
Other *rabi* food crops	17.5	7.4	25.8	17.2
Sugarcane	17.2	21.2	4.0	4.1
Rice	3.8	16.3	3.2	–
Cotton	8.9	2.9	8.9	40.9
Indigo	2.9	–	3.5	–
Maize	2.6	4.1	4.2	–
Other crops	7.0	8.7	10.1	7.3
Total irrigated area (triennial average in acres)	889,335	289,938	737,528	227,309

Source: Compiled from *IRR 1902–3/1904–5.*

GLOSSARY OF INDIAN TERMS

agaul: strain of sugarcane
Ahir: caste name; relatively low-caste cultivator–grazier
amin: suboverseer, subordinate official
anicut: weir
anna: one-sixteenth of a rupee
arhatia: commodity broker, commission agent
asarhi: indigo planted with the rains and cut in September
badni: crops pledged by grower at fixed terms in exchange for advances
bahi: account book
baisuri: noxious weed which favours dry conditions
bajra: species of millet
Baluchi: name of Baluchistan tribe
bangar: arable upland plain, as distinct from *khadir* (q.v.) or riverain land
bania: small-scale trader, moneylender
banjara: itinerant trader, cattle-dealer
bara: fields immediately surrounding village site
batai: division of crop between cultivator and landlord; payments may be in kind or in cash
bazar: marketplace
begar (begari): compulsory free labour services by tenant or low-status villagers to landlord or powerful villagers
Behea: type of sugar mill
bejhra: barley and *gram* (q.v.) mixed
bel: small-scale sugar factory
beldar: labourer on channel maintenance
beopari: itinerant local trader
beri: baskets of split bamboo, used for lifting canal water
bhaiachara: coparcenary tenurial form in which the revenue demand is apportioned among village proprietors on some principle other than ancestral shares, usually on the basis of cultivation
bhur: light, sandy soil
bhusa: straw

353

bigha: unit of area; the standard revenue *bigha* in the NWP was five-eighths of an acre

bohra: village banker

bokhar: fever

boro: rice grown during the dry season

Bowrea: caste name; nomad given to lawlessness

chakra: two-wheeled bullock cart

Chamar: caste name; leatherworker, labourer

chaprassi (chuprassee): attendant

chari: fodder

charpoy: wooden frame bed with webbing

chaudhri: headman of village, guild, or caste

Chauhan: name of low-status Rajput (q.v.) clan

chaukidar: watchman, patrol

chaupai: four-wheeled bullock cart

chaupal: village assembly hall

crore: ten million

dal: lift irrigation

daulhu: thin *desi* (q.v.) strain of sugarcane

desi: indigenous, local

dhan: common rice

dhaur: relatively high-yielding strain of sugarcane

dhenkli: well worked by hand using a pulley

doab: area between rivers running roughly parallel; plain

dofasli: double-cropped land, bearing two successive crops in the course of a year (normally a *kharif* (q.v.) and a *rabi* (q.v.) crop)

dumat: soil of stiff loam

durkjwast: *kolaba* (q.v.) application

fakir: religious mendicant

faslana: charge realised at harvest time

gadhwala: hired carrier

garoli: brushwood lining of *kachha* (q.v.) well shaft

gauhan: highly manured fields lying close to village site

ghair-maurusi: tenant-at-will, statutory tenant (lit. without hereditary land rights)

ghar: land in the form of a shallow trough

ghat: landing place

Ghoshi (Goshaen): name of Muslim caste of the Punjab; pastoralist, milkman

godown: warehouse

gojai: wheat and barley mixed

gram: pulse grown during *rabi* (q.v.)

Gujar: caste name; grazier

gul: village water-course

gur: cakes of unrefined sugar made by boiling *ras* (q.v.)

haq: right, fee
har: outlying land
harwaha: labourer formerly of cultivator status
hisab: loan account
ikrarnama: formal agreement
illakadar: suboverseer, subordinate official
jama: revenue assessment on holding
jamowa: indigo sown before the rains (with the aid of irrigation) and cut in time to allow a *rabi* (q.v.) crop to be sown
jarhan: transplanted rice
Jat: caste name; cultivator
jhil: shallow depression, pond
Jhojha: caste name; cultivator
jita: cultivator who exchanges labour with another cultivator
juar: species of millet
kachha: temporary; earthen; lacking substance
Kachhi: caste name; relatively low-caste cultivator
kachhi chini: refined, whitish floury sugar; also called *khand*
kapas: cotton not separated from the seed
karahi: boiling pan for *ras* (q.v.)
karinda: agent, manager
khadir: valley, flood plain area
khallassi: sailor
khand: refined, whitish floury sugar; also called *kachhi chini*
khandsari: small-scale sugar refinery; manufacturer of coarse sugar
kharif: growing season of crops harvested in autumn
khasra: official fieldbook compiled at time of village survey
khatauni: final statement of irrigation dues for village
khatti: grain storage pit
khetbat: unconsolidated or scattered land partitions
khudkasht: leased-in land cultivated by proprietor
khunti: indigo crop obtained from stumps of previous year's plants; also called *ratoon*
khushkharid: lit. 'good bargain'; agreement made at harvest time, as opposed to *badni* (q.v.) system
kiari: field compartment made up of low earthen ridges
kili: system of well irrigation employing two ramps per lift and allowing more than one pair of bullocks to be used at one time on the same lift
kist: revenue instalment
kolaba: irrigation outlet serving village or part of village
kolhu: traditional bullock-driven sugarcane-crusher
kothi: establishment, such as factory or storehouse
kumhar: potter
kunkar: alluvial formation of limestone

Kurmi: caste name; relatively low-caste cultivator

lagor: system of well irrigation employing one ramp and limited to one pair of animals per lift

lakh: one hundred thousand

lakkar: roller fashioned from a tree trunk, used in field preparation

lambardar: officially appointed representative of village

mahajan: merchant, moneylender

malbah: record of village cash payments

malbah-nahr: water-course account

Mali (Malee): caste name; low-caste gardener–cultivator

mandi: regular market

manjha: middle-quality land

maund: unit of weight, roughly 82 pounds

maurusi: occupancy status

mota: water-hoarding bed of clay

muddud: obligation of community or landholder to supply labourers for public works

mukaddam: headman; manager of village proprietary body

munji: fine rice

munshi: recorder–clerk

nadi: river

nagla: hamlet, village

Nahan: form of bullock-driven sugarcane press

naib-ziladar: deputy *ziladar* (q.v.)

nazrana: gift from tenant to landlord

nil: dye

nulla: drainage line

Oade: caste name; migratory earthworker

octroi: municipal levy on incoming goods and commodities

paimana: standard volumetric unit of canal water, regulated by module of the same name

painth: periodic market for group of villages

pakka: permanent; substantial; of solid material, such as masonry

paleo: single preliminary watering

pandit: learned Brahman

panni: coarse grass common in the Doab

parcha: statement of irrigation dues for individuals or irrigating groups

pargana: administrative subdivision of a *tehsil* (q.v.)

Pathan: name of Muslim caste

patta: engagement to pay revenue

pattidari: system of tenure in which the land is farmed in severalty and the revenue demand apportioned by kin group according to ancestral shares

patwari: village accountant

phaora: mattock used in canal excavation

pice: one-quarter of an *anna* (q.v.)

pounda: variety of sugarcane, mainly eaten raw

pur: leather bag, used for lifting well water

Purbeah: caste name; migratory earthworker from Oudh

rabi: growing season of crops harvested in spring

rais: notable, man of position

rajbaha: distributary channel of canal system

Rajput: caste name; warrior, landholder

rao: flow of water

ras: sugarcane juice

rati: well worked by hand using a weighted lever

ratoon: indigo crop obtained from stumps of previous year's plants; also called *khunti*

Rawa: caste name; cultivator

reh: saline deposits on or near the surface of affected fields

ryot: cultivator, peasant

Sani: caste name; relatively low-caste cultivator

saretha: variety of sugarcane distinguished by its comparatively high juice content

seer: unit of weight, one-fortieth of a *maund* (q.v.) or about two pounds

Shaikh: name of Muslim caste

shakar: unrefined sugar made by boiling *ras* (q.v.) and by stirring occasionally during cooling period

shroff: traditional banker, broker

sir: land held by a *zamindar* (q.v.) under title of personal cultivation

siwai: twenty-five per cent interest over a single six-month season

surnai: craft of inflated buffalo skin, used by canal officers to inspect headworks

surnaiwalla: swimmer who pilots a *surnai* (q.v.)

Taga: caste name; landholder of Rohilkhand

takavi: official advance to cultivator from public funds for agricultural purposes

talukdar: large landowner

tatil: periodic closure of a *rajbaha* (q.v.), or outlets on a section of a *rajbaha*, in order to conserve supplies

tatti: grass mat

tehsil: administrative subdivision of a district

tehsildar: subordinate official of revenue administration in charge of a *tehsil* (q.v.)

Thakur: caste name; warrior, landholder

thela: four-wheeled bullock cart

thokdar: appointee within village responsible for *guls* (q.v.) and *kolabas* (q.v.)

tor: flow or flush irrigation

tulai: weighment fees

tumri: elongated gourd forming a crude lifebelt placed under the chest

ukh: sugarcane

unt gari: camel cart with large rectangular cage

usar: barren land, usually affected by alkalinity

warabandi: list of days and part-days on which each cultivator is permitted to take canal water

zamindar: landowner, of varying size

zamindari: property of a *zamindar* (q.v.)

ziladar: subordinate official; head of local branch of canal administration

BIBLIOGRAPHY

UNPUBLISHED RECORDS

India Office Records, London
Government of NWP/NWP&O/UP, Department Proceedings:
  Public Works (Irrigation Branch) 1867–1930
  Revenue   1860–1930
  Sanitation   1900–1910
  Scarcity   1896–1921

Muzaffarnagar Collectorate Records, Muzaffarnagar
Revenue Records:
  Combined (Village) Assessment Statements, Pargana Khatauli, Muzaffarnagar
  (1917–18)

Records of the Commissioner of Meerut, Meerut
Files of the Scarcity Department
Revenue Records:
  Pargana and Tahsil Settlement Handbooks, Assessment Reports, and Rent-Rate
  Reports for Meerut Division

Uttar Pradesh Board of Revenue Library, Allahabad
Revenue Records:
  Tahsil Settlement Handbooks for Meerut and Agra Divisions

Uttar Pradesh Board of Revenue Library, Lucknow
Revenue Records:
  Pargana and Tahsil Settlement Handbooks, Assessment Reports, and Rent-Rate
  Reports for Meerut and Agra Divisions

Uttar Pradesh Secretariat Library and Records, Lucknow
Files of the Scarcity Department

Uttar Pradesh State Archives, Lucknow
Files of the Board of Revenue:
  Canal Shelf Series
  District Series

Uttar Pradesh State Regional Archives, Allahabad
Transferred Files of the Commissioner:
  Agra Division
  Meerut Division

PUBLISHED OFFICIAL RECORDS
British Parliamentary Papers

1862, XL, *Report on the Famine of 1860–1 in the North-Western Provinces of India*
1865, XXXIX, *Resolution by the Government of India Public Works Department Relative to the Canals of the North-Western Provinces*
1870, LIII, *Report on Irrigation Works in India*
1878–9, IX, *Report from the Select Committee on East India (Public Works)*
1880, LII, *Report of the Indian Famine Commission*
1881, LXXI, *Further Report of the Indian Famine Commission, with Proceedings, Evidence and Appendices*
1898, LXII, *Further Papers Regarding the Famine and Relief Operations*
1899, XXXI and XXXIII, *Report of the Indian Famine Commission, 1898*
1904, LXVI, *Report of the Indian Irrigation Commission, 1901–03*
1909, LXII, *Administration of Famine Relief in the United Provinces of Agra and Oudh, 1907–8*

Government of India

(A) Departments

Commercial Intelligence and Statistics Department (Department of Statistics up to 1902–3), *Returns of the Agricultural Statistics of British India*, Calcutta, annual, from 1884–5
Directorate of Marketing and Inspection, *Report on the Marketing of Wheat in India*, Simla, 1937
Home, Revenue, and Agriculture Departments, *Selections from the Records of the Government of India*, no. CLX of 1879, *The Wheat Production and Trade of India*, Calcutta, 1883
Imperial Council of Agricultural Research, *Report on the Cost of Production of Crops in the Principal Sugarcane and Cotton Tracts in India*, vol. III plus *Supplement*, New Delhi, 1939–40
Public Works Department, *Review of Irrigation in India*, Simla, annual, from 1899–1900
Public Works Department, *Triennial Review of Irrigation in India, 1918–21*, Calcutta, 1922

(B) Reports of Miscellaneous Committees, Commissions, etc.

Census of India, 1921, vol. XVI, *United Provinces of Agra and Oudh*, Allahabad, 1923
Indian Cotton Committee: Report, Calcutta, 1919; and *Minutes of Evidence*, 6 vols., Calcutta, 1920–6
Indian Irrigation Commission, 1901–03, Minutes of Evidence, United Provinces of Agra and Oudh, Calcutta, 1903
Report of the Indian Sugar Committee, Simla, 1920
Report of the Indian Taxation Enquiry Committee, 1924–5, Madras, 1926
Report of the Irrigation Commission of India, 3 vols., New Delhi, 1972
Report of the Irrigation Conference, Simla 1904, 3 vols., Calcutta, 1905–6
Royal Commission on Agriculture in India: Report, London, 1928; and *Evidence*, vol. VII, London, 1927

Government of NWP/NWP&O/UP

(A) Annual Reports (dates of commencement given in brackets; all published at Allahabad, unless stated otherwise in footnotes)

Irrigation Administration Report (1902–3)
Irrigation Revenue Report of the North-Western Provinces and Oudh (1871–2)
Papers on the Revenue Returns of the Canals of the North-Western Provinces (1860–1)
Progress Report of the Public Works Department (Irrigation Branch) (1870–1)
Report of the Board of Revenue on the Revenue Administration of the North-Western Provinces and Oudh (1870–1)
Report of the Department of Agriculture and Commerce (1877–8)
Report of the Department of Land Records and Agriculture (1894–5)
Report of the Government Experimental Farm, Cawnpore (1877–8)
Report of the Sanitary Commissioner for the North-Western Provinces and Oudh (1877)
Report on the Seasons and Crops of the United Provinces (1901–2)

(B) Settlement Reports, etc. (all published at Allahabad)

(a) Districts

AGRA
Report on the Settlement of the Agra District, by H.F. Evans, 1880
Final Report on the Settlement and Record Operations in District Agra, by R.F. Mudie, 1930

ALIGARH
Final Report on the Revision of Settlement in the District of Aligarh, by W.H. Smith, 1882
Final Settlement Report of the Aligarh District, by W.J. Burkitt, 1903
Final Settlement Report of the Aligarh District, by S. Ahmed Ali, 1943

BULANDSHAHR
Revised Settlement Report of the District of Boolundshuhur, by C.A. Daniell, 1869
General Report of the Settlement of the Bulandshahr District, 1865–6, by R.G. Currie, 1877
Final Report of the Settlement of Bulandshahr District, by T. Stoker, 2 vols., 1891
Final Settlement Report of the Bulandshahr District, by E.A. Phelps, 1919

CAWNPORE
Final Report on the Settlement of the Cawnpore District, by F.N. Wright, 1878
Final Report on the Seventh Settlement of the Cawnpore District, 1903–6, by H.K. Gracey, 1907

ETAH
Settlement Report of the Eta District, by S.O.B. Ridsdale, 1874
Final Report on the Etah District, 1904–5, by H.O.W. Robarts, 1905

ETAWAH
Report on the Settlement of Etawah District for the years 1868–74, by C.H.T. Crosthwaite, 1875
Final Settlement Report of the Etawah District, 1911–14, by E.S. Liddiard, 1915

FARRUKHABAD
Final Report of the Settlement of Farrukhabad District, by H.F. Evans, 1875
Final Settlement Report of the Farrukhabad District, Seventh Revision, 1898–1903, by H.J. Hoare, 1903

FATEHPUR
Final Settlement Report of the Fatehpur District, by A.B. Patterson, 1878
Final Settlement Report of the Fatehpur District, by C.L. Alexander, 1915

MAINPURI
Report on the Settlement of the Mainpuri District, by M.A. McConaghey, 1875
Final Settlement Report of the Mainpuri District, 1902–6, by W.J.E. Lupton, 1906
Final Settlement Report of District Mainpuri, by Zahurul Hasan, 1944

MEERUT
Settlement Report of the District of Meerut, by E.C. Buck, 1874
Final Settlement Report of the Meerut District, by R.W. Gillan, 1901
Final Settlement Report of the Meerut District, by C.H. Cooke, 1940

MUTTRA
Report on the Settlement of the Muttra District, by R.S. Whiteway, 1879
Final Settlement Report of the Muttra District, by H.A. Lane, 1926

MUZAFFARNAGAR
Report on the Settlement of the District of Moozuffurnuggar, by E. Thornton, 1841
Settlement Report of the District of Muzaffarnagar, including a Report on the Permanent Settlement of the Western Parganas of the District and also a Report on the Settlement of the Ganges Canal Tract, by A. Cadell, 1882
Final Report on the Settlement of the Muzaffarnagar District, by J. O. Miller, 1892
Final Report on the Settlement of the Muzaffarnagar District, by H.A. Lane, 1921

SAHARANPUR
Report on the Settlement of Saharanpur, by H. Le Poer Wynne, 1871
Final Report of the Settlement of the Saharanpur District, by L.A.S. Porter, 1891
Final Report on the Settlement Operations of the Saharanpur District, 1917–20, by D.L. Drake-Brockman, 1921

(b) Parganas/Tehsils
AGRA DISTRICT
Assessment Report of Pargana Fatehabad District Agra, by H.J. Frampton, 1929
Agra Pargana Handbook, by R.F. Mudie (?), 1930
Rent-Rate Report of Pargana Fatehabad, District Agra, by H.J. Frampton, 1931

MAINPURI DISTRICT
Handbook of Pargana Kishni, Tahsil Bhongaon, Mainpuri District, by W.J. Lupton, 1904
Rent-Rate Report of Karhal Tahsil, Mainpuri District, by W.J. Lupton, 1904
Rent-Rate Report of Pargana and Tahsil Shikohabad, Mainpuri District, by W.J. Lupton, 1905

(C) Miscellaneous Reports and Publications
Buck, E.C. 'Note on the Effect of the Canal on the Agricultural Conditions of the Pergunnah of Tirwa, Zilla Farruckhabad', 1 May 1871, *Revenue Reporter*, v, n.d., pp. 212–27
Crosthwaite, C.H.T. 'Report on the Average Rent-Rates for Part of Pergunna Phuppoond', 20 Sept. 1870, *Revenue Reporter*, v, n.d., pp. 1–20
District Census Statistics: Aligarh, Allahabad, 1895
Patterson, A.B. 'Report on the Presumed Rates to be Used in the Settlement of Tehseel Eglas', 27 Aug. 1870, *Revenue Reporter*, v, n.d., pp. 42–6
Report on Agriculture in the United Provinces, Naini Tal, 1926

Report of the Irrigation Rates Committee, United Provinces, Lucknow, 1939
United Provinces Provincial Banking Enquiry Committee, 1929–31: Report and *Evidence*, 4 vols., Allahabad, 1931
United Provinces Provincial Congress Committee: Agricultural Distress in the United Provinces, Allahabad, 1932

Government Reports and Published Papers Listed under Individual Authors

Atkinson, E.T. *Statistical, Descriptive and Historical Account of the North-Western Provinces of India*, 14 vols., Allahabad, 1874–86
Baird Smith, R. *Revenue Reports of the Ganges Canal, 1855–6*, Roorkee, 1856
——. *The Special Commissioner's Final Report on the Famine of 1860–61*, Calcutta, 1861
Baker, W.E., Dempster, T.E., and Yule, H. *Report of a committee assembled to report on the causes of the unhealthiness which existed at Kurnaul, and other portions of the country along the line of the Delhi canal, and also whether any injurious effect on the health of the people of the Dooab is, or is not, likely to be produced by the contemplated Ganges Canal, with appendices on malaria by Surgeon T.E. Dempster*, Calcutta, 1847; reprinted in *Records of the Malaria Survey of India*, 1:2, 1930, pp. 1–68, and in P.T. Cautley, *Report on the Ganges Canal Works*, vol. III, London, 1860, Appendix B.
Benett, W.C. *Report on Scarcity and Relief Operations in the North-Western Provinces and Oudh during 1877–8 and 1879*, Allahabad, 1880
Cautley, P.T. *Notes and Memoranda on the East Jumna or Doab Canal, North-Western Provinces, 1845*, Calcutta, 1845
Chatterjee, A.C. *Notes on the Industries of the United Provinces*, Allahabad, 1908
Clibborn, J. *Irrigation Works in India*, Roorkee, 1891
——. *Papers Relating to the Construction of Wells for Irrigation*, Roorkee, 1883
Crofton, J. *Report on the Ganges Canal, with Estimates and Plans*, 3 vols., Calcutta, 1865
Cutcliffe, H.C. *A Sanitary Report on Certain Districts in the Meerut Division*, Allahabad, 1868
Duthie, J.F. and Fuller, J.B. *Field and Garden Crops of the North-Western Provinces and Oudh*, 3 pts., Roorkee, 1882–93
Girdlestone, C.E.R. *Report on Past Famines in the North-Western Provinces*, Allahabad, 1868
Graham, J.D. *Report on an Enquiry into the Prevalence of Malaria in Kosi (District Muttra)*, Allahabad, 1910
——. *Report on an Enquiry into the Prevalence of Malaria in Meerut, 1911*, Allahabad, 1913
Gupta, R.B. *Agricultural Prices in the United Provinces*, Allahabad, 1933
Hadi, S.M. *The Sugar Industry in the United Provinces*, Allahabad, 1902
Hailey, H.R.C. 'Prices of Gur and Cane in the United Provinces', *Agricultural Journal of India*, IX, 1914, pp. 217–26
Henvey, F. *A Narrative of the Drought and Famine which prevailed in the North-Western Provinces during the years 1868, 1869 and the beginning of 1870*, Allahabad, 1871
Hill, A. 'The Advantages of Irrigation when the Supply Available is Used for Rabi or Cold Weather Irrigation', *Agricultural Journal of India*, IV, 1909, pp. 233–47
Holderness, T.W. *Memorandum on Food Production in the North-Western Provinces and Oudh*, Allahabad, 1891
Howard, A. and Howard, G. 'The Irrigation of Alluvial Soils', *Agricultural Journal of India*, XII, 1917, pp. 185–99
Hutton, C.H. 'Rainfall, Irrigation, and Subsoil Water-level of the Gangetic Plain in the United Provinces', *Agricultural Journal of India*, XIII, 1918, pp. 197–201
Leake, H.M. 'The Trend of Agricultural Development in the United Provinces', *Agricultural Journal of India*, XVIII, 1923, pp. 10–22

MacKenna, J. 'Notes on the Fodder Problem in India', *Agricultural Journal of India*, IX, 1914, pp. 38–58

Mathur, J.K. *The Pressure of Population: Its Effects on Rural Economy in Gorakhpur District*, Allahabad, 1931

Misra, B. *Overpopulation in Jaunpur*, Allahabad, 1932

Misra, R. *Economic Survey of a Village in Cawnpore District*, Allahabad, 1932

Molony, E.A. *The Manual of Irrigation Wells*, Allahabad, 1907

——. *Papers Relating to the Construction of Wells for Irrigation*, Allahabad, 1883

——. 'Rainfall, Irrigation, and the Subsoil Water Reservoirs of the Gangetic Plain in the United Provinces', *Agricultural Journal of India*, XII, 1917, pp. 84–9

Moreland, W.H. 'Conditions Determining the Area Sown with Cotton in the United Provinces', *Agricultural Journal of India*, I, 1906, pp. 37–43

——. *Conditions Determining the Area Sown with Crops in the United Provinces*, Allahabad, 1907

——. *Improvements in Native Methods of Sugar Manufacture*, Allahabad, 1905

——. 'The Sugar Industry of the United Provinces', *Agricultural Journal of India*, II, 1907, pp. 15–21

——. 'Wells in the Gangetic Alluvium', *Agricultural Journal of India*, IV, 1909, pp. 34–42

Nevill, H.R., Drake-Brockman, D.L., et al. *District Gazetteers of the United Provinces of Agra and Oudh*, 44 vols., plus *Supplementary Notes and Statistics*, 10 vols., Allahabad, 1903–21

Pande, J.K. and Gupta, R.B. *Prices of Cereals in the United Provinces*, Allahabad, 1938

Prasad, T. *The Organisation of the Wheat Trade in the North-West Region of the United Provinces*, Allahabad, 1932

Sayer, W. 'Sugarcane Cultivation in Non-tropical Parts of India', *Agricultural Journal of India*, XI, 1916, pp. 360–70

Subbiah, P. *The Cultivation of Maize*, Allahabad, 1901

Swinton, A. *Report of the Committee on the Ganges Canal*, Roorkee, 1867

Wright, F.N. *Memorandum on Agriculture in the District of Cawnpore*, Allahabad, 1877

SECONDARY WORKS

Amani, K.Z. 'Variability of Rainfall in Relation to Agriculture in the Central Ganga–Yamuna Doab', *The Geographer*, XIII, 1966, pp. 35–47

Anderson, A.G. 'The Rural Market in W. Java', paper presented at the Second New Zealand Conference on Asian Studies, University of Canterbury, Christchurch, May 1977

Anonymous. 'Canal Rent vs. Land Revenue', *Calcutta Review*, XLIX, 1869, pp. 1–36

——. 'English Capital and Indian Irrigation', *Calcutta Review*, XXII, 1859, pp. 172–85

——. 'Memoir on Lt.-General James Crofton', *Royal Engineers' Journal*, V, April 1909, pp. 174–84

Ansari, N. *Economics and Irrigation Rates: A Study in Punjab and Uttar Pradesh*, London, 1968

Anstey, V. *The Economic Development of India*, 4th edn, London, 1957

Aziz, A. 'Urban Gradients around Aligarh', *The Geographer*, XX:2, 1973, pp. 134–50

Bagchi, A.K. 'Foreign Capital and Economic Development in India: A Schematic View', in K. Gough and H.P. Sharma (eds.), *Imperialism and Revolution in South Asia*, New York, 1973, pp. 43–76

——. *Private Investment in India, 1900–39*, Cambridge, 1972

Baird Smith, R. *Italian Irrigation: A Report on the Agricultural Conditions of Piedmont and Lombardy*, 2nd edn, 2 vols., London and Edinburgh, 1855

Bharadwaj, K. *Production Conditions in Indian Agriculture: A Study Based on Farm Management Surveys*, Cambridge, 1974

Bhatia, B.M. *Famines in India: A Study in Some Aspects of the Economic History of India (1860-1965)*, London, 1967

Brown, J. 'A Memoir of Colonel Sir Proby Cautley, F.R.S., 1802–71, Engineer and Palaeontologist', *Notes and Records of the Royal Society of London*, XXXIV, 1979–80, pp. 185–225

———. 'Sir Proby Cautley (1802–71), a Pioneer of Indian Irrigation', *History of Technology*, III, 1978, pp. 35–89

Bruce-Chwatt, L.J. *Essential Malariology*, London, 1980

Bruce-Chwatt, L.J. and Glanville, V.J. (eds.). *Dynamics of Tropical Disease by G. Macdonald*, London, 1973

Buckley, R.B. *The Irrigation Works of India and their Financial Results*, 2nd edn, London, 1905

Buckman, H.O. and Brady, N.C. *The Nature and Properties of Soils*, New York, 1968

Byres, T.J. 'Land Reform, Industrialization and the Marketed Surplus in India: An Essay on the Power of Rural Bias', in D. Lehmann (ed.), *Agrarian Reform and Agrarian Reformism*, London, 1974, pp. 221–61

Cautley, P.T. 'Canals of Irrigation in the North-Western Provinces', *Calcutta Review*, XII, 1849, pp. 79–164

———. *A Disquisition on the heads of the Ganges and Jumna Canals, North-Western Provinces, in reply to strictures by Major-General Sir Arthur Cotton*, London, 1864

———. *A Reply to Statements made by Major-General Sir Arthur Cotton, on the projection of the Ganges Canal Works*, London, 1864

———. *Report on the Ganges Canal Works*, 3 vols. plus *Atlas*, London, 1860

———. *A Valedictory Note to Major-General Sir Arthur Cotton, respecting the Ganges Canal, with a postscript touching certain misrepresentations of a writer in the 'Times' on the same subject*, London, 1864

Chambers, R. 'Men and Water: The Organisation and Operation of Irrigation', in B.H. Farmer (ed.), *Green Revolution? Technology and Change in Rice-growing Areas of Tamil Nadu and Sri Lanka*, London, 1977, pp. 340–63

Chandra, B. 'Reinterpretation of Nineteenth-Century Indian Economic History', *Indian Economic and Social History Review*, V:1, 1968, pp. 35–75

Chayanov, A.V. *The Theory of Peasant Economy*, ed. D. Thorner, B. Kerblay, and R.E.F. Smith, Homewood, Ill., 1966

Christophers, S.R. 'Epidemic Malaria of the Punjab, with a Note on a Method of Predicting Epidemic Years', *Paludism*, II, 1911, pp. 17–26

Commander, S.J. 'The Agrarian Economy of Northern India, 1800–1880: Aspects of Growth and Stagnation in the Doab', unpublished Ph.D. dissertation, University of Cambridge, 1980

Connell, A.K. *The Economic Revolution of India and the Public Works Policy*, London, 1883

Corbett, A.F. *Climate and Resources of Upper India*, London, 1874

Cotton, A. 'Private Memorandum upon the Ganges Canal', printed as an Appendix in P.T. Cautley, *A Reply to Statements made by Major-General Sir Arthur Cotton, on the projection of the Ganges Canal Works*, London, 1864

———. *Public Works in India*, London, 1854

———. *Reply to Colonel Sir Proby Cautley's Disquisition on the Ganges Canal*, London, 1864

———. *Reply to Sir Proby Cautley's Valedictory Note on the Ganges Canal*, London, 1865

Cotton, A. and Cautley, P.T. *A Discussion, regarding the projection and present state of the Ganges Canal, and the measures required to make it reliably useful and profitable*, London, 1864

Covell, G. and Singh, J. 'Anti-malarial operations in Delhi, Part IV', *Journal of the Malaria Institute of India*, V, 1943, pp. 87–106

Crooke, W. *The North-Western Provinces of India: Their History, Ethnology and Administration*, London, 1897

Dacosta, J. *Facts and Fallacies Regarding Irrigation as a Prevention of Famine in India*, London, 1878

——. *The Indian Budget for 1877–8: Remarks on the Financial Position of the Government of India*, London, 1877

Danvers, J. 'The Public Works and Progress of India', *Asian Review*, I, April 1886, pp. 327–53

Deakin, A. *Irrigated India: An Australian View of India and Ceylon, and their Irrigation and Agriculture*, London, 1893

Dewey, C.J. 'The Agricultural Output of an Indian Province: The Punjab, 1870–1940', paper presented at the Institute of Commonwealth Studies, London, April 1972

——. '*Patwari* and *Chaukidar*: Subordinate Officials and the Reliability of India's Agricultural Statistics', in C.J. Dewey and A.G. Hopkins (eds.), *The Imperial Impact: Studies in the Economic History of Africa and India*, London, 1978, pp. 280–314

Dutt, R.C. *The Economic History of India in the Victorian Age*, 3rd edn, London, 1908

——. *Speeches and Papers on Indian Questions*, Calcutta, 1902

Epstein, T.S. *Economic Development and Social Change in South India*, Manchester, 1962

Ezekiel, H. and Mathur, P. 'Marketable Surplus of Food and Price Fluctuations in a Developing Economy', *Kyklos*, XIV, 1961, pp. 396–408

Fisher, C.M. 'Planters and Peasant: The Ecological Context of Agrarian Unrest on the Indigo Plantations of North Bihar', in C.J. Dewey and A.G. Hopkins (eds.), *The Imperial Impact: Studies in the Economic History of Africa and India*, London, 1978, pp. 114–41

Gadgil, D.R. *Economic Effects of Irrigation: Report of a Survey of the Direct and Indirect Effects of the Godavari and Pravara Canals*, Poona, 1948

Ganguli, B.N. *Trends of Agriculture and Population in the Ganges Valley*, London, 1938

Geertz, C. *Agricultural Involution: The Process of Ecological Change in Indonesia*, Berkeley, 1963

Gulhati, N.D. *Administration and Financing of Irrigation Works in India*, New Delhi, 1965

Gustafson, W.E. and Reidinger, R.B. 'Delivery of Canal Water in North India and West Pakistan', *Economic and Political Weekly*, VI:27, 3 July 1971, pp. A157–62

Harnetty, P. 'Cotton Exports and Indian Agriculture, 1861–70', *Economic History Review*, Series 2, XXIV:3, 1971, pp. 414–29

Howard, A. *Crop Production in India: A Critical Survey of its Problems*, Oxford, 1924

Jacob, L. 'Irrigation in India', *Contemporary Review*, CV, June 1914, pp. 800–8

Kanetkar, B.D. 'The Core of Price Policy for the Sale of Canal Waters', *Artha Vijnana*, V:2, 1963, pp. 97–110

——. 'Pricing of Irrigation Service in India (1854–1959)', *Artha Vijnana*, II:2, 1960, pp. 158–68

Kessinger, T.G. 'The Peasant Farmer in North India, 1848–1968', *Explorations in Economic History*, XII:3, 1975, pp. 303–31

Kuriyan, G. 'Irrigation in India', 2 pts, *Madras University Journal*, XV, 1943, pp. 46–58, and XVI, 1944, pp. 161–85

Lele, U. *Food Grain Marketing in India: Private Performance and Public Policy*, New York, 1971

Lipton, M. 'The Theory of the Optimising Peasant', *Journal of Development Studies*, VI:3, 1968, pp. 327–51

Macpherson, W.J. 'Economic Development in India under the British Crown, 1858–1947', in A.J. Youngson (ed.), *Economic Development in the Long Run*, London, 1972, pp. 126–91

McAlpin, M.B. 'Railroads, Prices and Peasant Rationality: India 1860–1900', *Journal of Economic History*, XXXIV:3, 1974, pp. 662–84

McGeorge, G.W. *Works and Ways in India*, London, 1894

Mellor, J.W., *et al. Developing Rural India: Plan and Practice*, New York, 1968

Millar, J.R. 'A Reformulation of A.V. Chayanov's Theory of the Peasant Economy', *Economic Development and Cultural Change*, XVIII:1, 1970, pp. 219–29

Moorti, T. and Mellor, J.W. 'A Comparative Study of the Costs and Benefits of Irrigation from State and Private Tubewells in Uttar Pradesh', *Indian Journal of Agricultural Economics*, XXVII, 1973, pp. 181–9

——. 'Cropping Pattern, Yields and Incomes under Different Sources of Irrigation', *Indian Journal of Agricultural Economics*, XXVII:4, 1972, pp. 117–25

Moreland, W.H. *The Agriculture of the United Provinces*, Allahabad, 1910

Morris, M.D. 'Towards a Reinterpretation of Nineteenth Century Indian Economic History', *Journal of Economic History*, XXIII:4, 1963, pp. 606–18

——. 'What is a Famine?', paper presented at a Conference on Indian Economic and Social History, St John's College, Cambridge, 1975

Musgrave, P.J. Review of *Agrarian Conditions in Northern India*, by E. Whitcombe, *Modern Asia Studies*, IX:4, 1975, pp. 552–5

——. 'Rural Credit and Rural Society in the United Provinces, 1860–1920', in C.J. Dewey and A.G. Hopkins (eds.), *The Imperial Impact: Studies in the Economic History of Africa and India*, London, 1978, pp. 216–32

Neale, W.C. *Economic Change in Rural India: Land Tenure and Reform in Uttar Pradesh, 1800–1955*, New Haven and London, 1962

Nicolls, L.R. 'Agricultural Engineering in India (I–XII)', *Engineering* XLV and XLVI, 1888

Poff, B.J. 'Land Management and Modernization in a Punjab Canal Colony: Lyallpur, 1890–1939', paper presented at the Institute of Commonwealth Studies, London, Feb. 1974

Puxley, H.L. *Agricultural Marketing in Agra District*, Calcutta, 1936

Ranis, G. 'Equity with Growth in Taiwan: How "Special" is the "Special Case"?', *World Development*, VI:3, 1978, pp. 397–409

Reidinger, R.B. 'Institutional Rationing of Canal Water in North India: Conflict between Traditional Patterns and Modern Needs', *Economic Development and Cultural Change*, XXIII:1, 1974, pp. 79–104

Repetto, R.C. *Time in India's Development Programmes*, Cambridge, Mass., 1971

Robinson, F.C.R. *Separatism among Indian Muslims, 1860–1923*, Cambridge, 1974

Schluter, M.G. and Mount, T.D. 'Some Management Objectives of the Peasant Farmer: An Analysis of Risk Aversion in the Choice of Cropping Pattern, Surat District, India', *Journal of Development Studies*, XII:3, 1976, pp. 246–61

Schultz, T.W. *Transforming Traditional Agriculture*, New Haven, 1964

Sen, A.K. 'Starvation and Exchange Entitlements: A General Approach and its Application to the Great Bengal Famine', *Cambridge Journal of Economics*, I:1, 1977, pp. 33–59

Siddiqi, A. *Agrarian Change in a North Indian State: Uttar Pradesh, 1819–33*, Oxford, 1973

Singh, K., Goel, B., and Murty, V. 'Estimation of the Availability of Bullock Power in Certain Tracts of India', *Agricultural Situation in India*, xxvi, 1971, pp. 483–7

Singh, K. and Hrabovszky, J.P. 'An Economic Analysis of Bullock Labour Use on Delhi Farms', *Indian Journal of Agricultural Economics*, xx:4, 1965, pp. 17–28

Skinner, G.W. 'Marketing and Social Structure in Rural China', 2 pts, *Journal of Asian Studies*, xxiv, 1964–5, pp. 3–43, 195–228

Stokes, E.T. 'Dynamism and Enervation in North Indian Agriculture: The Historical Dimension', in E.T. Stokes, *The Peasant and the Raj: Studies in Agrarian Society and Peasant Rebellion in Colonial India*, Cambridge, 1978, pp. 228–42

——. *The English Utilitarians and India*, Oxford, 1959

——. 'The First Century of British Colonial Rule in India: Social Revolution or Social Stagnation?', *Past and Present*, LVIII, 1973, pp. 136–60

Stone, I. 'Canal Irrigation and Agrarian Change: The Experience of the Ganges Canal Tract, Muzaffarnagar District (U.P.), 1840–1900', in K.N. Chaudhuri and C.J. Dewey (eds.), *Economy and Society: Essays in Indian Economic and Social History*, Delhi, 1979, pp. 86–112

——. 'Peasant Agriculture: An Analysis of Production Determinants in Pargana Koil (Aligarh, U.P.) 1900–01', paper presented at the Second New Zealand Conference on Asian Studies, University of Canterbury, Christchurch, May 1977

Strachey, J. *India: Its Administration and Progress*, 4th edn, London, 1911

Strachey, J. and Strachey, R. *The Finances and Public Works of India*, London, 1882

Thompson, W.P. *Punjab Irrigation*, Lahore, 1925

Thornton, W.T. *Indian Public Works*, London, 1875

Voelcker, J.A. *Report on the Improvement of Indian Agriculture*, London, 1893

Wade, R. 'The Social Response to Irrigation: An Indian Case Study', *Journal of Development Studies*, xvi:1, 1979, pp. 3–26

Watt, G. *A Dictionary of the Economic Products of India*, 6 vols. in 9 pts., Calcutta, 1889–93

——. *Pamphlet on Indigo*, Calcutta, n.d.

Whitcombe, E. *Agrarian Conditions in Northern India*, vol. I, *The United Provinces under British Rule, 1860–1900*, Berkeley, Los Angeles, and London, 1972

——. 'Development Projects and Environmental Disruption: The Case of Uttar Pradesh, India', *Social Science Information*, II:1, 1972, pp. 29–49

Wilson, H.M. *Irrigation in India*, Washington, D.C., 1892

Wittfogel, K. *Oriental Despotism: A Comparative Study of Total Power*, New Haven, 1957

Yule, H. *The Book of Ser Marco Polo, the Venetian, Concerning the Kingdoms and Marvels of the East*, 2 vols., London, 1871

——. *Hobson Jobson, a Glossary of Anglo-Indian Colloquial Words and Phrases*, London, 1886

INDEX

369

CAMBRIDGE SOUTH ASIAN STUDIES

These monographs are published by the Syndics of Cambridge University Press in association with the Cambridge Centre for South Asian Studies. The following books have been published in this series:

Lightning Source UK Ltd.
Milton Keynes UK
UKHW010723100322
399840UK00007B/382

9 780521 526630